Manufacturing Technology

Volume 2

Second Edition

R. L. Timings
S. P. Wilkinson

LONGMAN

Pearson Education Limited
Edinburgh Gate
Harlow
Essex CM20 2JE
England

and Associated Companies throughout the world

Visit us on the World Wide Web at:
http://www.pearsoned.co.uk

First published 1993
Second edition 2000

British Library Cataloguing in Publication Data
A catalogue entry for this title is available from the British Library

ISBN 0-582-357977

10 9 8 7 6 5 4
08 07 06 05 04

Set by 32 in Times $9\frac{1}{2}$/12 and Frutiger
Produced by Pearson Education Singapore (Pte) Ltd
Printed in Singapore

Contents

Part C   Management of health and safety at work

12   Health and safety at work — 377

Preface

Manufacturing Technology, Volume 2, was originally written not only to satisfy the requirements of the Business and Technician Education Council (BTEC) standard units for Manufacturing Technology at Levels 3 and 4, but also to satisfy the main requirements of those students following college-devised syllabuses. At the design stage of this text, most of the colleges and new universities offering HNC/HND in manufacturing engineering were sent a questionnaire requesting information concerning the topic areas and treatment they would like to see incorporated. The response was a record for questionnaires and the author was swamped with information and spoilt for choice. However, after analysing the returns and asking supplementary questions of the respondents, a specification was drawn up and became the synopsis for the first edition of this book.

The time has now come to bring out a new edition of this book and the original author, R. L. Timings, has been joined by Dr Steve Wilkinson of Leeds Metropolitan University to correct, expand and update the original text:

- There is a new Chapter 1, setting out the requirements for design for manufacture and assembly. This is cross-referenced with the processes discussed in Volume 1 and in the subsequent chapters of this book.
- Much more emphasis has been placed on aspects of computer-aided engineering and the computer-aided management of manufacture. This has resulted in the content of the previous chapter on assembly being subsumed within other chapters of this revised text to make room for the new material.
- Photoelectronic and laser-measuring techniques have been included together with touch-trigger and analogue probing. The use of probing is considered in connection with three-dimensional co-ordinate measuring machines and one-hit machining.
- The use of the Renishaw Ballbar for machine tool testing is also included.
- The section on Health and Safety has been expanded to take account of the Workplace Regulations 1992, Provision and Use of Work Equipment Regulations (PUWER), *the Management of Safety* (BS 8800), and *Environmental Management Systems* (BS EN ISO 14000) as they relate to manufacturing.

All these changes have necessitated much restructuring and the book is now in three closely related parts:

- **Part A** Technology of manufacture
- **Part B** Management of manufacture
- **Part C** Management of health and safety at work

Whilst still satisfying the original aims, this revised edition not only reflects changes in technology but, at the same time, the opportunity has been taken to:

- Make the language less formal and more reader friendly.
- Expand some of the explanations to improve their clarity.
- Correct some residual errors and omissions.
- Modernise the general format of the book.
- Introduce self-assessment exercise at key points in the text.
- Make more use of bulleted summaries.

Each chapter is now prefaced by a summary of the topic areas to be covered and finishes with a selection of practice exercises.

This revised text follows on naturally from the third edition of *Manufacturing Technology*, Volume 1. Students who have made a 'direct entry' onto their HNC/HND course will need to read Volume 1 in conjunction with Volume 2 for a full understanding of the processes described.

The broad coverage of *Manufacturing Technology*, Volume 2, not only satisfies the requirements of higher technician engineering students with reference to the latest EdExel modules, but also provides essential reading for undergraduates studying for a degree in manufacturing engineering, mechanical engineering or combined engineering. The book is also suitable for the new technology degrees that are tending to supplant the HNC/D courses and which lead to incorporated engineering status.

R. L. Timings
S. P. Wilkinson
1999

Acknowledgements

The authors wish to thank Mr Richard Duffill of Coventry University for his original contributions on metal cutting, advanced machining and automated assembly, and Mr Frank Race for his major contributions on the management of manufacture. Their contributions to the first edition helped to form a sound foundation for this new edition. The authors would also like to thank:

- Mr Tom Watters of GKN Birfield Extrusions for his time, advice and help concerning cold- and warm-forging techniques.
- Mr M. J. Nash of GKN Sinter Metals for his time, advice and help concerning sintered particulate metal manufacturing techniques.
- Dr Maurice Bonney, Professor of Production Management, University of Nottingham, for granting permission to use illustrations and data from the *Handbook of Industrial Robotics*.
- Mr D. Blizzard of BSA Machine Tools for his assistance in connection with single- and multi-spindle automatic lathes.
- Mitutoyo (UK) Ltd for their assistance in connection with three-dimensional measuring machines.
- Renishaw Ltd for their assistance in connection with probing and laser measurement together with their Ballbar system for machine calibration.
- Taylor Hobson for their assistance in connection with optical measuring equipment and systems.

We are grateful to BSI for permission to reproduce our Figs 11.6, 12.1, 12.3(b), 12.14, 12.6, 12.9 & 12.10, and Section 12.12 (pages 401–3) adapted from BS EN ISO 14000 and BS 8800. Extracts from British Standards are reproduced with the permission of BSI under licence number PD\1999 0788. Complete copies of the standards can be obtained by post from BSI Customer Services, 389 Chiswick High Road, London W4 4AL.

The authors and publishers are grateful to the following for permission to reproduce copyright material:

Hayes Shellcast Ltd for our Fig. 2.6; Fordath Ltd for our Fig. 2.9; Trucast Ltd for our Fig. 2.10; P. I. Castings Ltd for our Fig. 2.13; Kenrick Hardware Ltd for our Fig. 2.15; B & S Massey Ltd for our Fig. 3.3; National Machine Corp. USA for our Fig. 3.7; Sweeney and Blacksidge Ltd for our Fig. 3.18; Bulpitt Engineering Ltd for our Figs 3.34 & 3.35; GKN Sinter Metals Division for our Figs 4.1, 4.1 & 4.11; Colchester Lathe

Co. for our Fig. 6.6; H. W. Ward Ltd for our Fig. 6.11; BSA Machine Tools plc for our Figs 6.13–6.16; Cassell (publishers) for our Figs 6.17 & 8.15; Alfred Herbert Ltd for our Fig. 6.20; Stavely Machine Tool Co for our Fig. 6.43; Sylvac/Bowers for our Fig. 7.4; Taylor Hobson Ltd for our Figs 7.16–7.19 & 7.22; Renishaw plc for our Figs 7.32 & 7.37; Butterworth Heinemann for our Fig. 7.44; Visual Component Engineering Systems Ltd for our Fig. 9.4; HSE for our Fig. 12.3(a); Silvaflame Guards Ltd for our Figs 12.5.

How to use this book

There are many ways of using a textbook. You can read through it from cover to cover and try to remember all that you have read, but this is rarely successful unless you have a photographic memory, and few of us have. You may consider using it just as a reference book by looking up individual topics as and when you need that specific information. This can be useful as a reminder once you know the subject thoroughly, but until then it can lead to misconceived ideas. So, here are a few thoughts on how to maximise the benefit that you can get from this book.

The book is divided into three parts:

- Part A covers the technology of manufacture (Chapters 1 to 8).
- Part B covers the management of manufacture (Chapters 9 to 11).
- Part C covers health, safety and environmental issues (Chapter 12).

Although interrelated, it is not essential that Parts A, B and C are read sequentially. However, the chapters *within* each part should be read sequentially. It is assumed that the reader will already be familiar with *Manufacturing Technology*, Volume 1, as many topic areas build on the foundations laid in that book.

Each part consists of a number of chapters, each of which covers a major syllabus area. For example, **Chapter 1** covers design for manufacture and assembly. It is divided into sections. If you turn back to the Contents page you will find that:

- **Section 1.1** deals with an introduction to design for manufacture and assembly.
- **Section 1.2** deals with the philosophy of design for manufacture and assembly.
- **Section 1.3** deals with the implementation of DFMA, and so on.

Sometimes it is necessary to divide these sections up further. For example, Section 1.5 subdivides into:

- **Section 1.5.1,** which deals with complexity.
- **Section 1.5.2,** which deals with material suitability.
- **Section 1.5.3,** which deals with wall sections.

As previously stated, it is not essential to read each part of this book sequentially, but to obtain the maximum benefit from this book (and this may seem obvious!), start with the first chapter of the part you wish to study and work sequentially through that part. If you wish to work in some other order and be selective in your reading, then this should only be done in consultation with your tutor. Each chapter is prefaced with a list of the main topic areas that you are going to find in it.

As you work through the chapters, you will find **self-assessment tasks** at key points. If you have understood what you have read previously, you should be able to complete these exercises *without looking back* at the text. If you have to look back, then you are not yet sure of your ground and there is no point in moving on. Try again and, if you are still not sure, have a chat with your tutor to clear up your difficulty.

At the end of each chapter there is a selection of more extended **exercises**. Your tutor will guide you as to their use and check your responses to ensure that you have the background knowledge required to understand the next topic area to be studied.

No matter whether you work through the book from beginning to end systematically, or work through the foundation chapter and then move on to the specialist areas you require, you must complete the self-assessment tasks and the end-of-chapter exercises as you reach them. This will ensure that you will understand the next stage of your journey through your study of manufacturing technology.

Finally, once you have qualified, you can still obtain benefit from this book. The numbering of the sections and subsections, the comprehensive list of contents and the extended index, will all help these texts (including the companion volumes: *Manufacturing Technology*, Volume 1 and *Engineering Materials*, Volumes 1 and 2) to act as quick reference books during your future career in engineering.

Useful web addresses

DFMA

- www.dfma.com

CNC

- www.fanucamerica.com
- www.heidenhain.com
- www.cincinnati.com

Robotics

- www.fanuc-robotics.co.uk
- www.staubli.com
- www.abb.com
- www.adept.com
- www.epson.com

Measurement

- www.npl.co.uk
- www.renishaw.co.uk
- www.taylor-hobson.com
- www.mitulogo.com

Safety and standards

- www.open.gov.uk/hse/signpost.htm
- www.bsi.org.uk

Simulation

- www.byg.co.uk
- www.rob.sim.com
- www.promodel.com
- www.camtek.co.uk
- www.cadcentre.co.uk

Part A
Technology of manufacture

1 Design for manufacture and assembly

The topic areas covered in this chapter are:

- Philosophy of design for manufacture and assembly.
- Function of components.
- Processes for the manufacture of components.
- Handling of components (automatic or manual).
- Fitting or assembly of components (automatic or manual).
- Computer-aided design and manufacture.

1.1 Introduction

Some basic principles of design specification and the factors affecting the selection of manufacturing processes were introduced in *Manufacturing Technology*, Volume 1. Many different processes were described together with the materials used. The importance of specifying dimensional tolerances and surface finishes was also considered, together with their relationship with the process selected. Remember that it is bad design to specify dimensional and/or geometric tolerances that are closer than are required for a given component to achieve its specified fitness for *purpose*. Unnecessarily close tolerances that cannot be achieved by the selected process may force a change to a more expensive process and this may render the product financially uncompetitive. In Volume 1, you were also introduced to the fact that surface finish and dimensional tolerance must be matched. Figure 1.1 shows that close tolerances cannot be achieved by any process that leaves the component with a rough surface.

The relationships between the different departments within a manufacturing company were also described in *Manufacturing Technology*, Volume 1. This second volume develops these concepts further, both in breadth and in depth, and considers current thinking in the area of design for manufacture.

All designs start with a *design brief*. This may be generated by market research within the manufacturing company itself. Alternatively, a design brief may be generated by an existing or by a potential customer who may be seeking to develop a new product in order to satisfy an identified market demand. A design brief is a description of the client's needs. It will state the:

Fig. 1.1 *Relationship between surface texture and dimensional tolerance: (a) limits of size and process mismatched; (b) process suitable for limits of size*

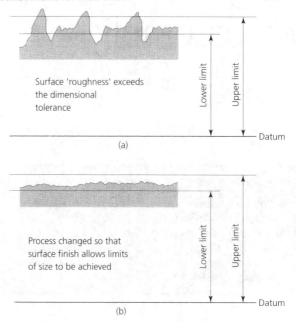

Surface 'roughness' exceeds the dimensional tolerance

Lower limit

Upper limit

Datum

(a)

Process changed so that surface finish allows limits of size to be achieved

Lower limit

Upper limit

Datum

(b)

- Client's requirements.
- Functional (operational) requirements.
- Maximum permissible project and product costs based upon the perceived market.
- Market details.
- Quality requirements.

The design brief is then analysed and *the key design features* are identified. The key design features will be concerned with:

- Aesthetic requirements.
- Contextual requirements.
- Performance requirements.
- Production parameters and constraints, including labour and material availability, available technology, health and safety requirements.

Alternative design proposals will then be drawn up and feasibility studies carried out in order that a realistic presentation can be made to the client. These initial concepts will then lead to detailed discussions with the client so that, finally, a detailed design specification can be drawn up together with the associated quality specification. The final design will then be produced conforming to these specifications. The final design will also need to conform to the requirements of national and international standards organisations, safety and environmental legislation and legally binding notes of guidance for the category of product concerned.

It would be most unfortunate if, at this stage, the design proved to be too costly to make or put demands on the available technology that could not be met. This has happened on more than one occasion in the past. Therefore the final design must obey the basic rules and requirements of *Design for Manufacture and Assembly*.

1.2 Philosophy of design for manufacture and assembly

Design is the first step in the manufacture of any product and is where most of the important decisions are made affecting the cost of a product. The philosophy of Design for Manufacture and Assembly (DFMA) examines techniques for reducing the product cost by improving 'manufacturability' and ease of assembly. *It is more important to improve 'manufacturability' and assembly than it is to increase automation.* For example, a study has shown that the least automated, domestic Japanese plant is the most efficient in the world. Yet this plant has only 34 per cent of all steps in the manufacturing sequence accomplished automatically. It also needs only half to a third the human effort of a comparable European plant. The European plant that is the most automated in the world requires 70 per cent more effort than this particular Japanese plant. The conclusions of the study show that there is a direct link between product *design* and *productivity*. Therefore, even if a product is designed so that it achieves fitness for *purpose,* that design is nevertheless *defective* if its 'manufacturability' has been ignored. No amount of automation can improve productivity more than the application of DFMA.

1.3 The implementation of DFMA

Traditionally, the attitude of designers was 'we designed it: you build it'. This has been termed an *over-the-wall approach* where the designer throws his designs over the wall to the manufacturing engineers who have to deal with the various design problems because they were not involved in the design process in the first place. To implement DFMA such attitudes, which unfortunately still exist all too widely in industry, must be overcome, and one means of overcoming this problem is to build teams. These teams are now called *Simultaneous* or *Concurrent Engineering* teams. The criteria examined by these teams for analysing manufacturability and assembly from proposed designs are:

- Functionality.
- Manufacturing issues.
- Handling and Assembly.

1.4 Functionality

Any industrial museum housing examples of engineering from the Victoria era will exhibit machines and engines that have most of their functional components embellished with features more usually associated with classical architecture as an indication of the artistry of the designer and craft skills of the maker. Fortunately for the designer, such exhibits were made in an era when highly skilled labour was available cheaply and competition was largely non-existent.

Fortunately or unfortunately, depending on one's point of view, such embellishment is now unwarranted and *functionality* is the key to modern design success. This technique reveals how non-functional components of a product can be designed out (eliminated) in order to improve *Design Efficiency*. For example, the bearing surfaces on a shaft are the *functional* features of the component, any other complex features or embellishments may be unnecessary, and will only add to the cost of manufacture. Components can be divided into two categories:

- 'Type A', which carries out functions vital to the performance of the product, such as drive shafts, etc.
- 'Type B', which is NOT critical to the function of the product, such as inspection covers, spacers, etc.

This technique was developed by Professors Boothroyd and Dewhurst in collaboration with Lucas Industries. They suggest starting with a major functional part such as rotor shaft and categorising mating components in a logical progression for the remainder of the parts. The idea is to eliminate as many 'Type B' components as possible by redesign.

$$\text{total design efficiency } (E) = \frac{\text{number of 'Type A' components} \times 100\%}{\text{total number of components}}$$

A new product should have a high design efficiency, with 60 per cent being a typical limit for 'good' designs. However, it is not unusual for the design efficiency to have a value as low as 30 per cent for complex products. Figure 1.2 shows an example of these types of components.

Fig. 1.2 *Functional, non-functional and auxiliary dimensions: F = functional; NF = non-functional; (A) = auxiliary or reference; R = redundant*

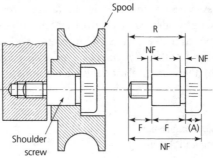

1.5 Manufacturing issues

Manufacturing issues are concerned with the analysis of the cost of using different materials and manufacturing processes. Manufacturing issues are also concerned with such matters as *part count reduction*. Can the *part count* be reduced by using alternative manufacturing processes to combine parts? For example, can plastic fasteners be moulded as part of a casing rather than using separate fasteners attached by means of additional nuts and bolts? Other considerations include the volume of material used. For example, if the strength of the product would still be adequate, is it possible to achieve a thinner wall by using an alternative process? The tolerances and surface finish achieved by a given process must also be considered. For example, could the design and quality specifications for a given component allow coarser dimensional and geometric tolerances and/or a rougher surface finish? If so, then it may be possible to achieve fitness for purpose by a cheaper process. Let us now look at some of these issues more closely.

1.5.1 *Complexity*

The complexity of components can range from simple cylindrical shapes to parts having very complex sections with thin walls. Also, different shapes require different processes. Depending upon the shape of the component, some processes are more suitable than others; for example, simple cylindrical shapes are best manufactured by turning. However, for very thin-walled, complex shapes, sand casting or machining such a component may prove extremely difficult to accomplish. Sand moulds would not have the necessary strength and accuracy. An investment (lost-wax) process may be more appropriate. Further, a thin-walled component may collapse during machining due to the forces exerted by the workholding devices or the cutting tool. These problems may also be overcome by redesigning the component so that it can be made out of sheet metal in a press. Alternatively, the best solution for a particular component may lie in die casting, forging, sintering, or plastic moulding.

1.5.2 *Material suitability*

The suitability of materials for different processes is another consideration. The material and the process must be carefully matched. Laminated plastics such as 'Tufnol' may be machined like a metal, whereas polystyrene is more suitable for moulding. Different metals also have widely different properties. For example: copper, zinc, lead and some aluminium alloys are easier to work by impact extrusion processes than steel and its alloys. Plain and alloy steels, copper, and aluminium alloys may all be hot forged. Brass alloys may also be worked hot but, in their case, the process is referred to as *hot pressing*. Surprisingly, aluminium alloys are more difficult to hot forge than alloy steels, the reason being that the hotter the metal, the more easily it flows under the forging press or hammer. Since aluminium alloys have a relatively low melting temperature, the temperature at which they can be forged is very much lower than the temperature for the ferrous metals (see section 3.4) so more force is required to make them flow. Aluminium alloys, copper and its alloys and low-carbon steel are all suitable for the manufacture of sheet metal pressings. Non-ferrous metals are generally easier to machine than most steels, especially stainless steel which is one of the most difficult steels to machine. While zinc alloys and aluminium alloys are the only metals suitable for die casting, some aluminium alloys and all the brass alloys, bronze alloys and the cast irons are suitable for sand casting. Materials and their properties are fully covered in *Engineering Materials*, Volumes 1 and 2.

1.5.3 *Wall thickness*

Different processes have different limits of wall thickness. For example, sand casting is more suitable for components requiring walls greater than 3 mm thick. Die-casting and investment-casting processes are better for walls thicker than 0.6 mm. Sheet metal pressings can be used for thin-walled components, i.e. less than 0.6 mm and so is impact extrusion.

1.5.4 *Tolerance and surface finish*

Chapter 7 deals with dimensional tolerance and surface finish in great detail. However, simple rules are useful – for example, keeping the tightest tolerance in only one plane, and not overspecifying the tolerance if it is unnecessary to the function of the part. Other processes such as lapping, honing, polishing or grinding can help to achieve a very tight tolerance but the cost implications of additional processing and the time required to achieve that tolerance and surface finish must be considered. It is better to select the process that can deliver the appropriate tolerance and surface finish suitable to the function and appearance of the part. For example, processes such as hot forging and sand casting produce a rough finish with dimensional tolerances over ± 0.3 mm, whilst die casting can produce a fine finish with dimensional tolerances of ± 0.08 mm. The plastic-moulding process can produce a super fine surface finish on the component with dimensional tolerances of ± 0.05 mm. (*Note:* The finish on the moulding reflects the quality of the finish in the mould cavity.) Presswork can also produce components with a good surface finish with a tolerance of ± 0.08 mm.

1.5.5 *Process cost*

Each process has different costs associated with it; for example, some processes such as plastic moulding have a lot of initial costs tied up in tooling and set ups. However, as soon as the machine starts operating many thousands of components can be produced at very high rates before the tools need to be replaced. The more components that are manufactured, the more quickly the high initial investment in mould tooling can be recovered.

Like plastic moulding, many metalworking processes require special dies and/or tools. These include processes such as impact extrusion, forging, die casting, press work and sintering. These processes are highly automated and can quickly recover the high initial tooling costs through high-volume production of the components. Computer numerical control (CNC) machining can now produce medium batch sizes more cost-effectively with automatic tool and workpiece set ups using touch trigger probes such as those produced by Renishaw (see Section 8.3.7). However, large-volume production is not always required, and for smaller quantities such as prototype work it may be better to sand cast and manually finish or machine a component.

1.5.6 *Waste*

Some processes are more wasteful than other processes. For example, casting and forging processes shape metal with little loss of volume, whereas machining components from the solid results in a large loss of volume and is very wasteful in terms of energy and material. The material that is cut away is called *swarf* and can only be disposed of as scrap. Therefore very complex components use less material if they are cast rather than machined from the solid. The choice of process is a matter of production economics. A balance has to be struck between the low tooling costs but high material cost for machining, and the high tooling costs but lower material costs of; say, die casting, forging, or extrusion. The availability of plant must also be considered. It is no good choosing a process, however ideal on paper, if it means that new and highly expensive plant has got to be purchased, unless the long-term volume of production warrants such an investment.

1.5.7 *Treatment*

The cost of decorative and/or corrosion-resistant surface treatment processes must also be considered. A balance has to be struck between the cost of the additional treatment of; say, a low-cost plain carbon steel or the higher cost of a material like stainless steel that does not require additional surface treatment. Heat-treatment processes may also be required and, again, process costs and material costs must be balanced against each other. Only the product designer in consultation with the customer can decide whether the lower cost of a case-hardened, plain carbon steel component can be substituted for the higher cost of a through-hardened alloy steel component.

1.6 Handling and assembly

The main issue when considering how parts can be handled and assembled is how far a redesign of a product can promote the reduction in the quantity of parts. Remember that a *lower part count* requires less time to be spent on handling and assembling the product. An example of this is the modern computer printer, where the part count is dramatically reduced when the latest models are compared with older models. In certain instances it was found that, when redesigning a product for automatic assembly, the resulting (simplified) product was so easy to assemble manually that it no longer justified the use of an automated process.

1.6.1 *Handling*

The orientation of parts and their suitability for manual or mechanical handling needs to be assessed. It is a poor design if a component is difficult to handle – that is, if the component is fragile and delicate, unnecessarily heavy, or requires some special handling device. The questions must be asked: Could the product be redesigned to ease manual assembly or allow for automated assembly? Does the product need to be repositioned a number of times in order to insert a component? Other factors that need to be considered include mechanical feeding; for example: Could a component be redesigned to make it symmetrical? This would allow it to be orientated in a number of directions and still be correctly positioned in the assembly. A number of feeding mechanisms exist and these are considered in Section 8.10.

1.6.2 *Assembly*

The task of assembling and testing the components of a product must also be assessed. For example, a poorly designed component may be difficult to grip, or the surface to be located may be obstructed by the gripping process. The features of good design include:

- Self-orientating parts that can be positioned correctly at the first attempt.
- No resistance to insertion.
- No restriction to access and/or vision.
- No additional joining process such as bolting, soldering or adhesive bonding.

To achieve these criteria, all components:

- Should be symmetrical, if possible, as shown in Fig. 1.3(a).
- Have a marked polar geometry and/or weight if they are not symmetrical, as shown in Fig. 1.3(b).

- Have the least number of important directions in order to assist orientation Fig. 1.3(c).
- Should be tangle proof, as shown in Fig. 1.3(d).
- Should be consistent in the dimensions used for feeding, orientation and location (Fig. 1.3(e)).
- Should be self-locating, e.g. taper lead on end of bolt (Fig. 1.3(f)).
- Should have a datum surface or point on which to build the assembly (Fig. 1.3(g)).

Fig. 1.3 *Design for ease of assembly*

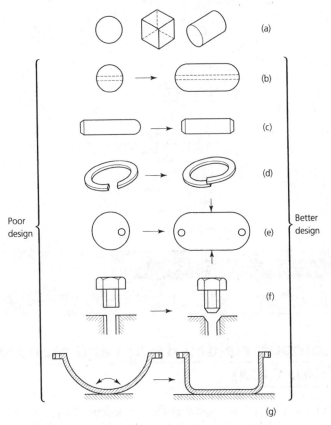

It is advisable that assemblies are designed for manual assembly first as this is a simple and low-cost process. However, the design should keep the possibility of automated assembly in mind, in case the volume of production and cost savings warrant the financial investment into automated assembly equipment required at a later date. With this in mind, it is important that assemblies:

- Should have location points.
- Should be designed so that they can be assembled from the top.
- Should be designed so as to avoid the necessity of turning the product over.
- Should be designed so that the product can be built as a series of subassemblies in order to allow intermediate checks, as shown in Fig. 1.4(a).

- Should incorporate standardised parts, such as fasteners in order to facilitate assembly and avoid the possibility of incorrect selection, as shown in Fig. 1.4(b).
- Should eliminate unnecessary components and complications, as shown in Fig. 1.4(c).

Fig. 1.4 *Elimination of unnecessary complexity and redundant components*

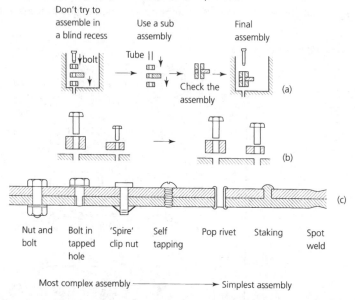

SELF-ASSESSMENT TASK 1.3

List and critically appraise the techniques that aid design for assembly.

1.7 Computer-aided design and manufacture (CAD/CAM)

In section 1.2 it was stated that the 'we designed it: you must make it' attitude must be abandoned if design for manufacture is to be encouraged. Although the analysis techniques considered are far more important than any computer application, it is apparent, however, that the integrative effects of some computer applications contribute towards the DFMA philosophy. For example:

- Geometry can be downloaded directly into the CAM package using different file formats; i.e. DXF (Data eXchange File).
- Modern computer systems make it easy to toggle between CAD and CAM applications so modifications can be made interactively – that is, if a manufacturing problem is encountered in the CAM package then a design modification can be made instantly. Design is an iterative process and CAD/CAM can greatly reduce the time between iterations (see case study).

- The use of CAD and parametrics to design a family of components that share common features will help to reduce the design time, and the number of set ups for the machine operator. In many instances, set-up time exceeds machining time for small batches of components. Parametrics can aid the automatic design of components from known features. For example, the operator would only need to enter length and diameter for simple turned components.
- Simulation of the machining process on the CAM package will help to analyse clamping positions, cutters used and suitable speeds and feeds. Also the automatic generation of cutter paths. For example, the use of automatic *area clearance cycles* reduces the time spent in programming CNC machines, hence concept models or trial runs can be produced very quickly.
- The nesting of components to make the maximum use of material can be automatically generated using CAM applications. For example, cloth in the clothing industry is cut using a profiling machine. The patterns are nested to get the maximum number of garments out of the minimum amount of material. Other applications using this technique include leather, steel plate, and nylon sheets. The computer-controlled profiling machine would use a knife in the case of cloth and a CO_2 laser in the case of wood and plasma or oxyacetylene flame cutting equipment in the case of steel.

1.8 Design for CNC

The previous section showed that the CNC program could be simulated before downloading onto the actual machine. As already stated, simulation also helps to analyse the types of cutter used and clamping positions. Here are a few simple rules for use when designing for CNC machining:

- Use one corner as a datum from which all dimensions are taken.
- Reduce the number of tool changes by using the same fillet radius throughout the component where possible.
- Do not specify a radius on a turned component merely to remove a sharp corner if a simple chamfer will do.
- Use automatic circular pocket cycles when generating holes of different sizes using one slot drill rather than a large number of drills, better still reduce the variety of hole sizes.
- Make as many machining operations as possible in one plane as this reduces the amount of clamping/unclamping operations (see case study).
- Avoid excessive depths (long series cutters required) and thin-walled section. This leads to weakness and lack of rigidity in the cutter and the component. Such cutter problems may cause component and/or cutter breakage, cutter deflection (inaccuracy), and poor surface finish due to cutter chatter.
- Consider using as many automatic (canned) cycles as possible when designing the component – that is, use the maximum functionality of the CNC machine when machining such features as slots, circular pockets and rectangular pockets. Also consider using automatic inspection by touch trigger probes. After all, every design must accommodate some form of inspection or gauging.

- Rationalise on material stock so that a range of designs can be produced on CNC machines from a minimum range of bar and billet stock. This has two main advantages:
  1. Minimising the number of fixtures required for the accommodation of different sizes.
  2. Reduced inventory by holding the minimum quantity of bar and billet (blank) sizes (see case studies).

1.9 Case studies

Let us now put the principles of DFMA into practice by examining some design case studies.

1.9.1 *Computer mouse*

This case study shows all aspects of DMFA mentioned above. A computer company redesigned its mouse using DFMA. First they carried out a market analysis in which they compared their mice with those of other competitors. They benchmarked (compared against a fixed standard) assembly times, part counts, the number of assembly operations, labour costs and the costs of each product. They then redesigned their mouse using DFMA and the result is shown in Fig. 1.5. In the new design, the assembly time (circuit board already assembled) was reduced from 130 seconds to 15 seconds. Other changes to the product structure also brought cost savings. Examples of the savings are listed below:

- The seven screws in the original mouse were reduced to zero with snap fits.
- The new mouse no longer requires any assembly adjustments, whereas the old mouse required eight.
- The entire number of assembly operations (including the circuit board) was reduced from 83 for the old mouse to 54 for the new mouse.

Other benefits included a dramatic reduction in the design cycle time, especially when making the injection-moulding tools.

1.9.2 *One-hit machining (Renishaw)*

Renishaw have designed a family of touch-trigger probes which can automatically set tooling and workpieces and also check the tolerances of components in-cycle (as shown in

Fig. 1.5 *Applications of DFMA to a digital mouse*

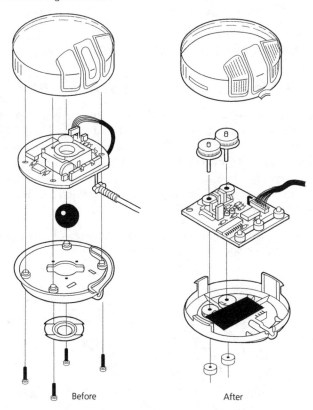

Before   After

Fig. 1.6). In the manufacture of the probe bodies they designed the components so that the majority of the casings and other parts could be manufactured from one size of bar stock. This meant that costs could be reduced because only a small quantity of bar was held. One other major benefit was the application of 'one-hit' machining. This is where the component is held in a 3-jaw chuck which could be automatically moved as the 4th axis of

Fig. 1.6 *'One-hit' machining*

the machine tool inorder to machine all sides and surfaces of the component without being unclamped at any stage. The benefits of using this technique are summarised as follows.

- Rationalised fixturing (only a 3-jaw chuck is required).
- Only one set up is required when using automatic touch-trigger probes.

1.9.3 *Electronic component assembly*

The electronic solid-state device shown in Fig. 1.7 can be placed in one of two positions because of the symmetry of its legs, as shown in Fig. 1.7(a). The designer of the device has included a notch and spot on the case to help the assembly operative to decide which way round it goes, as shown in Fig. 1.7(b). The designer of the printed circuit board completes the process by showing the outline of the device together with its notch and spot. By matching the device to the outline, decision making and potential error on the part of the assembly operative is reduced.

Fig. 1.7 *Designing to avoid assembly errors: (a) alternative positioning of integrated circuit with symmetrical 'legs'; (b) positioning an IC using identification marks*

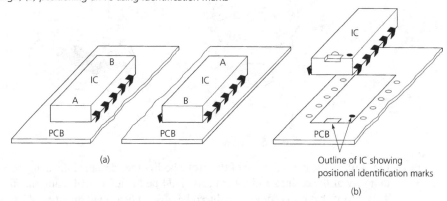

(a)

Outline of IC showing positional identification marks

(b)

1.9.4 *Design for maintenance*

At one time ease of assembly and access for repairs and routine servicing was given scant consideration, but with ever-increasing labour costs and the increasing use of automated assembly, it is now a major design consideration. During WW2 a truck widely used by all the armed services suffered from a maintenance problem that was the result of bad design. In order to change the engine oil filter one of the front wheels had to be removed. This type of vehicle was widely used in the North Africa campaign and the filters needed to be changed regularly because of the ingress of sand under desert conditions. More thought at the design stage would have made life a lot easier for the maintenance mechanics labouring in intense heat and often under enemy shell fire. All new designs should be appraised for ease of maintenance as well as ease of manufacture. For example:

- Can the fastenings be readily positioned and are they accessible to standard spanners, nut runners, screwdrivers, etc., and is there room to turn these devices?
- It must be remembered that special tools that are available in the factory may not be available to the service engineer in the field.

- Can key components, which have to be changed during routine servicing, be removed readily without having to strip off other components?
- It should not be necessary to dismantle a vehicle engine to reach a sparking plug, or to remove the engine to change an oil filter. At one time ease of assembly and access for repairs and routine servicing was given scanty consideration, but with ever-increasing labour costs and the increasing use of automated assembly, it is now a major design consideration.

1.9.5 *Radar scanner bracket*

A company which manufactured a mounting bracket for a radar scanner found that by using a 3D CAD/CAM system they were able to redesign the bracket so that it could be machined from a solid billet using a multi-axis milling machine, as shown in Fig. 1.8. Previously this component was manufactured by using a casting with functional features (holes and mating surfaces) that were then machined. The benefits resulting from redesigning the component to use a different process can be summarised as listed below.

- Less money is tied up in a billet of material, than an unfinished casting.
- No expensive casting pattern is required.
- Only one process is used to manufacture the component.
- Lead-time is reduced from receiving the initial order to receiving the finished article, as no patterns and dedicated fixturing will be required.
- Minor design changes can be made easily to suit customer requirements in the interests of 'customer-focused manufacture' (i.e. agile manufacturing).

Fig. 1.8 *Radar scanner bracket*

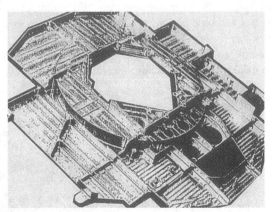

1.9.6 *Telephone design*

Owing to the liberalisation of the telephone system, telephone manufacturers found that the design life of a telephone had been reduced from 10 years to 18 months. Previously it had taken 2 years to place a new telephone on the market from the initial design inception to the assembly of the finished telephone. By using 3D CAD/ CAM techniques, as shown in Fig. 1.9, they were able to reduce this time and the number of design iterations to an acceptable level. The advantages they found were:

- The ability to manufacture concept models very quickly to test the market.
- The ability to manufacture short runs of telephone casing by using aluminium moulds.
- The ability to investigate mould flow by using computer-aided techniques to ensure the optimum number and size of gates in the mould.

Fig. 1.9 *Telephone body shell*

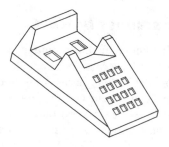

1.9.7 *Car design*

Most car manufactures have now adopted 3D CAD/CAM techniques, as shown in Fig. 1.10. The advantages over conventional design-and-make methods include:

- Reduction in the lead-times to manufacture press tools and injection moulds.
- Components can be redesigned to be lighter and stronger, by using *Finite Element Analysis*.
- The simulated testing of drag coefficients on the computer model.
- Checking the fit of the components on the computer model.
- Using a database of components, changes and redesigns can be made very easily and quickly.
- Concept cars can be manufactured very easily and quickly by using very large (almost robot like) multi-axis machining centres.

Fig. 1.10 *Surface mesh of a car body*

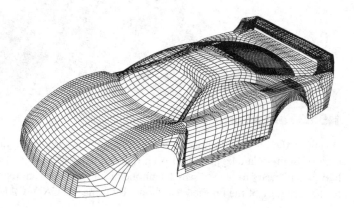

1.10 Chapter summary

This chapter briefly describes the benefits of DFMA and some of the techniques required for its implementation. Further information on this subject can be found on the websites given in the 'Useful web addresses' section on page xvii. The following chapters of this book develop the manufacturing concepts introduced in *Manufacturing Technology*, Volume 1, and introduce additional manufacturing processes together with the products and materials for which they are suitable. Also included is the application of computer-aided automation and manufacture to the assembly and production of components. These techniques will be set in the context of the management and costing of manufacture, again addressing the application of computer technology. Finally, safety must be addressed as an important issue in the use of any manufacturing technology. The location of topics within *Manufacturing Technology*, Volumes 1 and 2, relating to DFMA are listed in Table 1.1.

Table 1.1 *Topic areas related to DFMA*

DFMA Topic	Related subject	Volume	Chapter
Function of components	Tolerances	1	10
	Process selection	2	1
Manufacturing processes	Fabrication	1	7
	Press tooling	1	1
		2	3
	Casting	1	1
		2	2
	Moulds and dies	1	1
		2	5
	Forging	1	1
		2	3
	Drawing and hot extrusion	1	1
	Coatings and finishes	1	9
Handling of components	Robotics	2	1/8/9
Assembly of components	Assembly and joining methods	1	6
		2	1/8
Computer-Aided design and manufacture	CAD/CAM, CNC and Robotics	1	5
		2	1/8/9

EXERCISES

1.1 Examine the new mouse previously described and shown in Fig. 1.5 and categorise the benefits achieved under the following headings:
(a) improved function of components
(b) improved manufacturing process and part reduction of components
(c) improved handling of components
(d) improved assembly of components

1.2 What other benefits has the new mouse design accrued by using DFMS?

1.3 Compare different types of mouse design in terms of part count, assembly time, etc. (**Ask first** before you take them apart.)

1.4 Compare different types of computer printer, i.e. the old dot-matrix printer and the new ink jet printer, for part count, wall thickness, surface finish, the number of processes used, volume of material, etc.

1.5 Compare different scrap car cylinder heads and determine the number of set ups each type must have had during its machining process.

1.6 Redesign a garden wheel barrow (old model with steel fabrications) to reduce the number of parts and to reduce the cost of manufacture and assembly by selecting a different material and process.

1.7 Design a plastic toy car for a child. Use a number of redesigns to reduce the part count by determining alternative methods for fastening and joining. In addition:
   (a) which processes could be used to minimise the cost of the product if 100 000 were to be made?
   (b) determine assembly issues to find the best method of assembly. Don't forget child safety, i.e. choking on small parts

1.8 Compare different food and liquid containers by surface finish, wall thickness, strength and weight. Also determine which process was used to manufacture each container. Don't forget marketing issues, i.e. advertising labels (could they cover up a poor surface finish?).

1.9 State the benefits of using *Computer-Aided Engineering* techniques over manual methods.

1.10 Explain why a product should be redesigned for manual assembly before being redesigned for automated assembly?

2 Casting processes

The topic areas covered in this chapter are:

- Green-sand moulding and moulding sands.
- Casting defects.
- Shell moulding.
- Investment casting (lost-wax process).
- High-pressure die-casting.
- Low-pressure die-casting.

2.1 The casting process

The processes of casting metal to shape is the oldest and still one of the most widely used metal-forming processes. A casting is produced by pouring molten metal into a mould cavity and allowing it to solidify. The mould cavity is the shape of the required component. For ease of casting, the metal should have a high *fluidity* (ease of flow) and a high *fusibility* (low melting point). For example, cast iron, with its high carbon content, has a high fluidity and high fusibility compared with low-carbon steel. Therefore, cast iron can be readily cast into complex shapes, whereas steel can only be cast into simple shapes as it has a low fluidity ('treacle' consistency) and does not flow easily. Again, the high melting temperature of the steel means that the moulding sand has to be carefully selected or it will itself be melted on coming into contact with the molten steel.

Unlike other primary forming processes, the crystal structure of a casting is not homogeneous but varies from the core of a casting to its surface depending upon the rate of cooling. Such lack of uniformity and refinement in the grain structure can lead to planes of weakness in the casting, as shown in Fig. 2.1(a), if the design is not matched to the casting process. A generous radius at the corner of the casting, as shown in Fig. 2.1(b), helps to remove the plane of weakness.

In all casting processes the pattern has to be made oversize to allow for any contraction of the metal that may occur during the change of state from liquid to solid and also to allow for the continued contraction as the solid metal cools to room temperature. Note that cast iron is exceptional in that it *expands* as it solidifies and this helps it to take a sharp impression from the mould. However, after solidification, it contracts like any other metal. In addition to shrinkage allowance, a *machining allowance* also has to be provided when a surface is to be machined. Not only must sufficient additional metal be provided to ensure that the surface 'cleans up' during machining, but also that there is sufficient depth of metal for the nose of the cutting tool to operate below the hard and abrasive skin of the casting, as shown in Fig. 2.2.

Fig. 2.1 *Crystal structure of a casting: (a) inadequate corner radius leads to a plane of weakness; (b) increased corner radius helps to remove any plane of weakness*

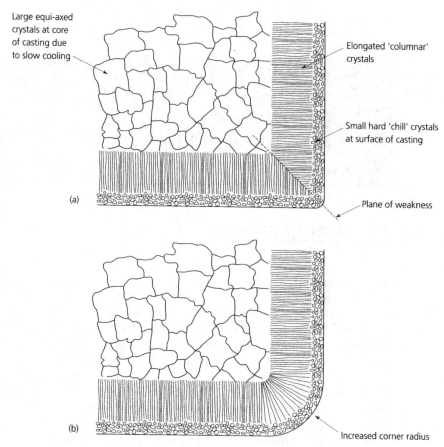

Large equi-axed crystals at core of casting due to slow cooling

Elongated 'columnar' crystals

Small hard 'chill' crystals at surface of casting

(a)

Plane of weakness

(b)

Increased corner radius

Fig. 2.2 *Machining allowance*

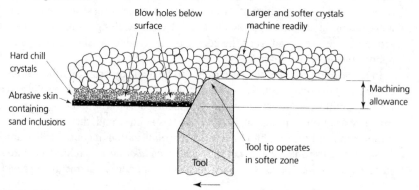

Blow holes below surface

Larger and softer crystals machine readily

Hard chill crystals

Machining allowance

Abrasive skin containing sand inclusions

Tool tip operates in softer zone

Tool

2.2 Green-sand moulding

Green-sand moulding was introduced in *Manufacturing Technology*, Volume 1. The main advantages of this process can be summarised as follows:

- Most metals and alloys can be cast in green-sand moulds.
- The size of the castings that can be produced in green-sand moulds ranges from small components for model engineers to those used in the manufacture of power station turbines.
- Castings produced by this process can be very simple or extremely complex, the latter requiring multiple internal cores, external cores, draw backs, loose pieces and other sophisticated foundry techniques to produce re-entrant surfaces.
- Quantities can range from single components to batch production.

The main disadvantage of the process is that the mould, which can only be used once, has to be made by highly skilled and highly paid moulders. This results in a process with relatively high unit costs. However, there are many large and complex components that cannot be made by any other casting process.

In Volume 1, you were introduced to green-sand moulding with a simple example using a split pattern. Often, the component shape demands more complex patterns and moulding techniques. Let's now look at some of them.

2.2.1 *Odd-side moulding*

This technique gets its name from the fact that the *parting surface is not flat*, but projects from one box or 'side' into the other box, as shown in Fig. 2.3. To make an odd-side mould the pattern is placed on a turnover board and packed up into position with sand. The moulding box is placed round the pattern and moulding sand is rammed in to form the drag (lower part of the mould). When the drag is complete the box is inverted, the board is removed, and the parting surface is carefully cut away with a trowel to expose the parting line of the pattern. The sides of the hollowed-out sand are carefully sloped from the parting line of the pattern to the parting line of the moulding box. Parting powder is then applied and the cope is rammed up and prepared, as previously described in Volume 1.

Fig. 2.3 *Odd-side moulding*

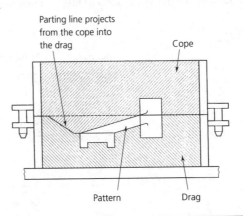

Parting line projects
from the cope into
the drag

Cope

Pattern

Drag

2.2.2 *Loose pieces*

Figure 2.4(a) shows a component with recesses that would prevent a conventional pattern from being withdrawn from the mould. This time a pattern has to be made with *loose pieces*, as shown in Fig. 2.4(b). The pattern is moulded upside down, as shown in Fig. 2.4(c). Upon completion of the cope and drag, the two parts of the mould are separated and the main body of the pattern is removed, leaving the loose pieces in the sand of the drag. The loose pieces are then carefully drawn back into the mould cavity one at a time and lifted out. The mould is then reassembled ready for pouring.

Fig. 2.4 *Moulding with loose pieces: (a) component requiring a pattern with loose pieces; (b) pattern with loose pieces; (c) removing the loose pieces*

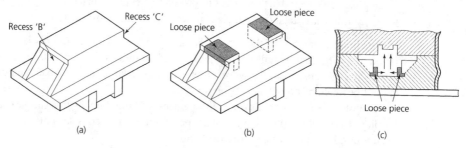

2.2.3 *Drawbacks*

A drawback is a body of sand that is used instead of an external core in order to facilitate the moulding of undercut parts of castings. An example is shown in Fig. 2.5. The drag is rammed up as usual and the sand that occupies the space that will eventually become the drawback is cut away with a trowel. After dusting with parting powder, the lifting plate and rod are inserted and the sand is rammed in again. After the mould has been completed, the drawback can be lifted out by means of the lifting plate, and the pattern can be withdrawn. In the example shown, the face CD of the drawback must be at least parallel to the face EF of the pattern. Preferably it should have a smaller angle to the horizontal to enable the drawback to clear the pattern as soon as lifting commences.

Fig. 2.5 *Use of a 'drawback'*

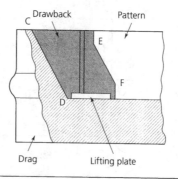

1. Explain why an ample machining allowance should be left on any cast surfaces that have to be finished by a machining process.

2. Explain why the corners of castings should be designed with ample radii.
   (a) Describe briefly what is meant by the term 'green-sand' moulding, and why the process is given this name.
   (b) Consult manufacturers' literature and list the content of a typical moulding sand and its recommended applications.

3. (a) Sketch a component that would need to be made using the odd-side moulding technique.
   (b) Sketch a mould that requires *either* the use of loose pieces *or* the use of one or more drawbacks to facilitate the removal of the pattern from the mould.

2.3 Moulding sands

Moulding sands used in green-sand moulds must possess the following properties:

- *Cohesiveness*: the ability to retain its shape after the pattern has been removed and during the pouring and solidification of the metal.
- *Refractoriness*: the ability to withstand the high temperatures associated with molten metals without itself melting or burning.
- *Permeability*: the ability to remain sufficiently porous after ramming so that the steam and other gases that evolved during pouring can escape. The permeability of the mould as a whole can be modified by the degree of ramming, as can the rigidity of the mould. A mould that is rammed too hard will lack permeability and gases will be trapped in the casting causing porosity. Also, a mould that is too rigid and has insufficient 'give' will cause the casting to crack as it shrinks on cooling. On the other hand, a mould that is too loosely rammed will lack rigidity and may collapse as the metal is poured.

Moulding sand usually contain from 5 to 20 per cent of colloidal clay to give it the property of cohesiveness, and it will usually contain up to 8 per cent moisture. Because of the moisture content, moulding sands are referred to as 'green' sands, despite the fact that their colour may range from red through dark brown to black.

2.4 Core sands

Cores are made from moulding sands to which an additional binding material has been added, or from sands that are free from clay and that are bonded entirely by the addition of synthetic binders. Typical binders are:

- *Water-soluble binders* such as the by-products of flour milling called glutin and dextrin. These are added to ordinary moulding sands.

- *Oil binders* such as linseed oil, soya-bean oil and fish oils added to clay-free sands. These oils undergo a chemical change during the baking of the core and set-off hard.
- *Resin binders* such as phenolic resins that are added to clay-free sands and are *cured* when the core is baked. The curing of thermosetting resins was discussed in *Manufacturing Technology*, Volume 1. When thermosetting resins are heated they undergo a chemical change (*curing*) that renders them hard and strong. They can never be softened again by heating. Such resins are also the basis of thermosetting plastics and produce very strong cores.
- *Carbon dioxide gas*: The core sand is mixed with sodium silicate which, after ramming into the core box, is exposed to carbon dioxide gas. This causes a chemical reaction in the sodium silicate that then provides a very strong bond. Less shrinkage occurs than for oven drying and more accurate cores can be produced. Greater productivity is also achieved and this offsets the expense of the materials and equipment used.

2.5 Casting defects

Although a long-established process, and despite its apparent simplicity, it is all too easy for defects to appear in castings made from sand moulds. Let us now look at some common defects and their causes.

2.5.1 *Blow holes*

These are smooth round or oval holes with a shiny surface, usually occurring just under the surface of a casting. Because they are not normally visible until machining is underway, their presence can involve the scrapping of a casting on which costly machining has already been carried out. They are caused by gases and steam being trapped in the mould, which may result from:

- Inadequate venting.
- Insufficient provision of risers.
- Excessive moisture in the sand.
- Excessive ramming reducing the permeability of the sand.
- Inadequate degassing of the molten metal immediately prior to pouring.
- A combination of one or more of these causes.

2.5.2 *Porosity*

This is also due to inadequate venting and inadequate degassing of the molten metal immediately prior to pouring. However, in this instance, the trapped gases do not form large bubbles just below the surface of the casting. Instead, a mass of pinpoint bubbles are spread throughout the casting rendering it porous and 'spongy'. Apart from being a source of weakness, porosity renders castings useless where pressure tightness is essential, as in fluid valve bodies and pipe fittings.

2.5.3 *Scabs*

These are blemishes on the surface of the casting resulting from sand breaking away from the surface of the mould. This may be caused by:

- Lack of cohesiveness due to insufficient clay content in the sand.
- Inadequate ramming adjacent to the pattern.
- Pouring too rapidly.
- An incorrectly proportioned in-gate. This can also result in the scouring away of the walls of the mould.

2.5.4 *Uneven wall thickness*

This can be caused by the moulder not assembling the core correctly or by accidentally displacing the core when reassembling the mould. Alternatively, the core may move in the mould due to ill-fitting or inadequately proportioned core prints. The buoyancy of the core sand can make it try to rise in the molten metal.

2.5.5 *Fins*

These are due to badly fitting mould parts and cores, allowing a thin film of metal to leak past the joints. Although fins can be removed by fettling after casting, this is an added and unnecessary expense and detracts from the appearance of the casting.

2.5.6 *Cold shuts*

These are usually caused by casting intricate components with thin sections from a metal that lacks fluidity and is also at too low a temperature. Consequently, individual streams of the metal may flow too sluggishly and either solidify before the detail in the mould is filled, or fail to merge when the streams of metal meet.

2.5.7 *Drawing*

This results from lack of risers or their incorrect positioning so that any thick sections are not fed adequately with molten metal as they cool and shrink. Since thick sections cool slowly compared with thin sections, they are last to solidify. This allows other parts of the casting to draw molten metal from thick sections rather than from the risers. The result is that the thick sections may have unsightly hollows that may not clean up when machined.

SELF-ASSESSMENT TASK 2.2

1. Describe the essential difference between a moulding sand and a core sand.

2. Discuss the relative advantages and limitations of the alternative binders and binding techniques that can be used when making sand cores.

3. Discuss the possible causes of the following casting defects and suggest possible solutions to overcome these problems: blow holes, porosity, cold shuts, scabs and drawing.

2.6 Shell moulding

The use of phenolic (Bakelite) resins as a core sand binder has already been introduced in Section 2.4. In *shell moulding* the resin-bonded sand is used to make the mould as well as the core. This process has a number of advantages over green-sand moulding for the quantity production of repetitive components. The shells are hard, strong and relatively light so that they can be safely handled and stored over long periods of time before use. They can be produced by hand or by automatic moulding machines more quickly and cheaply than green-sand moulds. The mould has good permeability so that sound castings of high definition and freedom from porosity can be produced. Figure 2.6 shows typical shells and cores ready for assembling together to form a mould.

Fig. 2.6 *Shells ready for assembly*

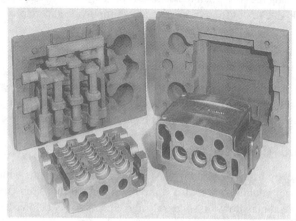

Figure 2.6 also shows a complete casting and a section through a casting made from such a mould. You can see that the traditional sand mould, rammed up in a moulding box, is replaced by a relatively thin, rigid shell of uniform wall thickness. Also, cores produced for the shell-moulding process are hollow and provide much improved venting compared with the solid, rammed cores used when green-sand moulding.

The shell is made in two or more parts to enable it to be stripped from the pattern. These shell parts are then assembled together with their associated cores to make a complete mould. High-strength, synthetic adhesives are used during assembly to bond the individual components of the mould together permanently. Let's now consider how a shell mould is made.

- For batch production a metal pattern having the profile of the required casting is used.
- As for green-sand moulding, the pattern is made oversize to allow for contraction of the metal as it cools.
- The metal pattern is heated in an oven to between 200 and 250 °C.
- The heated pattern is then removed from the oven, sprayed with a release agent and clamped over a trunnion-mounted dump box, as shown in Fig. 2.7(a).
- The dump box contains the sand and resin binder mixture that will eventually form the shell mould.

- The dump box is rotated on its trunnions so that it becomes inverted. This allows the resin–sand mixture to fall over the pre-heated half-pattern, as shown in Fig. 2.7(b).
- After some 30 seconds the resin–sand mixture becomes tacky and forms a soft 'biscuit' about 6 mm thick.
- The dump box is rotated back to its original position so that the surplus resin–sand mixture falls back to the bottom of the box ready for re-use, as shown in Fig. 2.7(c).
- The pattern with the 'biscuit' still adhering to it is transferred to an oven for about 2 minutes to cure the resin binder. Alternatively a silicate binder can be used and this is cured by the use of carbon dioxide gas. Either of these processes converts the 'biscuit' into a rigid shell.
- If the thickness of the shell needs to be increased it is returned to the dump box and a further layer of 'biscuit' is built up on it, after which it is returned to the oven to be cured.
- When the desired shell thickness has been reached, it is stripped from the pattern and placed on one side ready for assembly.

Fig. 2.7 *Making a shell mould*

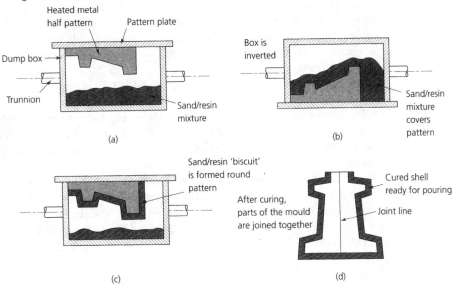

(a)

(b)

(c)

(d)

The various component parts of a shell mould may be assembled together by clamping, bolting or adhesive bonding to form the complete mould. Bonding with synthetic resin is the most usual method as it gives a clean casting free from flash lines. To strengthen large moulds, metal reinforcing rods are introduced into the 'biscuit' when the shell is built up in several layers.

Shell cores are made in split core boxes that are similar in appearance to those used in the making of conventional cores, except that the core boxes for shell cores are made from metal to enable them to be heated to the correct temperature to form the core 'biscuit'. Figure 2.8 shows the stages in making a shell core.

Fig. 2.8 *Making a shell core: (a) core box filled; (b) surplus sand and resin removed; (c) core removed for curing*

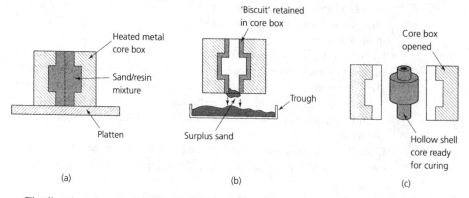

(a) (b) (c)

Finally, the assembled shell moulds are placed in a rack to prevent them from falling over whilst the metal is being poured. The moulds are usually placed over a sand bed for safety in case a mould fails and molten metal is released accidentally. Large moulds are often placed in strong metal boxes and packed with a coarse sand to provide reinforcement whilst not interfering with the venting of the mould. Alternatively, steel shot can be packed round the shell where greater rigidity is required and where some chilling of the mould is desirable for metallurgical reasons. A pouring cup is usually built into the mould, but risers are not generally provided because the mould is sufficiently permeable to allow the escape of any trapped gases.

The metal is poured in rapidly until the mould is full, because the heat of the metal starts to burn away the resin binder. By the time the metal has solidified the resin binder should have been completely burned away, leaving only a deposit of relatively loose sand around the casting. On tapping the residual shell with a metal bar it disintegrates and the mould and the core are easily removed, leaving a clean and accurate casting requiring a minimum of finishing.

For quantity production a *casting wheel* may be used. This is a horizontal carousel holding a number of shells that are filled as they pass under the pouring station. By the time a filled shell reaches the loading station, the metal has solidified and is sufficiently cooled to allow the filled mould to be removed and replaced with an empty shell. The casting wheel (carousel) rotates continuously.

2.7 Advantages of shell moulding

The shell-moulding process is widely used where castings with thin sections and high definition are required – for example, the finned cylinder of an air-cooled engine, as shown in Fig. 2.9. On a larger scale, the water-cooled cylinder block castings used for car and truck engines are also made by the shell-moulding process. Since the sand particles in a shell mould are only bonded at their points of contact, this gives the mould high permeability and allows castings of great complexity and varying wall thickness to be made without venting difficulties and without the necessity for a complex system of risers. The low mass of the shell allows it to heat up rapidly so that the metal is not chilled as it is poured. This results in:

- Rapid metal flow through the runners, allowing complete filling of complex moulds with thin sections and reducing the possibility of cold shuts (see Section 2.5).
- The absence of surface chilling and hardening of the cast metal. This improves the machining characteristics of the casting resulting in improved tool life.
- The metal cooling more slowly, resulting in improved grain structures and strength in thin sections.

Sections may be as thin as 1.5 mm or as thick as 50 mm, but thick sections are not recommended owing to the tendency for the shells to burst under the pressure of the molten metal. The surface finish of the castings is excellent and free from the fins and blemishes associated with the green-sand moulding process.

Fig. 2.9 *Typical shell mould showing the fine detail that can be achieved*

Dimensional tolerances can be as close as ±0.075 mm for small and medium-sized castings. Despite the fact that the shell and the resin–sand mix from which it is made is destroyed every time a casting is poured, this cost is more than offset by the rate of production of the mould and the casting, and the reduced finishing time required. Automatic moulding equipment for making shell moulds is lower in cost than that required for the automatic production of green-sand moulds. Unfortunately, the metal pattern equipment required for shell moulding is much more costly than the wooden or fibre glass patterns used with green-sand moulding, and it is not economical to use shell moulding for small quantity production and 'one-off' castings.

Both ferrous metals and non-ferrous metals can be cast in shell moulds provided a suitable sand–resin mixture is used. Excellent results can be obtained when casting aluminium and magnesium alloys providing an inhibitor is used in the sand–resin mixture to prevent the resin from reacting with the molten metal. Copper alloys can be used, provided the tin and lead contents are not too high, otherwise 'sweating' can occur at the surface of the casting. Grey and malleable irons can be cast in shell moulds with surface finishes approaching that of aluminium die castings if care is taken. Because of the effect of their high melting temperatures on the resin bond, plain carbon and alloy steels are not recommended for castings made by the shell-moulding process.

2.8 Investment casting

The investment casting (lost-wax) process is unique through its ability to produce castings of high integrity and excellent surface finish combined with virtually unlimited freedom of design. This is because the pattern does not have to be removed bodily from the finished mould but is melted out. There are two essential stages in producing the mould.

2.8.1 *Production of the wax pattern*

An expendable wax pattern is first produced in a pattern die. The wax pattern conforms exactly to the shape of the finished casting but its size is increased to allow for both the contraction of the wax pattern and the contraction of the metal casting. Pattern dies are constructed from various materials by a variety of techniques. The technique selected will depend upon:

- The shape and size of the component to be cast.
- The quantity of wax patterns required.

For small quantities of patterns, the pattern dies may be made from low-melting-point alloys and epoxy resins, For large quantities of patterns, multi-impression dies will be made from alloy tool steels for use in automatic injection machines. These latter dies have a life of 50 000 injections before maintenance is required. An average die life expectancy is about 10 000 injections. An investment casting pattern die is shown in Fig. 2.10 together with the wax pattern made from it.

The most commonly used pattern material is a blend of specially prepared pattern waxes, although polystyrene can be used on a limited scale. With few exceptions, it is normal practice to assemble together a number of individual patterns around a *sprue*, together with the necessary runners, risers (feeder heads) and a pouring cup. This is referred to as a *pattern assembly*, all of which is in expendable wax. Investment-casting patterns are produced to a high degree of accuracy and finish that is reflected in the mould cavity and the casting itself. Since the moulds have no natural permeability, provision for venting has to be designed and built into the pattern assembly.

Fig. 2.10 *Wax pattern and pattern die for investment casting*

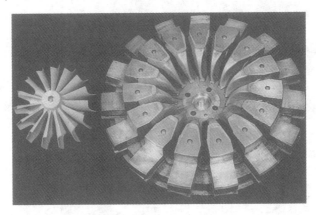

2.8.2 *Production of the mould*

Two basic processes are used for producing an investment mould: the *block-mould* or solid-mould process and the *ceramic shell-mould* process. Both techniques involve the application of a primary *investment* of the pattern assembly. This primary investment is in the form of a suspension of very fine refractory ceramic particles (200-mesh size) in a binder liquid. A *primary-stucco* of fine particles of dry refractory ceramic powder is applied to this wet primary coat or investment. This initial application of slurry and stucco constitutes the primary coat, which must be dried or chemically hardened prior to the application of secondary or 'back-up' layers.

In the *block-mould* process, a block or solid mould is produced around the coated pattern assembly, as shown in Fig. 2.11. An open-ended container or 'flask' is placed over a base plate to which has been attached the primary investment coated pattern assembly. A coarser refractory slurry, or secondary investment, is poured into the container and the base plate is vibrated to consolidate the mould and to facilitate the escape of entrapped air bubbles. The solid mould is allowed to dry slowly over a period of several hours to avoid cracking. After drying, the wax pattern is melted out at a temperature of 150 °C and the mould is fired at a temperature of approximately 1000 °C. This final firing develops the maximum strength of the ceramic bond and also removes any remaining traces of pattern material. The mould is now ready for pouring. The wax which formed the pattern, and which was melted out of the mould, can be re-used to make more patterns.

The *ceramic shell-mould* process has largely superseded the block-mould process except for certain specialist applications. This process involves the application of alternate layers of slurry and stucco to the primary coated pattern assembly, as shown in Fig. 2.12. Each successive coat is dried before the application of the next coat. This is similar to the building up of a resin–sand shell mould. A completed ceramic shell mould will normally consist of six to eight individual back-up coats, giving a final shell thickness of the order of 12 to 20 mm. The removal of the expendable pattern – 'dewaxing' – may be accomplished in a steam autoclave or in a 'flash firing' furnace. Both these 'dewaxing' processes are designed to impart a very high heat input at the pattern/mould interface. This causes immediate melting of the surface layers of the pattern which allows for the expansion of the

Fig. 2.11 *The block-mould process*

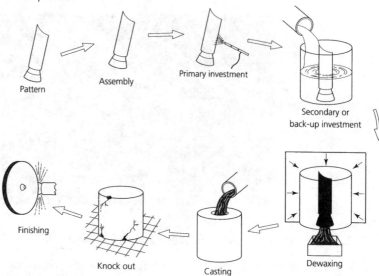

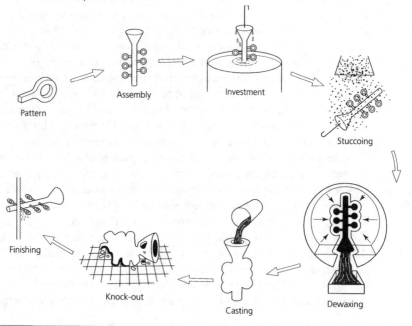

main body of the wax, as it – in turn – heats up and melts. The process is very economical since the wax, which has been melted out of the mould, can always be re-used. Following 'dewaxing', the mould is fired at approximately 1000 °C. The metal is cast into the mould as soon as firing is complete and, preferably, whilst the mould is still hot. In both the block-mould process and the shell-mould process the moulds have to be destroyed to recover the castings, and the broken mould material cannot be re-used.

Fig. 2.12 *The ceramic shell-mould process*

When investment casting ferrous metals, these are usually melted in indirect-arc and induction furnaces. Sophisticated alloys whose strength relies upon high purity are often melted in vacuum conditions, the ceramic mould being clamped or bolted to the furnace whilst the residual air is pumped out. When the metal has melted, the furnace and mould are inverted so that the metal flows directly into the mould. Pumping continues throughout the cycle so that any gases given off during melting and casting are also exhausted. Non-ferrous metals are usually melted in induction furnaces or in crucible furnaces prior to casting and they are then poured into the moulds in the normal way.

2.9 Advantages of the investment-casting process

The investment-casting process possesses the unique ability to transform liquid metal into a cast shape with a high degree of dimensional accuracy, surface finish, and a virtually unlimited freedom of design. No matter how complex the shape of the mould, the pattern is simply melted and poured out of it. These attributes offer to the design and value engineer the opportunity of achieving maximum material utilisation and a truly functional component. The increasingly heavy burden of machining and finishing costs necessitates a constant reappraisal of the established methods of producing both simple and complex metal shapes. An investment-cast component, requiring an absolute minimum of finishing, has rapidly emerged as a more viable proposition than a forged or a machined-from-bar component for an increasingly wide range of general and specialised engineering parts, such as shown in Fig. 2.13.

Fig. 2.13 *Examples of investment casting*

The limits of design possibilities for the process have been extended by the use of preformed ceramic cores, which allow the casting of highly complex and intricate internal shapes, including small-diameter holes of considerable length. The process itself permits the casting of parts accurate to within a few hundredths of a millimetre in a wide range of alloys, including those that cannot be forged and which are virtually unmachinable. The advantages of investment casting may be summarised as follows:

- A high degree of dimensional accuracy and surface finish of the 'as-cast' component, coupled with virtually unlimited freedom of design. A general tolerance of ± 0.013 mm per 25 mm can be readily achieved on small and medium-sized castings.
- The accurate reproduction of fine details, e.g. slots, holes and lettering.
- Maximum utilisation of the raw material and a truly functional component.
- Reduction of costly finishing operations to an absolute minimum. In many cases the component can be used 'as-cast' without further processing.
- Savings in capital expenditure on plant, equipment, labour costs, space utilised, storage, and in the interprocess movement of materials.

The only limitation on the size and weight of an investment-cast component is the physical limitation of being able to handle the pattern assembly and the resultant moulds. The increasing use of mechanical handling within the industry, together with continual improvements in process materials (i.e. pattern waxes, mould binders, and refractories) has steadily extended the upper limits in terms of both volume and mass. Whilst the majority of investment castings in current production fall within a mass range of 0.1 to 5 kg, castings up to 45 kg and more are in regular production. In terms of size, the maximum linear dimension in current production in the UK is of the order of 1.2 metre.

The ability of the process to produce castings with extremely thin-walled sections together with the inherent ability of the process to provide very good dimensional accuracy, finish and reproduction of fine detail is a further major attribute. Generally, the minimum wall thickness is 1.5 mm, but a wall thickness of 0.75 mm for steel castings and 0.5 mm for light alloy castings have been achieved over small areas. The minimum economical batch size depends upon the complexity of the component and, therefore, the cost of the pattern dies. Normal batch sizes range from as few as 500 for a simple component to 50 000 for a more complex component. With automated production lines, production runs of over a million components are quite common for the automobile industry.

2.10 Designing for investment casting

The general principles of good casting design apply as much to investment casting as to other casting processes. The volumetric shrinkage of the metal as it solidifies, the creation of localised 'hot spots' in the mould and the stresses set up in the casting as it contracts and cools in the solid state are inherent problems for any casting process. The adoption of the following design principles will greatly minimise these problems.

2.10.1 *Directional solidification*

The provision of 'feed' metal during solidification will be greatly facilitated by the avoidance of isolated, heavy sections. Changes in section thickness should be gradual. Ideally, thick sections should allow for the progressive feeding of a thinner section.

2.10.2 *Fillets and radii*

The avoidance of abrupt changes in section thickness and the use of a fillet or radius of the correct size will avoid the creation of local 'hot spots' in the mould and will help to reduce contraction stresses set up by unequal rates of cooling of different metal sections. Generous corner radii also help to eliminate planes of weakness in the crystal structure of the metal, and reduce the possibility of fatigue failure in service.

2.10.3 *Angles and corners*

Relatively sharp angles and corners can be satisfactorily investment-cast compared with other casting processes, but re-entrant angles will generally be cast more easily with the incorporation of a root radius.

2.10.4 *Shaped holes, bosses, etc.*

In general, the investment casting of irregularly shaped holes can be accomplished just as easily as casting a round shape. Holes with a high depth/diameter ratio should be designed so that preformed ceramic cores can be used.

2.10.5 *Freedom of design*

The absence of a mould joint or parting line removes the need, on the designer's part, to consider draught (taper) or any other technique to assist pattern withdrawal from the mould. This contributes to the improved dimensional accuracy of investment castings compared with other casting processes and presents the possibility of unrestricted shapes in cast form.

2.10.6 *Metals for investment casting*

Investment castings are now available in almost the complete range of engineering metals and alloys currently available for specialised and for general applications. These include alloys that are virtually impossible to shape by conventional metal-shaping techniques. Vacuum-melted and cast superalloys have become a major product of the industry, including the casting of high-purity titanium alloys.

The author is indebted to the British Investment Casters' Technical Association for their assistance in preparing the foregoing information on the process of investment casting.

SELF-ASSESSMENT TASK 2.4

1. Discuss the circumstances in which the investment-casting process would be chosen in preference to a green-sand or a shell-moulding process.

2. Discuss the cost relationships of green-sand moulding, shell moulding and investment moulding in terms of labour, moulding materials and capital investment.

3. (a) Compared with green-sand moulding and shell-moulding techniques explain how investment casting gives greater freedom to the component designer.
   (b) Discuss how the production of the wax pattern may still impose constraints upon the component designer.

2.11 Die casting (high-pressure)

The processes of gravity and pressure die casting were introduced in *Manufacturing Technology*, Volume 1. Figure 2.14 shows the principle of the hot chamber process for pressure die casting. The process derives its name from the fact that the injection equipment is constantly in contact with the molten metal. Automatic high-pressure die-casting machines can achieve up to 1000 'shots' per hour for small components and up to 200 'shots' per hour for large and complicated castings. When small components are being produced, multi-impression dies are frequently used to increase the rate of production and to enable the machine to operate economically at its full injection capacity. Figure 2.15 shows components as ejected from a multiple-impression die, together with a clipped spray to show how the die is designed to retain the spray in one piece to facilitate removal.

Fig. 2.14 *Hot chamber pressure die casting: (a) dies close; (b) metal is injected; (c) dies open, component is ejected*

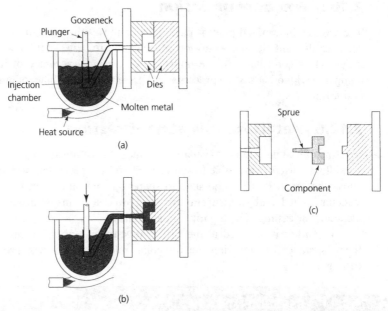

Note how the components are arranged symmetrically so as to balance the forces acting on the dies, and the ample provision for runners feeding the die cavities from the central sprue. Even after taking these precautions, it is apparent that some leakage of metal has occurred between the two halves of the die. Since the metal dies are impervious, vents have to be machined into them along the flash line to allow the air trapped in the cavities to escape as the metal is injected. Because the molten metal is being injected under very high pressure, the machine has to be fully guarded to contain any metal that may escape if a die failure occurs or if the dies fail to close properly. The guards have to be interlocked with the machine controls in such a way that the injection cycle cannot commence until the guards have been properly closed.

Fig. 2.15 *Multi-impression casting: (a) 'spray' of castings ready for clipping (note: peripheral web that improves metal flow during casting and gives the rigidity essential during flash clipping operation); (b) the flash after clipping (this is melted down and used again); (c) the eight components after clipping them from the spray*

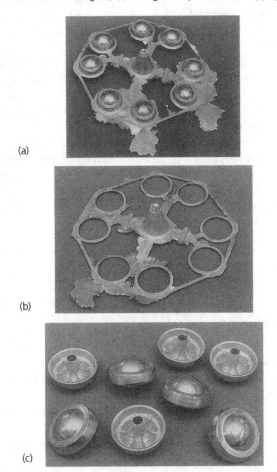

(a)

(b)

(c)

High-pressure die-casting machines and the dies themselves represent a large capital investment and it should be apparent that the process is only economic where large batches of components are involved. The minimum economic batch size varies between about 5000 components for simple parts to 20 000 components for complex parts, in order to ensure that the die costs and the cost of setting the more complex dies in the machine are both recovered. To benefit fully from the high-pressure die-casting process it is essential that the dies are expertly designed and manufactured by specialists in this field. Some of the more important factors are:

- The location and profile of runners and risers so that they do not lock the components into the dies.
- The joint lines of the dies have to be carefully selected so that the component can be easily ejected. The component itself has to be designed with the process in mind as no undercut sections (re-entrant surfaces) can be tolerated.

- Unlike sand moulds, metal dies are not self-venting and vents have to be built into the dies at the time of manufacture.
- The size of the casting is limited by the injection capacity of the machine; therefore, the component design should incorporate thin walls with reinforcing ribs to give the required strength but reduce the volume of metal required, rather than use heavy sections.
- The destructible sand cores used in gravity and low-pressure die casting are not suitable in the pressure die-casting process. Collapsible metal cores can be used but are so costly to make, maintain and use that they are only economical in special cases. Components should be designed to obviate the need for loose pieces and internal and external cores.
- The positive knock-out pins should be sufficient in number and positioned so as to eject the casting or spray of castings from the dies without causing distortion.
- The metal from which the dies are made must be selected to resist the operating temperature and the corrosive effect of the molten metal. Since alloy die-steels are expensive and difficult to machine, die-steel inserts are usually built into a cheaper plain-carbon steel yoke.

The advantages and limitations of the hot-chamber die-casting process can be summarised as follows.

Advantages

- Machining is minimised due to the high accuracy and surface finish of the castings.
- High rates of production can be achieved.
- Material consumption is low as clipped sprays and sprues can be recycled without loss of properties and without contamination.
- Complex shapes, as in decorative motifs and fine detail, can be faithfully reproduced.

Limitations

- Only low-melting-point alloys and metals that are chemically unreactive with the metal components of the injection equipment can be cast in hot-chamber machines. Therefore this process is usually limited to zinc-based alloys such as Mazak. Usually the cold-chamber process is used with the more reactive aluminium alloys. These also have higher melting temperatures. The process gets its name from the fact that the injection equipment is only subjected to the molten metal at the moment of injection. Figure 2.16 shows the principle of the cold-chamber process.
- The casting metal or alloy must have a short freezing range and the rapid solidification keeps the cycle time as short as possible.
- Plant costs, die costs and setting costs are sufficiently high to be prohibitive for small batches.
- The maximum size of the casting is limited by the capacity of the machine. Even the largest machines can only inject a few kilograms of metal for each 'shot'.

Fig. 2.16 *Cold-chamber pressure die casting: the metal is only poured into the chamber at the moment of injection*

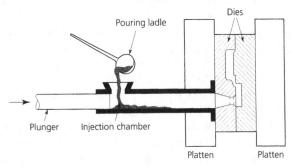

2.12 Die casting (low-pressure)

This process was developed during the Second World War for the production of air-cooled cylinders in aluminium alloy for the aircraft engine industry. Up to that time, castings of this size, and in alloys suitable for engine cylinders, had to be sand cast or gravity die cast. The low-pressure process is now widely used for the broad band of medium- and large-sized, high-quality die castings that need to be produced in large quantities. It should be noted that, with the improved materials now available, the high-pressure die-casting process can also produce high-quality castings but generally of far less substantial dimensions and in a more limited range of metals and alloys than is possible with the low-pressure process.

Figure 2.17 shows the principle of low-pressure die casting. You can see that the molten metal is retained within a crucible and furnace. The die, which is normally situated immediately above the crucible, is connected to the molten metal by means of a vertical riser tube. Since the crucible and the riser tube are made from non-metallic refractory materials they are completely impervious to molten aluminium and its corrosive properties.

Fig. 2.17 *Low-pressure die casting*

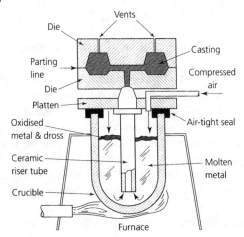

The riser tube is immersed nearly to the bottom of the crucible in order to ensure that all the metal fed into the die is taken from below any dross, metallic oxides and other impurities that always collect on the surface of molten metals. The crucible is sealed against leakage of the pressure air supply by gaskets, as shown. The riser is suspended from the top plate (platten) by a flange and is connected to the gate of the die by the riser cap. After the die is closed, the crucible is pressurised by compressed air. It is this pressure on the surface of the molten metal that causes it to flow against gravity into the die cavity to form the casting. Even when the die is full, the pressure is maintained so that feed metal is provided to avoid shrinkage voids until solidification is complete. The injection air pressure is of the order of 14–700 kPa depending upon the size of the casting. These are substantially lower pressures than those found in high-pressure die-casting machines.

During the initial die filling, the molten metal rises smoothly up the riser and through the die passageways at too low a velocity to cause turbulence within the molten metal. Because the feed is from the bottom of the die, the air within the die cavity can escape naturally upwards through the parting lines and vents. The result of this gradual die filling allows progressive solidification to start at the top extremities of the casting and slowly work its way down to the point of entry where the molten metal is kept under pressure until solidification is complete. After solidification, the pressure in the crucible is released, the dies are opened, and the casting is removed ready for the next cycle to commence. Little fettling of the casting is required.

A variation on this process is to use resin–sand shell moulds and shell cores where the aluminium alloy to be cast would attack a metal die material, or where the casting is too large and too complex to be produced economically in conventional metal dies. For example, motor vehicle engine cylinder blocks in light alloys are frequently produced in pressure-fed shell moulds. The more important advantages of the low-pressure die-casting process can be summarised as follows:

- The dimensional accuracy and surface finish is equivalent to, or better than, gravity die castings.
- Production rates are higher than for any casting process other than hot-chamber pressure die casting.
- The casting design is extremely flexible with the choice of metal dies or shell moulds and the option to use sand cores. Heavier sections can then be used with other die-casting processes.
- There is an uninterrupted metal feed without impurities since the metal is taken from the bottom of the crucible.
- Good mechanical properties result from the dies or moulds being filled under pressure and the pressure being maintained during cooling.
- All the aluminium alloys, including the heat-treatable alloys, can be cast by this process. Even alloys that would normally erode and corrode the metal components of other die-casting equipment, because of their high melting temperatures and chemical affinity, can be low-pressure die cast as all the components in contact with the molten metal can be made from non-reactive refractory materials.
- Operation and control are fully automatic, including ejection and removal of the finished casting.
- Rapid die changing enables short runs to be justified.

- Capital and running costs are relatively low compared with high-pressure die-casting equipment.

Thus it can be seen that this process has all the attributes of good-quality gravity die casting, but with reduced costs for fettling and finishing operations, coupled with the levels of automation usually associated with the high-pressure process. Since the process can be automated, operator fatigue and operator error can be eliminated.

SELF-ASSESSMENT TASK 2.5

1. Compare and contrast the advantages and limitations of the high-pressure and low-pressure die-casting processes.

2. Consult die-casting machine manufacturers' literature and discuss the relative advantages and limitations of the hot- and cold-chamber processes.

2.13 Justification of casting processes

Except for die casting and investment casting, the casting processes do not produce precision components. Castings produced by green-sand moulding and shell moulding not only require fettling, they also need to be shot blasted or 'barreled' to remove loose sand and improve the surface finish. Working surfaces and features with important dimensions need to be finish machined, which adds to the cost.

Nevertheless, casting is often the only process that can be used for manufacturing components of complex shape. In the case of green-sand moulding the cost of patterns is low compared with the manufacture of dies for forging and die casting. Further, casting is the only available process for manufacturing very large components such as the casing for power station turbines and the engines for large ships.

Nowadays, most manufacturers are looking at near finished size (NFS) strategies so as to minimise the need for expensive machining processes. To this end casting results in less waste of material than machining components from the solid. Also, surplus metal trimmed from castings, such as sprues, runners and risers, can be melted down and recycled. In the case of die casting and investment casting, very little finishing is required. The casting industry is constantly seeking ways to improve the accuracy and finish of its products whilst maintaining its cost advantages with other processes of greater precision.

EXERCISES

2.1 Outline, with the aid of diagrams, the principles of the process of shell moulding and compare the advantages and limitations of this process with the green-sand moulding process as described in Volume 1.

2.2 Outline, with the aid of diagrams, the principles of the process of the investment-moulding (lost-wax) process and compare the advantages and limitations of this process with the green-sand moulding process described in Volume 1.

2.3 Compare the advantages and limitations of any two of the following die-casting processes:
(a) gravity casting
(b) low-pressure casting
(c) high-pressure (cold-chamber) casting
(d) high-pressure (hot-chamber) casting

2.4 Describe typical components and materials associated with the following processes, giving the reasons for your choice:
(a) shell-mould casting
(b) investment casting
(c) high-pressure (hot-chamber) casting

2.5 Describe the essential differences between the die-casting process and the sand-casting process in terms of capital outlay and operating costs. Explain why die casting is more economical for the mass production of components in aluminium casting alloys.

2.6 The dies used for pressure die casting represent a considerable investment and their design is a specialist task. Summarise the main design factors and show, with the aid of diagrams, how the profitability of the process can be increased by the use of multiple-impression dies and describe the precautions that must be taken in the design of such dies.

3 Flow-forming processes

The topic areas covered in this chapter are:

- The principles of hot and cold working.
- Hot forging and forging plant.
- Open-die, closed-die and upset forging.
- Cold forging (cold or extrusion).
- Warm forging.
- Thread rolling.
- Presswork, press tools and presses.
- Flexible die pressing and stretch forming.
- Spinning and flow turning.

3.1 Hot and cold working

The forming of metal by plastic flow without loss of volume was introduced in *Manufacturing Technology*, Volume 1. These flow-forming processes were categorised into *hot forming* and *cold forming*. Hot-forming processes are those flow-forming processes carried out above the temperature of recrystallisation, whilst cold-forming processes are carried out below the temperature of recrystallisation. Let's now consider these processes in greater depth.

Metals other than lead and zinc will be cold worked if they are flow formed at normal ambient (room) temperature. During a cold-working process the grain of the metal becomes distorted and internal stresses are introduced into the grains. If the internal stresses are sufficiently great, and if the temperature of the metal is raised over a critical value, new crystals will start to grow at each stress point. The formation of these new seed crystals is called *nucleation*.

The more severe the cold working, the greater will be the distortion of the original grain structure, and the greater will be the internal stresses. The greater the internal stresses, the lower will be the critical temperature at which nucleation commences for any given metal. If the metal is maintained at this temperature, the new crystals will continue to grow until their boundaries collide, a new grain pattern will emerge, and the original, distorted grain structure will no longer exist. Since there are several stress points in each distorted crystal, and since a new crystal grows from each stress point, it follows that the new grain structure will consist of a larger number of smaller grains than existed originally; that is, the grain structure has been *refined*. The process of nucleation and crystal growth is referred to as *recrystallisation.* The principle of nucleation and recrystallisation, as outlined above, is shown in Fig. 3.1.

Fig. 3.1 *Recrystallisation: (a) before working; (b) after cold working; (c) nucleation commences at recrystallisation temperature; (d) crystals commence growth as atoms migrate from the original crystals and attach themselves to the nuclei; (e) after annealing is complete the grain structure is restored*

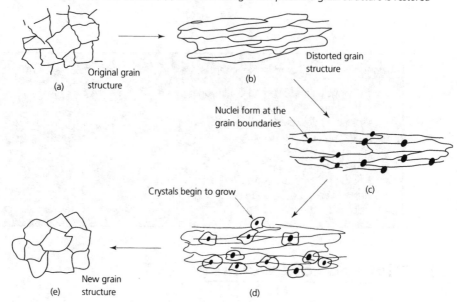

3.1.1 *Cold working*

As previously stated, this occurs when metal and alloys are flow formed below the temperature of recrystallisation. Since cold working results in distortion of the crystal lattice of the grains of the metal, flow becomes increasingly difficult and the metal *work hardens,* becoming harder and stiffer. If flow forming continues, the metal will eventually fracture. Once work hardened, the metal must be heated to its recrystallisation temperature and held at that temperature until recrystallisation is complete, before further flow forming can be undertaken. This heat-treatment process is variously described as subcritical annealing, process annealing, and interstage annealing. These heat-treatment processes are described in detail in *Engineering Materials,* Volume 1.

3.1.2 *Hot working*

As previously stated, this occurs when metals and alloys are flow formed above the temperature of recrystallisation. Since the process temperature is above the temperature of recrystallisation, the grains reform as fast as they are distorted and no work-hardening takes place. If the metal could be maintained at a constant forging temperature, there would be no limit to the amount of manipulation the metal could tolerate. In practice, of course, the metal is constantly cooling down from the moment it is taken from the furnace and flow forming should not continue below the minimum process temperature or the load on the forming equipment will become excessive, leading to structural failure of that equipment. Surface cracking of the work may also occur. On the other hand, the maximum process temperature must not be exceeded or the metal may become 'burnt' – that is, oxidation of the grain boundaries occurs and the metal is seriously weakened, which results in the work having to be scrapped.

3.2 Hot forging

When metals are forged to shape, their grain structure flows to the shape of the component and this greatly increases the strength of the workpiece, as does the grain refinement that also occurs during forging. The orientation of the grain is shown in Fig. 3.2. This compares a gear wheel machined from the bar with a gear wheel machined from a forged blank. Since metals break more easily along the lay of the grain than across the lay of the grain, the teeth of the gear wheel machined from the forged blank will be stronger than the teeth of the gear wheel machined from the solid.

Fig. 3.2 *Grain orientation*

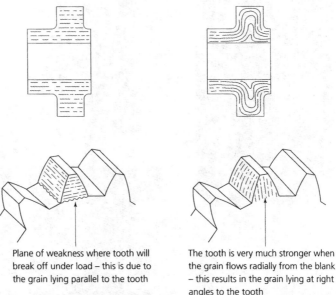

Plane of weakness where tooth will break off under load – this is due to the grain lying parallel to the tooth

The tooth is very much stronger when the grain flows radially from the blank – this results in the grain lying at right angles to the tooth

Forgings are widely used where a component is subjected to high impact loads, high fatigue (vibrating) loads, abrasive wear, high operating temperatures, and where high strength is required.

3.3 Forging plant

The energy required to cause the metal to flow may come from a manually wielded hammer for small components, from various sizes of power hammer for larger components, and from hydraulic presses for the largest components. Forging machines may be classified as:

- *Energy-restricted machines:* pneumatic hammers, steam hammers, drop hammers and screw presses.
- *Stroke-restricted machines:* crank and eccentric mechanical presses.
- *Load-restricted machines:* hydraulic presses.

Conventionally, the capacity of a hammer is specified by the falling mass of the moving parts or their equivalent if the downstroke is power assisted. Figure 3.3(a) shows a drop hammer or drop stamp. These are frequently used with closed-dies for the mass production of small and medium-sized components for the automotive industry but can also be used for open-die work for single components. Pneumatic hammers are made in capacities up to 4 tonnes, as shown for example in Fig. 3.3(b). Steam hammers are larger and are made in

Fig. 3.3 *Hot-forging plant: (a) drop hammer; (b) pneumatic hammer; (c) steam hammer; (d) hydraulic forging press*

(b)

(d)

(a)

(c)

capacities up to 12 tonnes, as shown in Fig. 3.3(c). Hydraulic presses are the largest and can range from 200 to 14 000 tonnes capacity, and an example is shown in Fig. 3.3(d). The size of this machine can be assessed by comparing it to the man in the bottom right-hand corner of the picture.

As a useful 'rule of thumb', the falling mass required to hot forge a low-carbon steel is $3.5 \, kg/cm^2$. Thus a 135 kg hammer can forge stock 63 mm square and a 5.5 tonne hammer can forge stock about 380 mm square. Hammering tends to concentrate the maximum flow towards the surface of the metal, whilst squeezing tends to modify the grain structure throughout the mass of the metal.

In the past, a team of operatives was required to manipulate the metal and drive the hammer or press. However, modern electronic control now allows a single operator to carry out all the functions, previously performed by the team, from an operating console in a comfortable, soundproof control cabin. A *manipulator* holds and positions the work under the hammer, whilst a *charger* loads and unloads the furnace. These functions are often combined when components of less than 20 tonne mass are being forged.

3.4 Open-die forging

These are hot-forming processes, similar to blacksmithing on a large scale, in which the metal is hammered or squeezed to shape whilst it is above the temperature of recrystallisation. Suitable temperature ranges are shown in Fig. 3.4, and the basic principles of the process were introduced in *Manufacturing Technology*, Volume 1. Open-die forging

Fig. 3.4 *Hot-working temperatures*

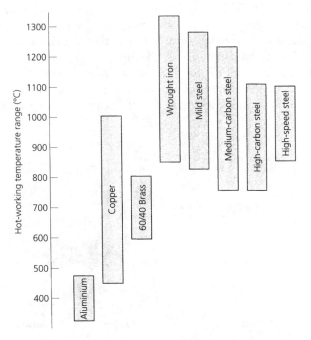

refers to any forging operations performed without the metal being completely constrained; that is, the forging operations are carried out between a hammer and an anvil with only the aid of standard forging tools such as swages and fullers. These tools and their uses were introduced in Volume 1. When applied to power hammers and presses, they are rigidly attached to the hammer (tup) and the anvil. The accuracy of the finished component depends upon the skill of the hammer operator and machining will be necessary to finish the component. Open-die forging requires highly skilled labour and is a relatively slow and expensive process compared with closed-die forging. It is used for making single components, or for small batches of similar components, of any size from the smallest to the biggest. It is a relatively expensive process and is used where the special properties imparted to forged metal needs to be exploited.

3.5 Closed-die forging

This is also a hot-forging process in which the metal is worked above the temperature of recrystallisation. However, in closed-die forging the metal is totally constrained within 'impressioned dies' and, after clipping off the residual flash, the forgings require the minimum of finishing. Closed-die forging can be used for the finishing of components roughed out by open-die forging or it can be used for the complete manufacture of forged components. The latter is the more usual and the process is used for the mass production of parts for the automotive industry.

Figure 3.5 shows a section through a pair of impressioned forging dies. To ensure complete filling of the die cavity, the blank is slightly larger in volume than the finished forging. As the die closes the surplus metal is squeezed out into the *flash gutter* through the *flash-land*. The flash-land offers a constriction to flow of the surplus metal and tends to hold it back in the die cavity to ensure complete filling. It also ensures that the flash is thinnest adjacent to the component being forged, resulting in a neat, thin flash line being left after the flash has been trimmed off. The flash-land should be kept as short as possible otherwise the dies may fail to close. The *rapping faces* ensure that the component is the correct thickness. The hammer driver knows when the dies are fully closed by the sound of the sharp 'rap' when the hardened die surfaces meet and the operation is complete. The dies are

Fig. 3.5 *Closed-forging dies*

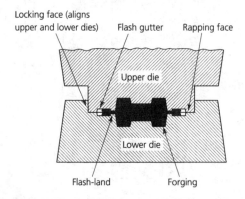

given a 7° taper, or 'draught', on all vertical surfaces so that the forging can easily be removed.

Except for very small and simple components, most forgings have to be produced in a sequence of operations. Figure 3.6 shows a set of forging dies and the component produced at the various stages in the sequence. Instead of using separate forging dies as shown in Fig. 3.6, components are frequently produced in multi-impression dies. The preforming impressions are sunk into the outer edges of the die and the finishing operations are sunk into the middle of the die where the blow is more solid and there is less chance of the dies tilting out of alignment.

Fig. 3.6 *Set of drop-forging dies*

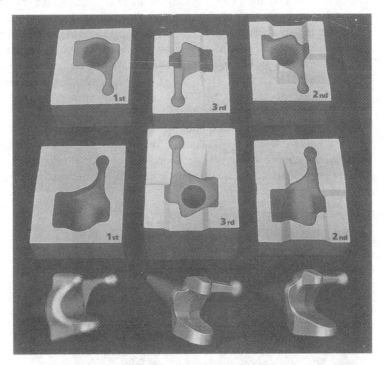

Die life depends upon many factors and may range from only a 1000 components up to 25 000 or more depending upon the size and complexity of the forging and the material from which the forging is being made. The composition of the die-steel and its heat treatment also has a significant effect upon the die life. Low-alloy die steels generally provide toughness and shock resistance but wear fairly quickly, whilst high-alloy die steels have better wear resistance and thermal fatigue resistance. The usual causes of die rejection are:

- *Abrasive wear* This occurs where pressures are highest and metal sliding is greatest, e.g. flash lands and other points where the reduction in vertical cross-section is greatest.
- *Thermal fatigue* Heat or 'craze' cracks on the surface of dies caused by frequent and high ranges of temperature change.

- *Mechanical failure* Catastrophic failure usually occurs as a result of overloading the dies; that is, using an oversize blank. This is particularly the case when precision forging without a flash land and flash gutter.

3.6 Upset forging

Figure 3.7 shows an upset forging machine and a set of dies. Unlike the processes described so far, the *header* moves horizontally and strikes the stock bar 'end on' in order to force the metal into the split dies. Figure 3.8 shows the principle of an upset-forging operation. The bar stock is fed into the machine after being heated to the correct temperature and gripped between the split dies. The dies are closed by a crank or by a cam-operated toggle mechanism. A crank or eccentric-operated header then moves the punch forward against the end of the bar stock and forces it into the cavity formed in the dies (or partly in the dies and partly in the punch). This process is used to manufacture high-tensile bolt blanks, internal combustion engine valve blanks, small cluster gear blanks and similar components. Large components are made individually from blanks that have been heated to their forging temperature. Small components such as nails and rivets are made from wire in the cold condition. The process of upset forging their heads and cutting them to length is entirely automatic. Rivets are annealed after heading, but nails and screw blanks are left in the original drawn condition to improve their stiffness and mechanical properties.

Fig. 3.7 *Upset-forging machine and dies*

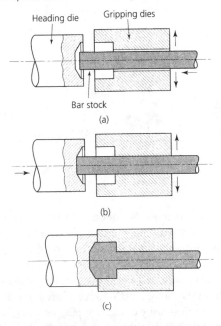

SELF-ASSESSMENT TASK 3.1

1. Compare and contrast the advantages and limitations of hot-forging processes with casting processes in terms of material properties, structure, process accuracy and cost.

2. Discuss the relative merits of steam/pneumatic hammers, drop hammers, and hydraulic presses for forging operations. State suitable applications for each class of machine.

3. (a) Explain why less force is required to flow form metal at its forging temperature than when it is cold.
   (b) Explain why more force is required to flow form aluminium alloy aircraft components than plain-carbon steel components of comparable size.

3.7 Cold forging (impact extrusion)

Figure 3.9 shows the principles of forward and backward extrusion processes as performed on mechanical and hydraulic forging presses. These cold-forging processes were developed for the manufacture of constant velocity universal joints and similar components where close tolerances, high levels of surface finish and maximised mechanical properties are required. In fact, the tracks in constant velocity joints produced by this process require no machining or further finishing after extrusion.

Fig. 3.9 *Cold forging (impact extrusion): (a) forward extrusion; (b) backward extrusion*

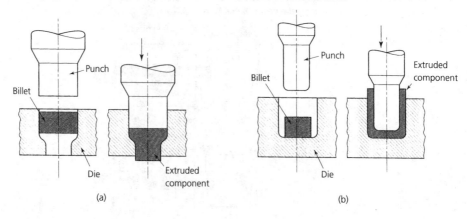

Punch

Billet

Punch

Extruded
component

Billet

Die

Extruded
component

Die

(a)

(b)

The following process description is for the manufacture of a constant velocity joint member from 527A19 carburising (case-hardening) grade steel. The bar stock is supplied in the bright condition with the outer skin removed by a peeling operation to ensure that no surface or subsurface defects are present. Defects that are too small to be detected by commercial ultrasonic testing can still cause component failure during extrusion. The billets are cropped from the bar to a high degree of accuracy. Each blank is automatically weighed as it is cut off and the setting of the cropping machine is automatically adjusted in order to keep the mass of the billets within very close tolerances. A 2.5 kg billet has to be kept within -0 g and $+20$ g. The billets are subcritically annealed at about 710 °C in a controlled atmosphere furnace. They are then barrelled to remove any sharp corners, pickled to ensure a chemically clean surface, treated with a phosphate bonder and lubricated with a metallic stearate soap. The phosphating process ensures that the lubricant adheres to the surface of the billet and also provides additional, extreme-pressure lubrication in its own right. The billets are then extruded in mechanical or hydraulic forging presses of 750–1000 tonne capacity depending upon the operation. The operation sequence is shown in Fig. 3.10.

Fig. 3.10 *Sections through cold forgings at each stage of the production of a constant-velocity universal joint member*

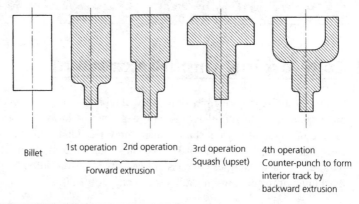

Billet

1st operation 2nd operation

Forward extrusion

3rd operation
Squash (upset)

4th operation
Counter-punch to form
interior track by
backward extrusion

Interstage annealing is carried out at subcritical temperatures (700–720 °C) between the squashing and backward extrusion operations. The extrusion is then normalised at 920 °C and again pickled, barrelled and lubricated with sodium strearate (PBL treatment) before the final cold calibration pressing operation which imparts an accuracy of $\leqslant 0.06$ mm across the tracks. No subsequent machining of the tracks is required.

The batch size is dependent on the die life of the most severe operation in the sequence. This is normally the backward extrusion operation and die life is limited to about 5000 components. At the other extreme, the die life of the calibration tools is about 500 000 components. Cold extrusion tools are extremely expensive and every care has to be taken to maximise their life. For example, a tool steel punch for the backward extrusion operation will often start to 'pick up' after between 1500 and 2000 components, whilst the same punch, after *nitriding*, can have a life of 30 000–35 000 components before recovery becomes necessary.

3.8 Warm forging

Warm forging is now taking over from cold extrusion since the forces required to form the metal are greatly reduced and the life of the tools is correspondingly increased. Both these factors decrease the operating costs of the process. As the name implies, the process is carried out below normal hot-forging temperatures but above the temperature of recrystallisation for the metal. The following sequence of operations is required to produce a constant velocity universal joint member similar to that described in the previous section.

The billets are again cropped to length and each billet is subjected to automatic weighing to ensure that any variation in mass does not exceed -0 g and $+20$ g on a billet of mass 2.5 kg. As previously described, the weighing station automatically adjusts the cropping machine to compensate for any variation in mass. After cropping, the billets are barrelled and degreased but, this time, they are not subjected to a phosphate treatment or a stearate soap lubricant treatment.

The billets are then given their first induction heating and are immediately immersed in a suspension of graphite (carbon) particles in water from a temperature of 120 °C. The suspension flash dries and leaves a protective film of graphite on the surface of the metal. This graphite film reduces oxidation of the metal to a minimum and provides adequate lubrication for the subsequent processing. The billets are then raised to the process temperature of 900–960 °C by further induction heating. Temperature control is 100 per cent and is performed automatically by infrared sensors. Any billets that are not at the correct temperature are automatically rejected and reheated. Correctly heated billets are fed automatically to a 1600 tonne, four-station transfer press. The sequence of operations is the same as in Fig. 3.10 but, this time, operations 1 and 3, and operations 2 and 4, are performed at the same time to balance the load on the press. The warm forged (extruded) components exit the press automatically into cooling trays. When cool, the components are returned to the PBS plant for the full treatment of pickling to remove any residual oxide film, barrelled, phosphate treated and lubricated with a sodium stearate soap. The soap reacts with the zinc phosphate coating on the component to form a film of zinc stearate at the surface of the component. This acts as an extreme pressure lubricant.

The treated components are then cold pressed to give them their final calibration. This is the same operation as that used during the cold-extrusion process and the finished accuracy is equally as good so that no machining of the tracks is required. Again, the highest wear occurs during the backward extrusion process. However, when warm forging, the die life is approximately 25 000 components compared with 5000 for cold extrusion. Output is approximately 15 finished components per minute. The combination of full automation and extended die life makes warm forging a much more economical process than cold extrusion. Components produced by this process have excellent mechanical properties and the microstructure and machineability can be controlled to suit the customer's specification direct from the press, thus avoiding the heat-treatment processes necessary after cold extrusion. Warm forgings are more accurate, have a much superior surface finish and have superior mechanical properties compared with components produced by conventional hot forging.

SELF-ASSESSMENT TASK 3.2

1. With the aid of sketches, describe the difference between forward and backward extrusion.

2. Describe how the steel blanks have to be prepared for *cold* forging in order to minimise die wear.

3. Describe how the steel blanks have to be prepared for *warm* forging in order to minimise die wear.

4. Discuss the economics of cold and warm forging compared with hot forging.

3.9 Thread rolling

Thread rolling is a flow-forming process carried out below the temperature of recrystallisation – that is, it is a cold-working process. The thread is flow formed to shape and no cutting occurs. This results in a thread in which the grain orientation follows the thread form and this imparts greater strength to the thread than is possible with a machine-cut thread.

The workpiece blank is rolled between a pair of flat dies or roller dies upon which is formed the impression of the thread to be produced. This process is usually used for screw threads under 25 mm diameter. The pressure between the dies and the blank causes plastic deformation of the metal to take place at the surface of the blank. This results in depressions to form the thread spaces and coining up of the metal to form the thread flanks and crests. Before rolling, the blank will be approximately the mean diameter of the thread. The rolling action results in displacement of the metal above and below this surface. In practice the blank diameter will vary slightly from the mean, depending upon the properties of the metal being formed. Although the dies are expensive, they have a long life and accurate threads can be produced in large quantities.

3.9.1 *Flat dies*

The principle of thread rolling using flat dies is shown in Fig. 3.11. In this example only one die is moved, but in some machines there is a counter-movement of both dies. Compensation of die and machine wear is achieved by the use of adjustable wedges. The blanks are usually fed automatically from a hopper and the die length is arranged so that the blank rotates about four times during one pass. The dies are made from high-quality alloy tool steel and have a life of up to 500 000 components. Rolling speeds are of the order of 30–75 m/s, depending upon the properties of the material being rolled. Use of an extreme pressure lubricant is essential to avoid early die wear.

Fig. 3.11 *Thread rolling (flat dies)*

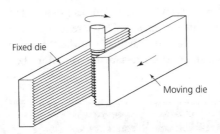

3.9.2 *Roller dies*

These are used for rolling small-diameter threads on capstan and automatic lathes. They could also be used on CNC turning centres. Since the work is supported by three roller dies located 120° apart, slender work can be threaded without difficulty. The highly polished rollers have helical grooves and are, in effect, screw threads. They are self-feeding and no lead screw is required. The length of the screw that is rolled is not limited by the face width of the roller dies. In this, they are superior to flat dies which cannot roll threads that are longer than the width of the dies. Figure 3.12 shows the general arrangement of a thread-rolling die head and its similarity with a thread-cutting die head is apparent.

Fig. 3.12 *Thread rolling (circular dies)*

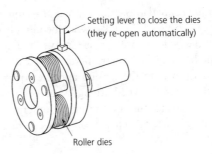

In operation, the die head is set and this closes the roller dies to give the correct thread size. When the thread is complete the die head opens automatically, and the roller dies retract clear of the work. This enables the die head to be withdrawn over the finished screw without having to stop or reverse the machine. As for flat die rolling, a copious flow of extreme pressure lubricant must be maintained to prevent early die wear. The thread rollers themselves are mounted on needle roller bearings to reduce friction to a minimum.

3.9.3 *Tapping*

Internal screw threads can also be flow formed in suitably malleable materials using lobed taps. The lobed tap can be screwed into a predrilled hole in the same way as a metal-cutting tap. However, greater axial force is required to start the tap. For flow tapping, the predrilled hole is the mean diameter of the thread and not the core diameter as in thread cutting. Again an extreme pressure lubricant must be used.

SELF-ASSESSMENT TASK 3.3

With the aid of manufacturers' literature, compare the economic and technical advantages and limitations of thread-rolling compared with thread-cutting processes. State how preparation of the workpiece differs in terms of size for thread rolling and thread cutting.

3.10 Press bending

Vee-bending, as shown in Fig. 3.13, is the simplest of the bending operations and was introduced in *Manufacturing Technology*, Volume 1. When the metal is bent, the metal on the outside of the bend is stretched (in tension) and the metal on the inside of the bend is shortened (compressed). Somewhere between these two extremes there is a layer of metal which is neither in tension nor compression – it has not changed in length. This layer lies on the *neutral axis*. It is important to understand the effects of bending on the material being processed. Initially, when bending is only slight, only elastic deformation occurs and if the bending pressure is removed the material will spring back to its original shape. As the bending pressure is increased, the stress produced in the outermost layers of the material (on both the tension and the compression sides) eventually exceeds the yield strength. Once the yield strength has been exceeded plastic deformation occurs. This permanent strain only occurs in the outermost regions of the material which are furthest from the neutral axis. Nearer to the neutral axis only elastic strain will be present.

When the bending force is removed, those regions of the material that have been subjected to plastic deformation will try to retain their permanent set. However, those regions of the material adjacent to the neutral axis which have been subjected only to elastic deformation will try to recover. To some extent this elastic recovery will overcome the permanent set and a degree of *spring-back* will occur. To allow for this the tools have to be designed with *over-bend* – that is, if a right-angled bend is required the tools have to be made to bend the metal to a more acute angle than 90°.

Fig. 3.13 *Vee bending*

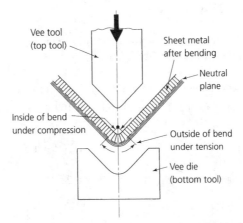

When materials are bent, allowance has to be made for the change of length that occurs on either side of the neutral axis. The length of the neutral axis does not change during bending. Because there is a slight difference between the amount of compressive strain and the amount of tensile strain when a metal blank is bent, the neutral axis does not lie on the centre of the material but is slightly offset. However, the difference between the position of the neutral axis and the centre line of the material is negligible for thin sheet metal blanks and the centre-line bend allowance is used to simplify the calculation. Example 3.1 shows how centre-line bend allowance can be calculated.

Fig. 3.14 *Centre-line bend allowance*

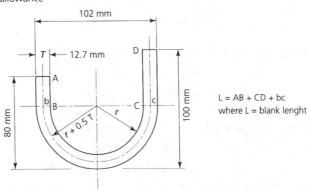

EXAMPLE 3.1

Calculate the length of the blank required to form the 'U'-bracket shown in Fig. 3.14. Assume that the neutral axis $= 0.5T$, where T, the thickness of the blank, is 12.7 mm.

The length of the blank required is equal to the sum of the flats AB and CD and the length of the mean line arc bc. Thus:

$$L = AB + CD + bc$$

Now bc represents a semicircular arc whose mean radius R is equal to the *inside radius r* plus half the thickness T of the blank.

The outside diameter of the semicircle $= 102$ mm   (see Fig. 3.14)

$$\begin{aligned}
\text{The inside diameter of the semicircle } &= 102 - 2T \quad (\text{where } T = 12.7 \text{ mm})\\
&= 102 - 25.4\\
&= \mathbf{76.6\,mm}
\end{aligned}$$

$$\begin{aligned}
\text{From which the inside radius } r &= 76.6 \div 2\\
&= \mathbf{38.3\,mm}
\end{aligned}$$

$$\begin{aligned}
\text{Thus the mean radius } R &= 38.3 + (0.5 \times 12.7)\\
&= 38.3 + 6.35\\
&= \mathbf{44.65\,mm}
\end{aligned}$$

$$\begin{aligned}
\text{Bend allowance bc} &= \pi R \quad (\text{for a semicircle})\\
&= 3.142 \times 44.65\\
&= \mathbf{140.3\,mm}
\end{aligned}$$

$$\begin{aligned}
\text{Length of flat AB} &= 80 - (102 \div 2)\\
&= 80 - 51\\
&= \mathbf{29\,mm}
\end{aligned}$$

$$\begin{aligned}
\text{Length of flat CD} &= 100 - (102 \div 2)\\
&= 100 - 51\\
&= \mathbf{49\,mm}
\end{aligned}$$

$$\begin{aligned}
\text{Total length of blank } L &= AB + CD + bc\\
&= 29 + 49 + 140.3\\
&= \mathbf{218.3\,mm} \quad (\text{before bending})
\end{aligned}$$

3.11 Deep drawing

The process of deep drawing is used to manufacture cup-shaped components from sheet metal. By way of introduction to the process, consider the component shown in Fig. 3.15(a). This shallow cup-shaped component, which is drawn from relatively thick sheet metal, can be produced in tools of the type shown in Fig. 3.15(b). A circular blank is forced through the die by the punch and the metal flows to shape. The punch and die clearances are arranged so that the metal is lightly 'pinched' between them. This 'irons out' any creases which tend to form during the drawing operation. This simple operation is called *cupping* and is limited to shallow components of the type shown.

Fig. 3.15 *Cupping: (a) cup-shaped component; (b) cupping tool*

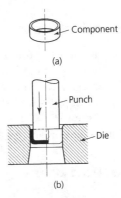

Component

(a)

Punch

Die

(b)

Let's now look at what is likely to happen when an attempt is made to cup a component having a greater depth to diameter ratio, particularly if the metal is thinner. Puckering occurs round the edge of the blank, as shown in Fig. 3.16(a). This may be sufficient to prevent the metal flowing smoothly through the die and the punch may then tear the bottom out of the component. Even if the tensile strength of the metal allowed it to be forced through the die, it would be impossible to 'iron out' all the pucker marks. The solution to this problem is to prevent the metal from puckering in the first place. This can be achieved by adding a *blank-holder* to the tools, as shown in Fig. 3.16(b).

Fig. 3.16 *Deep drawing: (a) edge wrinkling (puckering); (b) use of blank holder*

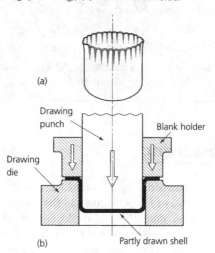

(a)

Drawing
punch

Blank holder

Drawing
die

(b)   Partly drawn shell

3.12 Deep-drawing tools

Deep-drawing tools vary widely in design, not only because of the range of components produced but also because of the variety of methods adopted for applying the necessary force to the blank holder. Let us now consider some of the alternatives.

3.12.1 *Double-action press*

In addition to the centre ram which is driven by the crank and carries the punch, there is also a cam- or toggle-operated outer ram that carries the blank holder. This outer ram is designed to descend first and grip the blank with a constant pressure whilst the centre ram (and punch) descends and completes the drawing operation as previously described. To control the blank-holder pressure accurately, the outer ram is counter-balanced by pneumatic cylinders. An example of this type of press is shown in Fig. 3.17.

3.12.2 *Single-action press*

It is also possible to deep draw components in single-action power presses providing they have sufficient length of stroke. Some presses have a variable throw crank so that the strokelength can be varied for different operations. An example of a single-action press is shown in Fig. 3.18. There are various techniques for applying blank-holder pressure when deep drawing in a single-action press and Fig. 3.19 shows two commonly used methods. The blank-holder pressure in Fig. 3.19(a) is provided by compression springs, whilst in Fig. 3.19(b) it is provided by a rubber buffer. The tool shown in Fig. 3.19(b) is often referred to as an *inverted tool* because the die is attached to the ram of the press and the punch is fixed to the bottom bolster. In the inverted type of tool the punch is often referred to as the *pommel*.

Fig. 3.17   *Double-action power press (reproduced courtesy of Taylor & Challen Ltd)*

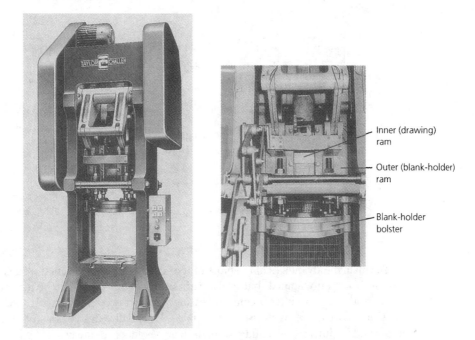

Inner (drawing) ram

Outer (blank-holder) ram

Blank-holder bolster

Fig. 3.18   *Single-action power press (reproduced courtesy of Sweeney & Blackridge Ltd)*

Fig. 3.19 *Drawing tools for a single-action press: (a) drawing tool with spring-loaded blank holder; (b) inverted drawing tool*

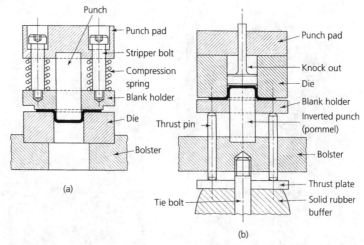

(a)

(b)

The main disadvantages of the types of tool shown in Fig. 3.19 is that the blank-holder pressure is not only limited, but it also varies throughout the stroke of the press as the springs or the rubber buffer become increasingly compressed. An alternative technique for use with inverted tools is to use a *die cushion*. Essentially, this consists of replacing the rubber buffer with a cylinder and piston, the blank-holder pressure being applied by feeding compressed air to the cylinder. The use of a die cushion has the following advantages:

- The blank-holder pressure can be adjusted for optimum performance by adjusting the air pressure being fed to the piston and cylinder.
- Providing the air pressure remains constant, the blank-holder pressure is constant throughout the stroke of the press.

Fig. 3.20 *Pneumatic die cushion*

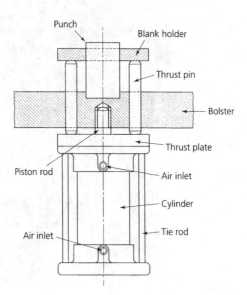

Figure 3.20 shows how a simple die cushion is used. The piston is fixed to the bolster by the suspension rod, and the force applied to the blank holder is transmitted by push rods engaging the top of the cylinder. In this instance the piston is stationary and the cylinder moves. Other designs use a more conventional fixed cylinder and moving piston. The choice of die cushion depends upon the type of press and the operation being performed.

3.13 Metal flow when drawing

Figure 3.21(a) shows a segment from a circular blank which is to be drawn through a circular die. The radius of the blank is R_1 and the radius of the die is R_2. During the passage of the blank across the die face, the arc length L_1 is reduced to L_2 and the shaded area of the metal has to be 'lost'. Since the volume of metal does not change between the blank and the component, this loss of surface area is compensated for by local thickening of the metal in the finished component, as shown in Fig. 3.21(b). This compression of the blank material sets up a hoop stress in the material outside the die, causing it to buckle and pucker. To prevent this happening, an opposing tensile force must be applied to the blank rim. This tensile control force is provided by the blank holder as previously described.

The very action of control immediately sets up a further condition that exerts an influence on the drawing operation. This is the tendency of the metal under the blank-holder to thicken uniformly around its annular rim and progressively between the die mouth and the outside edge of the blank, as shown in Fig. 3.21(b). In addition to the thickening of the metal and the compressive hoop stress at the rim, there is also a state of tension in the walls of the component as the punch forces the blank through the die. These three factors together decide the diameter ratio between the blank and the finished component in any one drawing operation. Figure 3.21(c) shows the stresses in a partly drawn shell.

Fig. 3.21 *Metal flow when drawing: (a) need for blank-holder control; (b) increased thickness due to plastic strain; (c) stresses in a drawn shell*

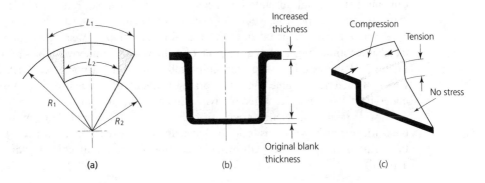

3.14 Material characteristics for deep drawing

Materials for deep drawing require the following properties:

- *High ductility*, to allow extensive plastic flow to take place.
- *Low work-hardening capacity*, to prevent the component cracking during drawing and to reduce the number of times the component has to be re-annealed (see Section 3.17).
- *High tensile strength*, to withstand the drawing stresses which tend to tear the walls of the shell as the metal is drawn from under the blank holder.
- *Uniform grain size and orientation*, to allow uniform flow to take place and to reduce *earing* to a minimum (see Fig. 3.22).

Fig. 3.22 *'Earing' due to grain orientation*

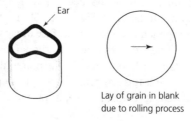

Ear

Lay of grain in blank
due to rolling process

Stress–strain analysis as applied to drawing processes is still an inexact science and the ability to predict the forces involved is still largely empirical. Although the characteristics of a material can be determined from its tensile test results, a more specialised test for determining the suitability of a material for deep drawing is the *Erichson Cupping Test*. This test simulates a drawing operation on a sample piece of material, as shown in Fig. 3.23(a). The circular sample is gripped between the pressure ring (blank holder) and the die face. After the pressure ring has been tightened down on the specimen it is slackened back by a constant amount (0.05 mm) to give the metal freedom to draw but not to pucker. The spherically ended punch is then pressed into the specimen, drawing it into the die to form a cup. The forward movement of the punch is continued until a crack appears at the base of the cup. The depth of the cup at fracture is read off the scale of the machine and is a measure of the drawing quality of the material. Figure 3.23(b) shows Erichson's standard curves giving the depth of cup for various thicknesses of sheet and various materials. Cupping tests are not absolute but merely compare a sample from an untried batch of material with a sample from a batch of material known to be satisfactory.

Account must also be taken of the critical relationship between the grain size of the metal after annealing and the amount of cold work it underwent before annealing. Figure 3.24 shows this relationship graphically for a low-carbon steel sheet. It can be seen that the critical value for cold working is about 10 per cent deformation for this material. If it is annealed after receiving this amount of cold working, a very coarse grain will result. This coarse grain not only weakens the component but also impairs the surface finish, which is said to have an 'orange peel' appearance. When heat treatment is to follow deep drawing, this critical amount of cold working must be avoided.

Fig. 3.23 *Erichson cupping test: (a) principle cupping test – the punch forces the spherical indenter into the blank, stretching it into the die – the more ductile the material, the greater will be the depth of the cup 'D' before the crack appears; (b) Erichson test curves*

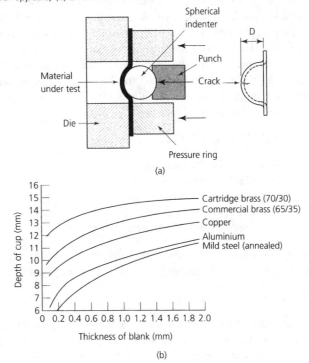

(a)

(b)

Fig. 3.24 *Effect of cold working on grain size*

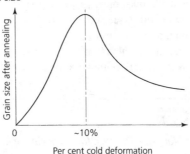

3.15 Blank-holder force

The blank-holder force is critical not only for the avoidance of wrinkles and puckering, but also for the extent to which it affects the elongation of the surface regions of the drawn component. As the force exerted by the blank-holder ring increases, so does its braking effect on the material being drawn between the blank holder and the die face. If the braking effect is insufficient, wrinkling occurs, but if the braking effect is too great, the bottom will be torn out of the drawn component by the punch.

When using double-action presses that have mechanically operated blank holders, the setting of the blank-holder force is usually determined empirically. This is done by drawing the shell progressively deeper using blanks of the same diameter, whilst progressively lowering the blank holder until the surface of the drawn shell is smooth and free from wrinkles. The pressure of the blank holder on the blank increase during the drawing operation as the edge thickness of the blank increases under the effect of the hoop compression forces set up (see Section 3.13). The blank holder force may be calculated using the following empirical formula:

$$P_B = \pi/4 \left[D^2 - (d_1 + 2r)^2 \right].p$$

where P_B = blank-holder force (N)
D = blank diameter (mm)
d_1 = diameter of die orifice (mm)
r = drawing edge radius (mm)
p = specific blank-holder pressure (MPa)

Since the drawing edge radius and metal thickness are small in relation to the blank diameter and the punch diameter (d), an approximate formula may be used:

$$P_B \approx \pi/4 \, (D^2 - d^2).p$$

The magnitude of the specific blank-holder pressure (p) varies with the material used, the punch diameter and the blank material thickness. The smallest blank-holder pressure that should just prevent wrinkling is given by the empirical formula:

$$p \approx 0.0025 \left[(\beta_0 - 1)^2 + 0.005 \, (d/s) \right].\sigma_B$$

where p = specific blank-holder pressure (MPa)
β_0 = drawing ratio (see Section 2.18)
d = punch diameter (mm)
s = material thickness (mm)
σ_B = tensile strength of material being drawn (MPa)

EXAMPLE 3.2

Calculate the blank-holder force for a drawing operation, given the following data:

Tensile strength of material (σ_B) = 370 MPa
Blank thickness   = 0.8 mm
Punch diameter   = 240 mm
Drawing ratio   = 1.75 : 1
Blank diameter   = 420 mm

$p \approx 0.0025 \left[(\beta_0 - 1)^2 + 0.005 \, (d/s) \right].\sigma_B$
$\approx 0.0025 \left[(1.75 - 1)^2 + 0.005 \, (240/0.8) \right] \times 370$

$$\approx 0.0025 \times 2.0625 \times 370$$

$$\approx \mathbf{1.9\,MPa}$$

$$P_B \approx \pi/4\,(D^2 - d^2).p$$
$$\approx \pi/4\,(420^2 - 240^2) \times 1.9$$
$$\approx \pi/4 \times 118\,800 \times 1.9$$
$$\approx \mathbf{178\,kN}$$

Sometimes it is not possible for the press to exert sufficient force on the blank-holder to prevent wrinkling, or there may be problems with lubrication of the blank itself. In these circumstances the tool may be designed with an entry bead to the die, as shown in Fig. 3.25.

Fig. 3.25 *Blank restraint: (a) entry bead; (b) braking beads*

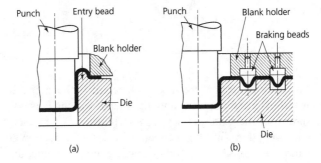

3.16 Die-edge and punch radius

The die-edge radius and the punch-corner radius are important factors in the design of the drawing tools. If the die-edge radius is too small, the blank is elongated and stressed too severely as it is drawn into the die. This results in rupture of the component wall and/or excessive thinning of the component. Alternatively, if the radius is too large, wrinkling can occur as the metal flow is uncontrolled for too great a distance between leaving the blank holder and becoming pinched between the punch and the die, particularly at the end of the draw. The die radius should normally lie between 5 and 10 times the blank thickness. When deep-drawn shells have comparatively thin walls (large d/s ratio) the higher value should be used.

The corner radius of the punch is also important and should be given careful consideration at the component design stage, especially when the drawing ratio is large. A generous punch radius helps to maintain the thickness of the component wall and, hence, maintain the wall strength. The punch-corner radius should never be less than the die-edge radius and, wherever possible, it should be between 0.1 and 0.3 times the punch diameter.

3.17 Multi-stage drawing operations

Where the depth to diameter ratio of the drawn cup is too great to be produced in one operation, it is necessary to produce the component in a series of draws. Figure 3.26(a) shows a first draw and a redraw tool. It can be seen that the punch in the first drawing operation is shaped so as to leave a taper of approximately 15° in the bottom of the cup. The reason for this is shown in Fig. 3.26(b). Not only does the taper provide location in the redraw tool but it allows the metal to flow smoothly past the blank holder (pressure sleeve) that provides the necessary braking to prevent wrinkling.

Fig. 3.26 *Multi-stage drawing tools: (a) first draw; (b) redraw*

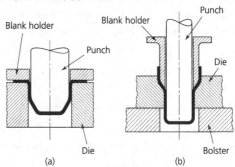

Alternatively, reverse drawing can be used, as shown in Fig. 3.27. In this instance the cup is turned inside out. The amount of cold working is greatly increased without undue pressure from the blank holder. This technique is often used with single-action presses when drawing thin-walled components. Further, the additional cold working of a reverse draw tool reduces the likelihood of coarse grain developing during subsequent annealing. The cold working will have exceeded the critical value (see Section 3.14, Fig. 3.24).

Fig. 3.27 *Reverse drawing tool*

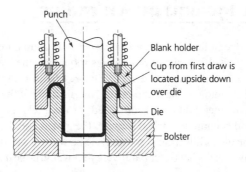

3.18 Drawing ratio

Where multi-stage drawing is required, the percentage reduction at each stage must be calculated carefully, as the allowable reduction becomes less at each stage. Figure 3.28

Fig. 3.28 *Stages in drawing and redrawing*

Stage	Per cent reduction	D/d ratio	Component (dimensions in mm)
Blank	–	–	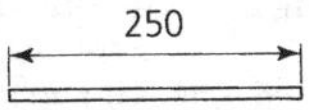 250
First draw	34	$\beta_0 = 1.51$	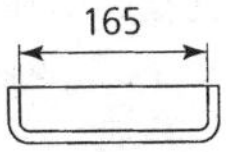 165
First redraw	24	$\beta_1 = 1.32$	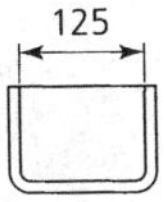 125
Second redraw	20	$\beta_2 = 1.25$	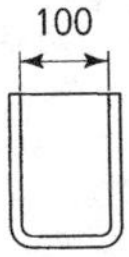 100
Third redraw	17.5	$\beta_3 = 1.21$	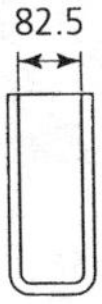 82.5

shows details of the stages in drawing and redrawing a cup from a given blank. The changes in percentage reduction and depth to diameter ratio (drawing ratio) at the successive stages can be clearly seen.

The drawing ratio (β_0) was introduced in the formulae used in Section 3.15:

$$(\beta_0) = D/d$$

where D = blank diameter (mm)
    d = punch diameter (mm)

This expression refers only to the first drawing operation and, for subsequent drawing operations β_1, β_2, etc., the ratios of the cup diameters before and after drawing are taken; that is:

$$\beta_1 = d/d_1 \quad \text{and} \quad \beta_2 = d_1/d_2, \text{etc.}$$

The largest ratio that can be drawn in a single operation is limited by the following factors:

- The tensile strength of the material being drawn.
- The tool dimensions and design.
- The blank-holder pressure.
- The material thickness.

- The friction affecting the flow of the material being drawn between the faces of the blank holder and the die (also referred to as the 'braking effect'). This will depend upon the lubrication of the blank, the surface finish of the blank material, the hardness and surface finish of the blank holder and die faces.

Figure 3.29 shows the largest drawing ratio for cylindrical forms as a function of the punch diameter and the material thickness. This figure may be used for different sheet materials drawn in steel tools of good design, using normal lubrication and appropriate blank-holder pressure.

Fig. 3.29 *Deep-drawing ratios for cylindrical shells: D = blank diameter; β_0 = maximum initial drawing ratio; d = punch diameter; A = good drawing quality material; s = sheet thickness; B = average drawing quality material*

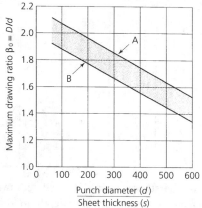

The drawing quality of the blank material also plays a significant part in determining the drawing ratio. As a rough guide, materials with good drawing characteristics follow the upper line (A), whilst general-purpose materials with poorer drawing characteristics follow the lower line (B). Materials with a rough surface finish need a reduced drawing ratio, particularly in the case of thin-walled components that have a high d/s ratio. The use of a good drawing lubricant can increase the drawing ratio by up to 15 per cent. These specialist lubricants not only allow an increased drawing ratio but reduce die wear and operating costs. The phosphate coating of low-carbon steel sheets, as used in the drawing of car body panels, has produced outstanding results, not only providing a key for conventional lubricants but also acting as an extreme pressure lubricant in its own right. Heavily sulphurised extreme pressure lubricants are also used in deep drawing, however, they tend to attack and discolour copper and copper-based alloys. More recently, use has been made of plastic spray coatings containing polytetrafluoroethylene (PTFE) which has the lowest known coefficient of friction. This coating can be stripped off after processing by the use of suitable chemical solvents.

3.19 Drawing force

It is not possible to determine a precise drawing force since this varies as the drawing operation proceeds. Since the drawing force increases proportionally to the drawing ratio, the force having to be exerted by the punch is greatest at the start of the draw when the blank diameter is at its maximum. However, as the blank is drawn into the die, the diameter of the blank remaining under the blank holder progressively decreases and the drawing ratio also decreases. Both these effects result in the force on the punch decreasing as the operation proceeds. Unfortunately, this is the reverse of what happens in a mechanical press where the force on the punch increases as the crank approaches its bottom dead centre.

Empirical formulae have been developed to estimate the maximum drawing force for a given set of circumstances (Example 3.3). One which is relatively simple yet gives a reasonable result is:

$$F_{max} \approx n.\pi.d.s.\sigma_u$$

where F_{max} = punch force (N)
d = punch diameter (mm)
s = material thickness (mm)
σ_u = UTS of blank material (MPa)
n = a coefficient based upon the drawing ratio and lying between unity for general purpose materials and 1.2 for very good drawing materials. Since this is only an approximate estimation of the punch force required, the coefficient n can be ignored for all practical purposes.

EXAMPLE 3.3

A cylindrical component is to be drawn to a diameter of 150 mm from 1 mm thick material. Calculate the approximate drawing force, given that the ultimate tensile strength of the material is 370 MPa.

$F_{max} \approx n.\pi.d.s.\sigma_u$ where $\pi = 3.142$; $d = 150$ mm; $s = 1.0$ mm; $\sigma_u = 370$ MPa
$\approx 3.142 \times 150 \times 1.0 \times 370$
$\approx$ **174 kN**

3.20 Blank size

The determination of the blank size and shape for drawing operations is largely a matter of trial and error based upon experience, particularly for non-circular and asymmetrical components. The drawing tools are made first and trial blanks are cut out by hand. These are gradually adjusted until the desired component shape, free from 'earing', is arrived at. A copy of this final blank is used as a template for the manufacture of the blanking tool. Figure 3.30 shows a rectangular, drawn component with corner radii. The 'earing' that occurs with the blank shape shown in Fig. 3.30(b) is due to crowding of the metal at the corners where it will be in compression. To obtain a reasonably flat top to the component and to reduce the amount of 'ironing' required between the punch and the die, the blank shape should be modified to that shown in Fig. 3.30(c). Remember that the need for

Fig. 3.30 *Blank profile for rectangular box: (a) drawn rectangular box showing earing; (b) incorrect blank profile causing crowding at the corners and earing; (c) correct blank profile with corner relief*

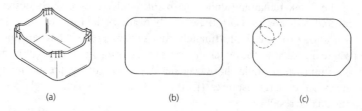

(a)   (b)   (c)

excessive 'ironing' of the component results in premature and unnecessary die wear. Where even slight 'earing' is unacceptable, the drawn shell may have to be clipped or machined.

The approximate blank size for cylindrical components can be calculated by equating the surface areas of the circular blank and the finished, cup-shaped component (Example 3.4). Assuming that no change in metal thickness occurs and that the corner radius of the bottom of the cup is minimal, the following expression may be used.

$$\text{surface area of blank} = \pi.D^2/4 \qquad\qquad (3.1)$$

$$\text{surface area of cup} \quad \approx (\pi.d^2/4) + (\pi.d.h) \qquad\qquad (3.2)$$

Equating (3.1) and (3.2) $\pi.D^2/4 \approx (\pi.d^2/4) + (\pi.d.h)$

therefore $D \approx \sqrt{(d^2 + 4d.h)}$

where D = blank diameter
        d = punch diameter
        h = height of cup

EXAMPLE 3.4

Calculate the approximate blank diameter for a cup-shaped component 250 mm diameter and 200 mm high.

$D \approx \sqrt{(d^2 + 4d.h)}$ where $d = 250$ mm; $h = 200$ mm; D = blank diameter
      $\approx \sqrt{(250^2 + 4 \times 250 \times 200)}$
      $\approx \sqrt{(62\,500 + 200\,000)}$
      $\approx \sqrt{262\,500}$
      $\approx \mathbf{512\,mm}$

Press tool design is largely a matter of experience and the calculations are largely for guidance and verification. No matter how carefully the tools are designed and set, faults frequently show up on 'try-out' whilst the tools are being 'proved' and later, during service, as wear sets in.

1. With the aid of sketches, show how 'puckering' can be prevented during deep drawing:

   (a) When using a double-action power press.

   (b) When using a single-action power press.

2. Calculate the approximate blank-holder force for a drawing operation, given the following data:

   Tensile strength = 320 MPa
   Blank thickness = 1.5 mm
   Punch diameter = 150 mm
   Drawing ratio   = 2 : 1
   Blank diameter = 350 mm

3. Calculate the approximate force required to deep draw a cylindrical component to a diameter of 100 mm from 0.8 mm thick material given that the ultimate tensile strength for the material is 350 MPa.

4. Calculate the approximate blank diameter for a cup-shaped component 100 mm diameter and 50 mm high.

3.21 Flexible die pressing

Figure 3.31(a) shows the principle of forming sheet metal pressings using a rigid punch and a flexible, rubber block die. This technique can be used when:

- The production run is small (under 1000 components).
- Only shallow drawing or forming is required.
- The material is soft and thin (e.g. aluminium panels for aircraft skins).
- Tooling costs must be kept to a minimum – the punch may be cast from reinforced epoxy resin.

When confined as shown in Fig. 3.31(a), the rubber acts like a fluid and transmits the pressure uniformly in all directions. Alternatively, a rubber diaphragm may be used as shown in Fig. 3.31(b). The pressing forces are transmitted to the diaphragm by fluid pressure. It is essential that the fluid pressure is kept constant throughout the operation and that the deformation of the diaphragm is restricted to prevent excessive wear leading to bursting. With both these techniques it is essential to avoid sharp corners and sudden changes of section.

Fig. 3.31 *Flexible-die pressing: (a) rubber block forming; (b) rubber diaphragm forming*

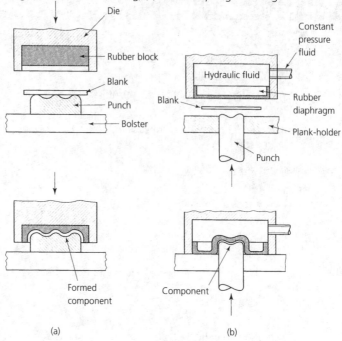

(a)　　　　　　　　　　　　　(b)

3.22 Stretch forming

The principle of this technique is shown in Fig. 3.32. The sheet is placed in a state of tension and wrapped around a former. Normally, when a sheet is bent, one side is in tension and the other in compression; the side in compression will start to buckle and pucker. In stretch forming, prestressing keeps both sides of the sheet in tension and prevents puckering from occurring. This process was developed so that large, thin sheets of metal could be formed economically in small quantities.

Fig. 3.32 *Stretch forming: (a) blank is pretensioned; (b) blank is stretched over former while still tensioned*

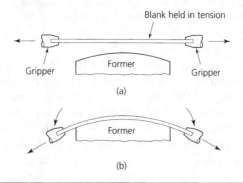

3.23 Spinning and flow turning

Metal spinning is used to produce symmetrical, round, hollow components from circular blanks with a surface area similar to that of the finished component. Unlike deep drawing where very expensive tooling is required, spinning only requires the blank to be spun over a simple former. By the very nature of the process, only vessels that are surfaces of revolution can be formed by spinning. This is done manually by a skilled operator using specially shaped stick tools. An example of a basic set up is shown in Fig. 3.33. Since the metal flow depends upon the pressure exerted by the operator, the process is limited to thin metal that is soft and ductile. It is also possible to internally spin metal from preformed cups on special internal spinning formers. Considerable material savings can be made using these methods as a skilled operator is capable of thinning or thickening up the material as and when required. The burnishing action of the spinning tools results in an exceptionally good surface finish that reduces the need for such finishing operations as polishing. The working of the metal also improves its mechanical properties and the high quality of the surface finish improves the fatigue life of the component.

Fig. 3.33 *The metal-spinning process: the tee-rest and holder may be adjusted to a variety of positions as work progresses*

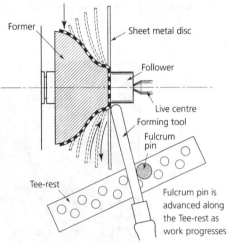

Because of the high level of skill required by the operator, the unit labour cost of production is high compared with pressing. However, owing to the very low tool costs and negligible setting costs involved in the spinning process compared with pressing, the overall unit cost of production becomes competitive for the production of single components and for small quantity batches. Spinning is also useful for producing components where the material strength and/or the blank thickness is inadequate to resist the forces encountered when deep drawing.

The principle of flow turning is shown in Fig. 3.34. You can see that the stick tool, used for manual spinning, is replaced by a roller and that some extrusion of the metal takes place resulting in wall thinning. The products manufactured by flow turning are as diverse as

Fig. 3.34 *Shear spinning*

hollow explosive charge units for under-water seismatic exploration, rocket motor tubes, parabolic nose cones, pressure cylinders, brake cylinders, funnels, champagne buckets and wine coolers, thick-based thin-walled hollow ware, accurate thin-gauged seamless tube, and cylinder liners. Traditionally, flow-turned components were limited to cylindrical and conical shapes but with the advent of computer-controlled flow-turning lathes more complex shapes can be produced. The principle of *shear spinning* is shown in Fig. 3.35.

Fig. 3.35 *Flow turning*

Both flow turning and shear spinning allows a greater degree of deformation of the material to be achieved than is possible by any other process, because deformation of the material only occurs at the point of contact between the rollers and the blank and the remaining material remains stress free. Therefore, in many instances, components can be produced in one operation that would require a number of operations with interstage annealing when produced by pressing. Owing to the severe working of the material the crystal structure is elongated, increasing the strength and hardness of the material in the finished component. Thus the possibility of manufacturing a given component from a thinner blank can be considered with a corresponding reduction in the cost of material required and a reduction in the mass of the component. Components up to 685 mm diameter and up to 600 mm in length can be spun. The benefits of flow turning are: low tooling costs and short lead times, close dimensional accuracy, considerable savings in materials, and the effect of the process on the microstructure of the metal results in improved strength and hardness and a surface finish that requires little if any further attention.

SELF-ASSESSMENT TASK 3.6

Compare and contrast the economic and technical advantages and limitations of flexible die forming, stretch forming and spin forming.

3.24 Advantages, limitations and justification of flow-forming processes

3.24.1 *Advantages*

- Economical use of materials. There is no loss of volume, except for conventional hot forging where the flash has to be clipped from the forged component.
- Improved mechanical properties owing to controlled grain refinement and orientation.
- Cold-forging and warm-forging operations produce components with a high geometrical and dimensional accuracy and a good surface finish. Such components require little further finishing or machining for completion.
- Many flow-forming processes lend themselves to full automation where large quantities of components are required.

3.24.2 *Limitations*

- Compared with casting, the stock material is in a highly processed condition and, therefore, more costly.
- Flow-forming processes are capital intensive as they have to be performed on large and expensive machines.
- Flow-forming processes generally require complex and expensive tooling which can only be justified for large batch or continuous production.
- The complexity of the tooling can result in lengthy production lead times.
- Down time for changing and setting can be relatively lengthy, thus frequent tool changing is uneconomical and flow forming is generally restricted to medium and large-quantity batch production. The exceptions are open-die forging using standard smithying tools and spinning. However, these latter processes require the use of highly skilled and expensive operators. Note that the use of preset and modularised tooling substantially reduces the down time for tool changing.

3.24.3 *Justification*

When comparing flow-forming processes with such alternatives as casting or machining from the solid, a number of factors have to be considered. Flow-formed components have better mechanical properties than castings and components machined from the bar. Compared with machining, there is much less waste of material; for example, when turning a bolt from a hexagonal bar, roughly half the material is wasted as swarf. Compared with casting the finish and accuracy of flow-formed components is superior with the exception of components produced by die casting and investment casting. In addition, die casting can only be used with relatively low-strength metals and alloys. Therefore, on an overall cost and performance basis, flow forming can be justified as follows, despite the high cost of plant and tooling:

- Low loss of volume – economical use of stock material.
- Improved mechanical properties. For some components, such as wing spars for aircraft, precision forging is the only viable process which can provide components with the necessary strength/weight ratio and service reliability.
- High accuracy and good finish reduces or eliminates machining and finishing operations, particularly when precision forging, so that the overall component cost is competitive with, say, a lower cost sand casting which has to be fettled by grit or shot blasting and extensively machined.
- High rates of production – especially in automated plants – leads to low unit production costs providing that the costs can be defrayed over an economically large number of components.

EXERCISES

3.1 Compare and contrast the advantages and limitations of *hot working* and *cold working* in terms of finish, material properties, accuracy, grain orientation and process costs.

3.2 With the aid of sketches, describe the process of *open-die forging* and *closed-die forging* and compare and contrast the advantages and limitations of these two processes.

3.3 Compare and contrast the advantages and limitations of *cold forging* (cold extrusion) and conventional *hot forging* in terms of finish, accuracy material properties, grain structure, and process costs.

3.4 With the aid of sketches, describe the process of *warm forging* and explain the advantages of this process compared with *hot forging* and *cold forging*.

3.5 With the aid of sketches, describe the process of thread rolling and discuss its advantages compared with conventional thread cutting.

3.6 (a) Draw a section through a 'cut and cup' deep-drawing press tool to produce a cup with the internal dimensions of 50 mm diameter and 20 mm deep from brass strip 1.00 mm thick.

(b) Calculate the approximate blank diameter.

3.7 Compare and contrast the advantages and limitations of flexible-die forming with rigid-die forming for sheet metal components.

3.8 Describe, with the aid of sketches, the process of stretch forming sheet metal components and compare the advantages and limitations of this process with flexible-die forming.

3.9 Compare and contrast the advantages of metal spinning with deep drawing in terms of finish, accuracy, process costs and the material properties of the finished component.

3.10 With the aid of sketches, explain the essential differences between shear spinning and flow turning and give examples of components where each process would be appropriate, with reasons for your choice.

4 Sintered particulate processes

The topic areas covered in this chapter are:

- Basic principles of sintered particulate processes.
- Powder production, selection and blending.
- Compaction of preforms.
- Sintering and sintering furnaces.
- Post-sintering treatments.
- Components for sintered processes.
- Die design.

4.1 Basic principles

As the name implies, sintered particulate processes are those processes by which components are formed by the compaction and sintering of particles of metals and metallic compounds. The study of these processes and the materials so formed are also referred to as powder metallurgy. Basically, the manufacture of components by means of sintered particulate processes involves a sequence of three operations.

1. The production of a controlled blend of metals, non-metals and metallic compounds, in powder form, which will provide the required properties when formed and sintered into the finished product.
2. The consolidation, or compacting, of the powder mixture in a die to form the unsintered or 'green' shape of the component. This will have the same shape as the finished component, but will allow for between +1 per cent growth and −1 per cent shrinkage by volume for most processes.
3. The furnace sintering of the 'green' compact at high temperatures in a reducing atmosphere for up to one hour. Sintering is a heat treatment process carried out at between 60 and 90 per cent of the melting point of the majority constituent powders. It causes the bonding together of the individual powder particles at their points of contact to provide a homogeneous solid. This usually results in a slight reduction in volume accompanied by a corresponding consolidation and increase in density. However, under some circumstances, there may be slight growth and corresponding

decrease in the density of the completed compact. Sintering is discussed more fully in Section 4.5.

Note that where close control of size and mechanical properties are required, some post-sintering operations may have to be performed. These are discussed in Sections 4.7 and 4.8. Figure 4.1 shows the basic production processes for making sintered components.

Fig. 4.1 *Basic production processes (reproduced courtesy of GKN Bound Brook Ltd)*

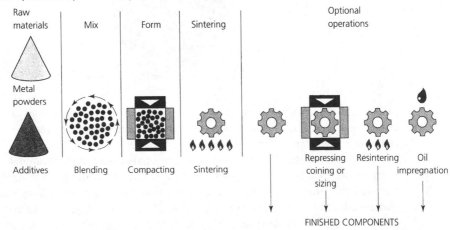

From the foregoing outline, it might well seem that the process of manufacturing components by the sintering of metal powders must be complex and costly and only applicable to highly specialised components. Nothing could be further from the truth.

Sintered components range from gears, pulleys and sprockets for washing machine to highly stressed aircraft components: from small spacing washers and oil-impregnated bearing bushes to aircraft undercarriage components. The process is also used for producing hard 'tips' for cutting tools. Since the process can be readily and highly automated both for the actual manufacture and for in-process quality control, the common factor between all the examples quoted is batch size. Providing the batch size warrants the cost of making and setting the compaction tools, the process can compete on a cost basis with casting, forging, pressing and machining. This is because the accuracy and finish of the sintered component often removes the necessity for further processing. As a rough guide a minimum batch size for a small and simple component would be 5000 components per batch with a minimum of four batches per year.

The component size is limited by the forming pressure and the maximum available capacity of the compaction press, where the forming pressure is dependent upon the projected area of the workpiece and the characteristics of the powder being compacted. For example, a 150 mm diameter sprocket wheel made from an iron–carbon powder mixture would require a compacting press capable of exerting a pressure of $925\,\text{MN/m}^2$, compared with a press capable of exerting $1850\,\text{MN/m}^2$, that would be required if the wheel was made from an alloy steel powder. In the UK, the largest diameter component normally produced is approximately 250 mm diameter with a mass not exceeding 2.5 kg. By introducing 'cut-outs'

(holes) into the component, as shown in Fig. 4.2, the projected area of the component can be reduced. This, in turn, reduces the compaction pressure required for a given component diameter.

It has already been mentioned that one of the main cost advantages of making components by a sintered particulate process, lies not in the process itself but in the fact that the finish and accuracy is so good that subsequent machining can be largely eliminated as the following example shows. Figure 4.3 shows a conveyor chain-link that was originally made from a forging. After the forging had been clipped to remove the flash and sand blasted to remove the scale, it had to be milled on the faces A and B and then cross-drilled. The time came when the milling machine and tooling required replacement, but the value of the component was insufficient to warrant such an investment.

Fig. 4.3   *Conveyor chain-link*

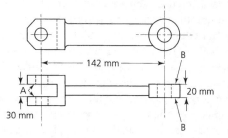

Fortunately, the links were – and still are – required in sufficient quantities to warrant manufacture by a sintered powder process. After slight modifications to the component to aid compaction, the links are now made to an accuracy and finish which has eliminated the need for milling the faces A and B. The only post-sintering machining required is the cross-drilling of the holes. Although the cost of the sintered links is higher than the cost of the original forgings, the elimination of the milling and surface finishing processes has resulted in the overall unit cost being lower. Further, the capital outlay on new milling equipment was avoided.

1. Describe the essential difference between sintered particulate processes and casting and forging processes.

2. Compare the advantages and limitations of sintered particulate processes with casting and forging processes.

4.2 Powder production

Sintered particulate processes can be applied to most metallic elements and many metallic compounds. The most notable exception being aluminium. This metal has an affinity for steel and, therefore, tends to cold-weld itself to the die walls during the compaction process (see Section 4.5.1). The basic properties of the individual powder particles depend not only upon their composition but also upon the process by which they were manufactured. These basic properties affect their behaviour during consolidation. The following processes can be used to produce the metal powders:

- The chemical reduction of metal oxides by carbon or by hydrogen gas results in a 'spongy' type of powder with good compaction and sintering properties. For example, sponge iron powder is widely used for the manufacture of ferrous components. The purity depends upon the quality of the metal oxides and the reducing agent used.
- Metal powders can also be produced by atomisation of liquid metal using water or an inert gas as the quenching (cooling) agent. Atomisation is akin to the metal spraying technique used for building up worn components. However, the atomised metal is sprayed into the quenching agent instead of onto a worn component. Atomisation can be used with most metals and is the most widely used commercial process for producing alloy powders.
- Metallic compound particles are usually produced in atritor mills where these materials are ground together.

4.3 Selection and blending of powders

High compressibility is a desirable feature in all powders, since it allows more efficient use of the pressing (compacting) equipment with corresponding reductions in pressing forces and die wear. A die lubricant such as a metal stearate is usually added to improve the particle flow in the die and reduce die wear. Unfortunately, the lubricant has to be removed before sintering to prevent contamination and this additional operation adds a slight cost to the process. Under uni-directional compaction forces – which exist in press and die compaction – the addition of a lubricant improves the uniformity of the compaction throughout the mass of the component; reduces friction between the particles and the die wall and eases ejection of the compacted component.

4.3.1 *Single powders*

Refractory metals such as molybdenum, tantalum and tungsten are very expensive and difficult to melt and cast by conventional means due to their high melting points. It is substantially cheaper to produce components from these materials by sintering powder compacts, provided there is sufficient demand to warrant the relatively high tooling costs for any given product. In these instances, single powders of a given metal are used and no blending is required.

4.3.2 *Metal/metal-compound two phase systems*

A typical example of such a system is the cobalt/tungsten carbide mixture widely used for cutting tool tips. When these powders are sintered, hard brittle particles of tungsten carbide form, dispersed within a softer but tougher matrix of metallic cobalt as shown in Fig. 4.4

Fig. 4.4 *Metal/metal-compound system*

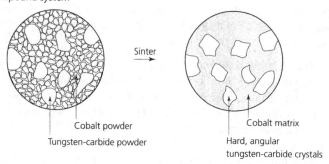

Sinter

Cobalt powder
Tungsten-carbide powder

Cobalt matrix
Hard, angular
tungsten-carbide crystals

4.3.3 *Alloy mixtures*

Sintered particulate processes permit alloys to be achieved between metals that are either completely or partially immiscible in the liquid and solid phases. If such metals are melted and cast together they will not give a satisfactory structure but will separate out into individual layers. However, by mixing together particles of the alloying elements, and sintering them, the required alloy will be formed by dispersion as shown in Fig. 4.5.

Fig. 4.5 *Alloy system formed by dispersion*

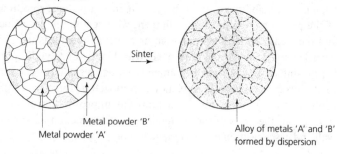

Sinter

Metal powder 'B'
Metal powder 'A'

Alloy of metals 'A' and 'B'
formed by dispersion

4.3.4 *Blending*

Manufacturers of sintered metal products buy in the powder in drums ready for blending and mixing. Weighed quantities of the powders are loaded into double cone mixing machines together with a lubricant and any other necessary additives. Mixing must be sufficiently thorough to ensure homogeneity but, at the same time, mixing must not be excessive or the particle size and compacting characteristics of the powder will be adversely affected, as will be the mechanical properties of the component produced from the powder.

When the blend has been thoroughly mixed it is tested for two important criteria. The apparent density (AD) and the flow rate (FR). The processing and properties of a sintered component are dependent upon the characteristics of the powder, particle size, size distribution, and particle structure, shape and surface condition. The apparent density of the powder (mass of a given volume) strongly influences the strength of the compact prior to sintering. A powder with irregular particles and a porous texture will have a low apparent density and this will give rise to a large-volume reduction during pressing. This, in turn, gives rise to a greater degree of cold welding during compaction and greater green strength. It also leads to more efficient sintering.

However, irregular particles with a low apparent density also have a low flow rate and will not readily fill the die cavity and this may lead to bridging-density variations and slower rates of production. The greater reduction in volume associated with irregular particle shapes usually requires greater compaction pressures resulting in greater tool loads and the need for more powerful presses.

The ease and efficiency of packing the powder in the dies depends largely upon the range of particle sizes in the powder, with the voids between the large particles being filled with progressively finer particles. Whilst an excess of 'fines' increases the density and improves the mechanical properties after sintering, too many fines reduces the flow rate and increases the compaction pressure necessary for a given size of component. Excessive mixing tends to break up the particles and increases the percentage of fines present in the blend.

The purity of the particles in a powder is also critical, particularly where the mechanical properties of the sintered component have to be closely controlled. Most powder grains have a thin oxide film on their surfaces, but these films are ruptured during compaction to leave clean and chemically active metal surfaces that are easily cold welded.

SELF-ASSESSMENT TASK 4.2

1. With the aid of manufacturers' literature list the more important powders available, their method of production and typical applications.

2. Discuss the need for and the precautions that must be taken during the selection and blending of powders prior to compaction.

3. Explain what is meant by the terms 'apparent density' (AD) and 'flow rate' (FR) and their effect on the processing and properties of a sintered component.

4.4 Compaction of the preform

Compaction of the particles into the required preform shapes can be performed by:

- Closed-die pressing.
- Cold isostatic pressing (CIP).
- Hot isostatic pressing (HIP).
- Roll compaction.
- Extrusion.
- Slip casting.
- Injection moulding.

The majority of preforms are compacted in closed dies using mechanical or hydraulic presses. Of the other processes listed above, only isostatic pressing is used to a significant extent commercially. Closed-die pressing and isostatic pressing techniques will now be considered in greater detail.

4.4.1 *Closed-die pressing*

The principle of compaction by closed-die pressing is shown in Fig. 4.6. A controlled amount of powder is automatically fed into the die cavity, as shown in Fig. 4.6(a). Counter-punches close on the powder and compact it into the required shape, as shown in Fig. 4.6(b). Finally the lower punch ejects the preform from the die, as shown in Fig. 4.6(c). As the punches close on the powder, compaction proceeds through a number of overlapping stages:

1. Slippage of the powder particles with little deformation.
2. Elastic compression of the particle contact points.
3. Plastic deformation of the particle contact points to form contact areas of greater size.
4. Bulk elastic and plastic deformation of the entire mass of the preform.

During die compaction, high shearing stresses are generated on those powder particles adjacent to the punch faces and the die walls, which results in more efficient compaction at the surfaces of the preform than within its body. Therefore counter-punches are used, as previously mentioned, to give a better density distribution to the preform. Figure 4.7 shows a comparison of the densification mechanism of a single punch compared with a counter-punch system. Under good conditions, up to 90 per cent of the theoretical maximum density can be achieved in closed-die compaction. However, the die-wall friction restricts the length to width ratio of a closed-die preform to 2.5 : 1 even when a lubricant is used.

It has already been stated that there are problems with the compaction of aluminium and aluminium alloy powders because of their chemical affinity for steel. If aluminium or aluminium alloys have to be used in a powder particulate process, then the compaction dies have to be manufactured from materials, such as tungsten carbide, for which the aluminium does not have an affinity. Such tools are much more expensive than tools made from alloy die steels and the batch size has to be sufficiently large to warrant the high tooling costs.

Fig. 4.6 *Closed-die pressing: (a) load powder; (b) powder compaction; (c) component ejection*

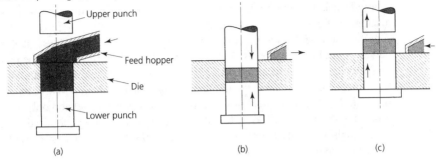

(a)   (b)   (c)

Fig. 4.7 *Densification: (a) single-punch system; (b) counter-punch system*

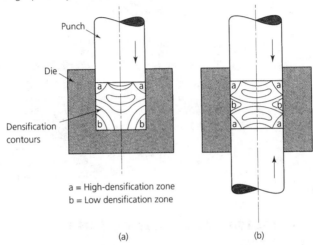

a = High-densification zone
b = Low densification zone

(a)   (b)

4.4.2 *Cold isostatic pressing (CIP)*

This technique was developed for the pressing of fine, non-ductile particles and is now being used widely for compacts whose shape precludes the use of rigid dies. Figure 4.8 shows the principle of cold isostatic pressing (CIP), where pressure is applied to flexible tooling at ambient temperatures. The main advantage of CIP is that more complex component design geometry can be used and components of increased size can be produced. Density levels are higher and more uniformly distributed than for rigid dies. Unfortunately, tolerances are less tightly held with CIP and depend upon such variables as:

- Uniformity and particle size of the powder fill in the tooling;
- Type of powder.
- Compaction ratio.
- Rigidity of the tooling.

Close tolerances and good surface finishes cannot be produced because a flexible membrane is used to form the external surfaces of the compact. However, the internal surfaces are formed by a rigid mandrel so that they can be held to tight tolerances, as can the ends of the compacts under certain conditions. Usually, with CIP, external tolerances cannot be held to better than 3 per cent of the linear dimensions, and closer control requires machining after compaction or sintering.

With 'wet-bag' tooling, as shown in Fig. 4.8(a), flexible rubber, neoprene, PVC, or polyurethane bags may be used. After filling, the tooling (bags) are sealed and placed in a pressure chamber. Several bags may be pressed in a single pressure chamber at the same time using a high-pressure, incompressible fluid such as water. With 'dry-bag' tooling, as shown in Fig. 4.8(b), the bag or tool is fixed in a pressure vessel and only one component is produced at a time. However, tool design is simpler, loading and ejection are easier and, with a high-durability membrane, high-volume production runs are possible with outputs up to 1500 components per hour.

Fig. 4.8 *Cold isostatic pressing: (a) wet-bag tooling; (b) dry-bag tooling*

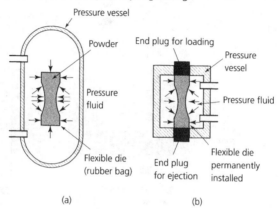

(a)   (b)

4.4.3 *Hot isostatic pressing (HIP)*

Hot isostatic pressing combines the processes of pressing and sintering. Although the associated equipment involves high capital expenditure, HIP has many advantages in terms of improved metallurgical control of structure and quality. Full theoretical densities can be achieved and, since the compacting pressure is applied at lower temperatures than those associated with conventional sintering, the components are free from massive secondary phases. Densification by HIP not only eliminates porosity, which improves the mechanical properties of the finished component, but it can also reduce scatter in the material properties resulting in closer quality control. Closer control of grain size in the micro-structure of the finished compact allows subsequent processing operations such as rolling, hot-forging, and heat treatment to be successfully carried out. In these cases the particulate process is used to create alloys between immiscible elements, and the compact is only a blank for further processing.

Hot isostatic pressing is successfully applied to a wide range of high-alloy systems where the high temperatures that would be required for conventional sintering would not only be uneconomical but would result in degradation of the grain structure of the finished compact. HIP is used to process tool steels, nickel-based super-alloys, 'hard metals' and some complex alloys based upon titanium and beryllium that cannot be processed economically by other, more conventional, techniques.

4.4.4 *Injection moulding*

This process is more widely used in the USA and Europe than in the UK. The metal powder blend is added to an organic binder to form a material suitable for injection moulding. The powder–polymer mixture is then injected into split dies, preheated to remove the binder and, finally, sintered. Volumetric shrinkage during sintering is very high, some 20 per cent, but this is consistent and predictable and can easily be allowed for in the design of the moulding dies. This process allows shapes to be produced that are impossible with conventional compaction in rigid dies.

SELF-ASSESSMENT TASK 4.3

1. Explain why sintered components have to be preformed by compaction and how this process influences the properties of the finished product.

2. Discuss the factors that influence the choice of the compaction technique.

4.5 Sintering

This is a heat-treatment process carried out at between 60 and 90 per cent of the melting temperature of the major powder present. However, in a polyphase system, one or more of the minority materials may become molten and will infiltrate between the particles of the solid material by capillary attraction. The sintering process occurs in three stages.

1. The sintering process begins with rapid neck growth between particles that remain essentially discrete. This neck growth is shown in Fig. 4.9. You can see that the initial cold weld caused by powder compaction is extended in area during the first stage of sintering. As the weld neck grows there is little change in the density of the compact. Thus, at this stage in the sintering process, it is possible to control the porosity of the component. For example, porous bearings would receive no further sintering beyond this stage.

2. During this stage, optimum densification with recrystallisation and particle diffusion occurs. This is the most important stage in the sintering process for the majority of components. At the same time, the volume of the compact may be reduced by up to 1 per cent with a corresponding increase in density and reduction in permeability of the structure. As previously stated, in some circumstances it may even expand by up to 1 per cent with a corresponding reduction in density. The structure and properties of the sintered compact can be affected significantly by the temperature and atmosphere of the furnace and the process time. The properties of the sintered material can also be modified by the physical and chemical treatment of the powder prior to pressing, as well as by the introduction of reactive gases into the furnace atmosphere during sintering.

3. During the final stage, the spheroidisation of isolated pores occurs with some further increase in densification. These are the result of surface tension effects and continuing diffusion that, although slower than at stage 2, produces components that are homogeneous. Such components have mechanical and physical properties equal to, or

in some instances superior to, similar components produced by conventional melting and casting processes.

Fig. 4.9 *Formation of a 'weld neck'*

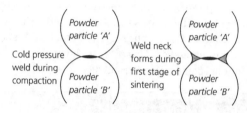

4.6 Sintering furnaces

Sintering furnaces may be heated by gas, oil or electricity. The components may be carried through the furnace on a wire-mesh belt made from a non-reactive, high-melting-point alloy such as inconel. Sintering is a relatively slow process taking up to 60 minutes at the sintering temperature. The furnace atmosphere is given reducing characteristics in order to avoid the possibility of oxidation. For example, metal carbides are usually sintered in an atmosphere of hydrogen at a temperature of 1350 °C. Alternatively, 'cracked' ammonia gas can be used. Although very much cheaper than hydrogen, it contains 25 per cent nitrogen and, therefore, can only be used with compact materials which are not susceptible to the formation of nitrides which would degrade their properties. The least costly and most widely used atmospheres for the commercial quantity production are made from hydrocarbon gases such as methane (natural gas). Such an atmosphere is quite suitable for components made from the less exotic materials. The hydrocarbon gas is 'cracked' by heating it at a high temperature with insufficient oxygen to complete combustion. This gives an atmosphere containing up to 45 per cent hydrogen. Unfortunately contaminants such as nitrogen, carbon dioxide and water vapour will be present.

Let us now consider the different 'chambers' into which the furnace is divided.

4.6.1 *Preheating chamber*

This chamber is used to burn off any lubricant that has been added to the powder as an aid to pressing. This 'de-waxing' stage is critical: should any residual lubricant be present when the compact is raised to the full sintering temperature, the lubricant will boil off causing porosity and blisters on the surface of the component. It will also contaminate the furnace atmosphere.

4.6.2 *Sintering chamber*

In this chamber the compact is raised to the full sintering temperature in an appropriately controlled atmosphere. The process time is controlled by the speed of the conveyor belt. Automatic temperature control to very fine limits is essential, as is uniformity of temperature within the sintering chamber, which must be devoid of hot or cold spots.

4.6.3 Carbon control chamber

When sintering compacts made from such materials as ferrous metal powders, decarburisation can occur at the high sintering temperatures encountered in the previous chamber. This carbon can be restored by the incorporation of a carbon control chamber operating at a lower temperature (carbon control shelf) following the sintering chamber and containing a carbon-rich atmosphere such as carbon monoxide.

4.6.4 Cooling chamber

It is essential to maintain atmospheric control until the components have cooled to below the temperature at which they will react with the normal atmospheric gases outside the furnace. The rate of cooling also affects the properties of the component, as in any other heat-treatment process. The rate of cooling can be controlled by the speed of the conveyor belt and by the rate at which the controlled atmospheric gases for this chamber are circulated through the heat exchanger, where they are cooled. A typical temperature-process chart for a ferrous powder compact is shown in Fig. 4.10.

Fig. 4.10 *Process temperature profile*

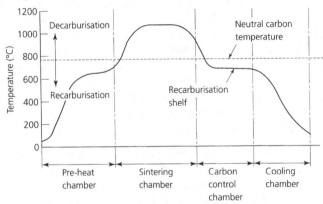

Time and distance through furnace

SELF-ASSESSMENT TASK 4.4

1. Describe the reason for sintering a compact and the structural changes that take place during this process.

2. Discuss the reasons why a sintering furnace has a reducing atmosphere, and the means of providing such an atmosphere.

3. Explain why it is necessary to preheat the compacts before they are sintered.

4.7 Post-sintering treatment (sizing)

The dimensional tolerances of sintered components can be held within 0.1 mm per 25 mm in the transverse direction and within 0.2–0.25 mm per 25 mm in the direction of pressing when compacted by the closed-die process. Nevertheless, since sintering is carried out at high temperatures, some distortion and dimensional inaccuracy will occur. When closer tolerances are required, sizing will be necessary after sintering. This is a re-pressing process, similar to coining, which is performed by squeezing the sintered component in dies using mechanical or hydraulic presses. The dies are similar to those used for the initial compaction, but with no allowance for densification. High rates of production are achieved for the sizing process, especially if automatically fed presses are used. After sizing, the component tolerance for linear dimensions can be within 0.02 mm per 25 mm in the transverse direction and 0.15 mm per 25 mm in the direction of pressing.

4.8 Post-sintering treatment (forging)

For this process the preform can be compacted by closed-die or isostatic processes. The compact is then sintered to a density of 85–90 per cent of the theoretical maximum. The sintering process improves the strength of the compact and reduces the included oxygen content from about 2000 ppm to below 200 ppm, thus improving the ductility and impact strength of the compact to that approaching wrought metal.

The preform is then used as a 'billet' in a conventional hot-forging operation in closed dies. This improves the densification of the sintered compact up to the theorectical maximum and a high-quality forging of great precision is produced. The mechanical properties achieved are comparable with conventional forging but the weight control and distribution are very much better, eliminating balancing operations for engine components used in the automobile industry. Other advantages of this process are:

- More economical use of material since no flash is produced.
- Increased productivity since flash clipping is eliminated.
- Tolerances can be held to closer limits than with conventional hot forging.
- Forging flow patterns are not introduced into the materials which have isotropic properties.

Unfortunately, the process is capital intensive and only economical for the continuous production of large numbers of the same component, or where the strength and quality achieved outweigh considerations of cost, as in some highly stressed, key components in the aircraft and automotive industries.

4.9 Components for sintered particulate processes

The most important consideration in component design geometry is that the component must be capable of ejection from a solid die. Thus, conventional casting features such as undercuts, reverse tapers, cross-holes, threads, and re-entrant angles are not feasible when compaction takes place in closed dies. However, a more flexible approach to component design is possible when isostatic pressing is used. Unfortunately, production costs are higher when using isostatic pressing. Figure 4.11 shows some typical components that may be produced by powder particulate processes. It can be seen that such features as holes, slots and teeth can be produced readily in the direction of pressing; however, the minimum size is limited by the rigidity and strength of the tool elements. Similarly, whilst parts with multiple steps can be produced, thin walls must be avoided; and where changes in the profile occur, feather-edges in the tooling must also be avoided.

Fig. 4.11 *Typical powder metal components (reproduced courtesy of GKN Bound Brook Ltd)*

4.9.1 *Structural materials*

Structural materials such as bronze, carbon steels and alloy steels have already been introduced. These are the most widely used materials and are found in components where mechanical strength is the main requirement.

4.9.2 *Porous bearings and filters*

The controlled porosity that can be achieved in sintered components is put to good use in porous, oil-impregnated bearings and filters. Mixtures of tin and copper powders form bronze alloys by diffusion during processing. Alternatively, pre-alloyed bronze powders may be used. Tin–copper mixtures tend to grow during sintering, whilst pre-alloyed powders tend to shrink. By using a combination of these two techniques, volume changes during sintering can be avoided. Sometimes graphite is added to act as a lubricant in its own right. For example, sintered bronze bearings and bearing shells have the advantage of being porous and are capable of being impregnated with a lubricant. Such bearings are invaluable in 'sealed for life' installations and other applications where the regular lubrication of conventional bronze bearings is not convenient. The porosity of the bearing is approximately 20 per cent by volume. The rate of oil supply automatically increases with temperature and, therefore, with speed of rotation of the shaft. Thus an impregnated bearing system is self-adjusting. Non-porous, steel-backed bearings for heavy-duty applications are also made by the sintering technique.

Sintered metal filters and diaphragms are also made, and are stronger than their ceramic counterparts. As well as separating solids from liquids, they can be used to separate liquids that have different surface tensions. In this way they can separate water from jet engine fuel, as well as filter out solid impurities.

4.9.3 *Friction materials*

Sintered metal friction materials are used for heavy-duty brake and clutch linings and disc brake pads. The metal matrix contains non-metallic friction materials such as carborundum, silica, alumina and emery. (Asbestos is no longer included for safety reasons.) Such friction materials can operate at temperature up to 800 °C where a copper matrix is employed, and up to 1000 °C where an iron matrix is employed. These are well above the operating temperature for resin-bonded friction materials and even at lower operating temperatures the wear rate of sintered friction materials is lower than that of resin-bonded materials.

4.9.4 *Electrical and magnetic components*

These often require materials with composite properties that can only be achieved by particulate processes. For example, commutator or slip-ring brushes require a combination of the high contact conductivity of silver or copper with the strength, heat resistance and resistance to arc erosion of tungsten, and the lubricating properties of graphite.

High-performance 'hard' magnetic materials, as well as high-performance 'soft' magnetic materials, can be made by the sintering process. Such magnetic components have the equivalent magnetic properties of cast alloys, but with a finer grain and better mechanical properties. Further, the high finish and dimensional accuracy of sintered magnets and cores, compared with similar cast components, eliminates much of the need for machining. This is important since some of the high-performance 'hard' magnetic materials are almost impossible to machine and in some instances can only be ground.

1. Compare the advantages and limitations of manufacturing components by post-sintered forging with conventional forging processes.

2. Discuss how a component's design geometry affects its suitability for manufacture by sintering.

3. Discuss how the controlled porosity of components manufactured by sintered particulate processes can be exploited in the manufacture of bearings and filters.

4.10 Die design

The high wall pressures generated during compaction (approximately 1500 MPa) necessitates the use of solid dies and punches for closed-die techniques, as shown in Fig. 4.12. As has already been mentioned, counter-punches are used to achieve uniform compaction. The powdered materials are very abrasive and frictional conditions in the die are very severe. For this reason the die is often made from tungsten carbide set in a high-tensile steel yoke to prevent it from bursting, Split-die or double-die techniques have been developed to permit components with simple offsets and projections to be manufactured in rigid dies. Figure 4.13 shows such a component, whilst Fig. 4.14 shows details of the double die required for its manufacture.

Fig. 4.12   *Solid-die compaction tool*

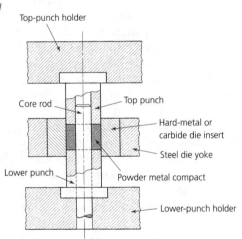

Fig. 4.13   *Component requiring split-die tooling*

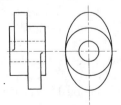

Fig. 4.14 *Split-die tooling*

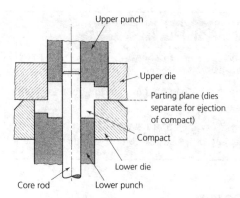

4.11 Advantages, limitations and justification of sintered particulate processes

Like any other processes, sintered particulate processes have a number of unique advantages and limitations that may be summarised as follows:

4.11.1 *Advantages*

- Sintered particulate processes have the ability to form components from refractory metals such as molybdenum, tantalum and tungsten, whose high melting points preclude normal casting processes.
- Materials whose phases are normally immiscible, and which will not form solutions under normal melting and casting processes, can be made to form alloy systems by the intimate mixture of particles of the materials prior to compaction and sintering.
- The porosity and permeability of components can be controlled during the compaction and sintering processes so as to produce such parts as porous metal bearings and high-pressure filter elements.
- Alloy systems can be controlled to very fine limits compared with conventional melting and casting. This is achieved by accurate mixing of the constituent particles prior to compaction and sintering. Alloys such as controlled expansion nickel alloys, high-speed steels, super-alloys, and 'hard' metals can be produced in this way.
- The micro-structure of the alloy can also be accurately controlled where grain orientation is important, as in permanent magnet materials requiring high domain alignment.
- Components made by sintered particulate processes can be produced to very close tolerances and high standards of surface finish compared with conventional melting and casting techniques. They can be made in a wide variety of sizes and shapes and, if correctly designed, need undergo little or no subsequent finishing following compaction and sintering.
- Sintered particulate processes lend themselves to quantity production and to process automation. This helps to reduce the production costs and makes such processes highly competitive with conventional forming processes.

- The quality control of sintered particulate products can be readily automated.
- The raw material, being in powder form, is more easily stored and handled than bar stock.

4.11.2 *Limitations*

- The process is capital intensive, requiring expensive compaction equipment and sintering furnaces.
- Compaction dies are expensive when rigid-die techniques are used. Lead times are also extensive, compared with the manufacture of casting patterns, whilst the dies are being made.
- The powdered materials are relatively expensive, although this is, to some extent, offset by economical usage. There is little waste compared with the clipped flash of forgings and the swarf produced during machining. Further, any scrap produced during setting up the compaction process can be ground down again into powder and re-used to a limited extent (a maximum of 10 per cent of the total mix) providing the compacts have not been sintered.
- The energy costs of the sintering process are high.
- There are limitations of size owing to the high compaction pressures required.
- There are limitations on the component design necessary to avoid problems of compaction die design and manufacture.
- Long runs are required to recover the tooling costs.

4.11.3 *Justification*

Despite the high level of the investment cost in the plant and equipment necessary for manufacture by sintered particulate processes, and despite the relatively high cost of the process itself, sintered particulate processes offer so many advantages over casting and forging that they are now used increasingly for a wide range of applications. Nowadays there is a rapid move to minimum metal removal strategies, where machining is looked upon only as final sizing and finishing operations, or eliminated altogether if possible. The good surface finish, accuracy and volumetric consistency of the sintered particulate processes go a long way to achieving such a strategy. For this reason one luxury car manufacturer has adopted sintered particular processes for the manufacturer of the engine connecting rods this not only reduced the amount of machining required but also improved the dynamic balance of the engine. This improved balance was achieved by the inherent consistency of the process that resulted in all the connecting rods having virtually identical masses.

EXERCISES

4.1 By means of a flow diagram, outline the basic principles of component production by powder metallurgy techniques.

4.2 Describe, with the aid of sketches, the tooling for and the production of a plain bronze bush by the powder particulate process.

4.3 Describe:
  (a) how the metal powders are produced
  (b) how the powders are selected and blended

4.4 With the aid of sketches, describe in detail any **three** of the following compaction processes:
  (a) closed-die pressing
  (b) cold isostatic pressing
  (c) hot isostatic pressing
  (d) roll compaction; extrusion
  (e) slip casting
  (f) injection moulding

4.5 With the aid of diagrams, discuss the essential requirements of a sintering furnace.

4.6 Explain what is meant by 'post-sintering treatment' and how these treatments affect the finished accuracy and properties of sintered components.

4.7 With the aid of sketches, discuss the principles of design for a component of your choice.

4.8 Discuss the economics of the production of a component of your choice by a powder particulate technique compared with conventional forging or casting followed by machining.

5 Plastic moulding processes

The topic areas covered in this chapter are:

- Plastic moulding materials.
- Polymerisation.
- Forms of supply.
- Positive die moulding, flash moulding and transfer moulding.
- Moulding conditions.
- Injection moulding.
- Extrusion moulding.
- Selection of a moulding process.

5.1 Plastic moulding materials

The moulding process selected depends upon two main factors: the geometry of the component to be moulded and the material from which it is to be made. There are two main groups of plastic moulding materials: *thermoplastic materials* that soften every time they are heated, and *thermosetting plastic materials* that undergo a chemical change during moulding and can never be softened again by heating.

5.1.1 *Thermoplastics*

Since this group of plastics can be softened every time they are heated, they can be recycled and reshaped any number of times. This makes them environmentally attractive. However, some degradation occurs if they are overheated or heated too often and recycled materials should only be used for lightly stressed components. Some typical thermoplastic materials are listed in Table 5.1.

5.1.2 *Thermosetting plastics*

This group of materials differs from the thermoplastic materials in that polymerisation (the forming of polymers) is completed during the moulding process and the material can never be softened again. Polymerisation during the moulding process is called 'curing'. Some typical examples of thermosetting plastics (thermosets) are given in Table 5.2.

Table 5.1 *Typical thermoplastic materials*

Type	Material	Characteristics
Cellulose plastics	Nitrocellulose	Materials of the 'celluloid' type are tough and water resistant. They are available in all forms except moulding powders. They cannot be moulded because of their flammability
	Cellulose acetate	This is much less flammable than the above. It is used for tool handles and electrical goods
Vinyl plastics	Polythene	This is a simple material that is weak, easy to mould and has good electrical properties. It is used for insulation and for packaging
	Polypropylene	This is rather more complicated than polythene and has better strength
	Polystyrene	Polystyrene is cheap, and can be easily moulded. It has a good strength, but it is rigid and brittle, and crazes and yellows with age
	Polyvinyl chloride (PVC)	This is tough, rubbery, and practically non-flammable. It is cheap and can be easily manipulated; it has good electrical properties
Acrylics (made from an acrylic acid)	Polymethyl methacrylate	Materials of the Perspex type have excellent light transmission, are tough and non-splintering, and can be easily bent and shaped
Polyamides (short carbon chains that are connected by amide groups – NHCO)	Nylon	This is used as a fibre or as a wax-like moulding material. It is fluid at its moulding temperature, but tough, and has a low coefficient of friction at room temperature
Fluorine plastics	Polytetrafluoro-ethylene (PTFE)	This is a wax-like moulding material; it has an extremely low coefficient of friction. It is relatively expensive
Polyesters (when an alcohol combines with an acid, an 'ester' is produced)	Polyethylene-terephthalate	This is available as a film or as 'Terylene'. The film is an excellent electrical insulator

Table 5.2 *Typical thermosetting plastic materials*	
Material	Characteristics
Phenolic resins and powders	These are used for dark-coloured parts because the basic resin tends to become discoloured. These are heat-curing materials
Amino (containing nitrogen) resins and powders	These are colourless and can be coloured if required; they can be strengthened by using paper-pulp filters, and used in thin sections
Polyester resins	Polyester chains can be cross-linked by using a monomer such as styrene; these resins are used in the production of glass-fibre laminates
Epoxy resins	These are also used in the production of glass-fibre laminates and as adhesives

5.2 Polymerisation

The difference between thermoplastics and thermosetting plastics is the way in which polymerisation occurs. In *thermoplastics* polymerisation occurs by the *addition of monomers*. For example, to manufacture such a plastic material, an *alkane* such as methane, ethane or propane has first to be converted into its corresponding *olefin*. This olefin is referred to as a *chemical intermediate* since it lies between the monomer and the *polymer*. A single atom of an olefin is referred to as a *monomer* (mono- means one or single), so the next stage in the process is to combine several monomers together to form a much larger molecule called a *polymer* (poly- means many). In the form of a polymer, the olefin takes on the properties of a plastic material. Some examples are shown in Fig. 5.1.

Fig. 5.1 *Simple thermoplastic polymers*

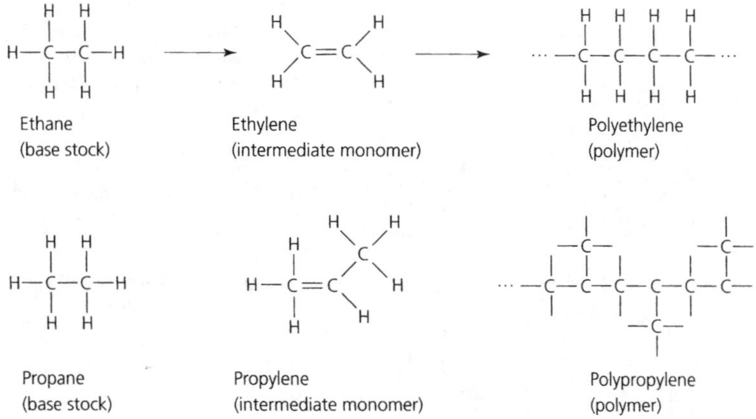

Ethane
(base stock)

Ethylene
(intermediate monomer)

Polyethylene
(polymer)

Propane
(base stock)

Propylene
(intermediate monomer)

Polypropylene
(polymer)

There are some simple basic rules governing the number of monomers that may be brought together to form a polymer. For example, at room temperature ethylene, which is made up of single molecules (monomers), is a gas. A polymer of 6 ethylene monomers is a liquid. A polymer of 36 ethylene monomers is a grease. A polymer of 140 ethylene monomers is a wax. A polymer of 500 or more ethylene monomers is the plastic polyethylene. The maximum number of monomers that can be brought together in a single polymer is limited in practice to about 2000. To enlarge the polymer further would result in little increase in strength but a considerable increase in hardness and brittleness.

In *thermosetting* plastic materials, polymerisation usually occurs through *condensation*. In this process, the plastic material reacts within itself – or with some other chemical (hardener) – when heated to or above a critical temperature. When this happens the plastic material releases ('condenses out') some small molecules such as water. This results in a permanent chemical change (polymerisation) taking place. The plastic material is said to be 'cured'. This change is irreversible and never again can the material be softened by heating. The loss of the water molecules causes shrinkage, and this has to be allowed for in the moulding process. Also the moulds have to be designed with vents so that the water vapour, generated as steam during the moulding process, can escape. The principle of condensation polymerisation is shown in Fig. 5.2.

Fig. 5.2 *Curing of thermosetting plastics*

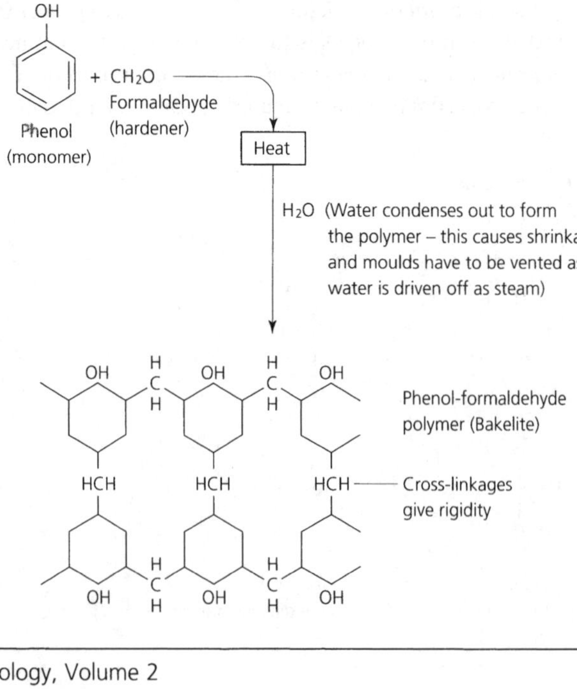

5.3 Forms of supply

Both thermoplastic and thermosetting plastic moulding materials are normally available as powders or as granules packed in bags or in drums.

Thermoplastic powders and granules are homogeneous materials consisting of the polymer together with the colouring agent (pigment), lubricant and die-release agent.

Thermosetting plastic materials are unsuitable for use by themselves, and the thermosetting plastics come in powder or granule form mixed with additives to make them more economical to use, to improve their mechanical properties, and to improve their moulding properties. A typical thermosetting plastic moulding material could consist of:

Resin powder or granules	38%	by weight
Filler	58%	by weight
Pigment	3%	by weight
Mould-release agent	0.5%	by weight
Catalyst	0.3%	by weight
Accelerator	0.2%	by weight

The low-cost filler not only bulks up the powder or granules and makes the material more economical to use, but also has a considerable influence on the properties of the mouldings produced from a given resin. For example, fillers improve the mechanical strength, electrical insulation properties and heat resistance. Fillers also reduce shrinkage during moulding. Typical fillers are:

- *Glass fibre*   High strength and good electrical insulation properties.
- *Wood flour*   Low cost, high bulk, low strength.
- *Calcium carbonate*   Low cost, high bulk, low strength.
- *Rock wool*   Heat resistance. (*Note:* Asbestos was used but is no longer recommended as it represents a health hazard.)
- *Aluminium powder*   Wear resistance and high strength.
- *Shredded paper*   Good strength but inclined to absorb moisture.
- *Shredded cloth*   Higher strength but also inclined to absorb moisture.
- *Mica granules*   Heat resistant with good electrical insulation properties.

The pigment is added to colour the finished product. Some pigments are also used to prevent ultraviolet degradation of the polymer when it is exposed to sunlight. This also applies to thermoplastics (for example uPVC for window frames). The mould-release agent prevents the moulding from sticking to the mould (most plastics are also good adhesives). It also acts as a lubricant and helps the moulding material to flow during the moulding operation. The catalyst promotes the curing process during moulding and ensures uniform properties throughout the moulding. The accelerator speeds up the curing time and, therefore, the moulding process, thus reducing the process cost. Figure 5.3 shows the stages in manufacturing a typical thermosetting plastic material, such as paper-filled melamine formaldehyde, up to the stage where it is packed in bags or drums for delivery to the moulding company. Polymeric (plastic) materials and adhesives are dealt with more fully in *Engineering Materials*, Volumes 1 and 2.

Fig. 5.3 *Manufacture of a typical thermosetting plastic moulding material*

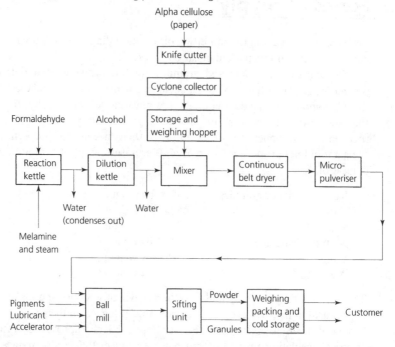

5.4 Positive-die moulding (thermosetting plastics)

In the positive-die moulding process the two-part mould is fitted into a hydraulic press of the type shown in Fig. 5.4. The press may be of the upstroke type in which the hydraulic ram and cylinder are housed in the base as shown, or the downstroke type with the cylinder and piston mounted above the upper bolster. In this latter type the upper bolster is free to move and the lower bolster is fixed. Large presses are usually of the upstroke type. The plattens are provided with steam or electric heating elements so that the moulds fastened to them can be heated to the curing temperature of the thermosetting plastic being processed. Modern presses are fully automatic in operation and this ensures constant moulding conditions at each stroke. This, in turn, ensures that mouldings of consistent quality are produced. The three factors which need to be preset are:

- The moulding pressure (closing force on the die).
- The time that the mould is closed (curing time).
- The temperature of the mould.

In addition, the moulding powder or granules may be added automatically and this helps to ensure correct filling of the moulds and uniform product quality.

Fig. 5.4 *Moulding press*

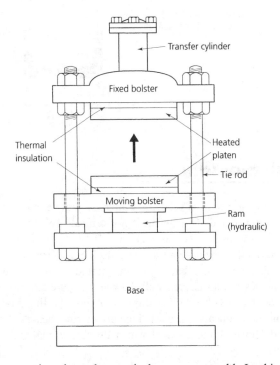

Figure 5.5 shows a section through a typical pressure mould. In this type of mould a predetermined amount of thermosetting plastic material, in powder or granular form, is placed in the heated mould cavity. The mould is then closed by the press whilst the moulding material cures under pressure. The mould is then opened and the moulding is ejected. In positive moulding a vertical flash is produced in the direction of the moulding pressure. This surplus material is extruded between the upper and lower mould through the slight gap (0.01–0.03 mm) which has to be left for venting purposes (remember that steam is produced during condensation polymerisation). 'Flash' is the name by which any surplus material is known, and must be trimmed from the moulding. The mould shown in Fig. 5.5 allows for complete closure to ensure constant thickness. The disadvantage of this type of mould is that the amount of moulding powder fed into the mould has to be very accurately controlled. Too little and the cavity will not be filled: too much and the mould will not close. In practice, very slight overfilling is aimed at to ensure that the mould is always filled. This very small amount of surplus material escapes through the steam vents. Where the

thickness of the product is not critical there are no closing faces on the two parts of the mould, and the product thickness is controlled solely by the amount of material fed into the mould and the moulding pressure.

Fig. 5.5 *Positive mould*

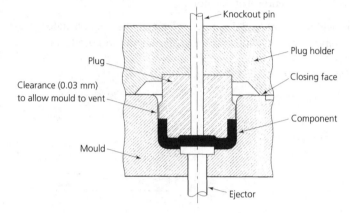

5.4 Flash moulding

The flash-type mould is used for simple, shallow components where compaction of the moulding powder during curing is not critical. To ensure complete filling, an excess of moulding powder is placed in the cavity. When the mould is closed, any excess material is squeezed out into the flash gutter. The 'flash-land' forms a constriction which tends to hold the moulding material back into the mould cavity during curing in order to reduce shrinkage losses and uneven wall thickness. Figure 5.6 shows a section through a typical flash mould. A disadvantage of this type of mould is the heavy flash that has to be removed from the moulding and the wasted material which, being a thermosetting plastic, cannot be recycled. However, the ease with which any surplus material can escape reduces the load on the press.

Fig. 5.6 *Flash mould*

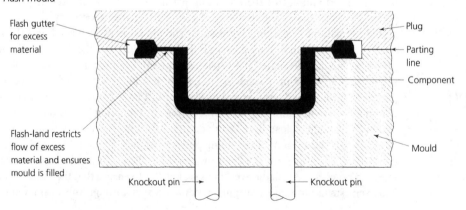

5.6 Transfer moulding

This technique uses a more complex three-part mould, as shown in Fig. 5.7. This type of mould is more costly than those previously described and transfer moulding is only used where:

- The moulding is of complex shape with many changes in wall thickness so that uniform filling of the cavity would be difficult in a more simple mould type.
- Multiple impression moulds are used, so that a number of components can be made at each stroke to ensure that the press is employed to its full capacity and, therefore, economically.

When transfer moulding, the powder is preheated and plasticised in the upper (loading) chamber of the mould. It is then forced under pressure into the mould cavity or cavities via the *sprue*. The sprue is removed after the mould is opened and the moulding has been ejected. Since the sprue does not form a useful part of the moulding, and cannot be recycled if a thermosetting material is used, it represents wasted material. However, this small waste

Fig. 5.7 *Transfer moulding*

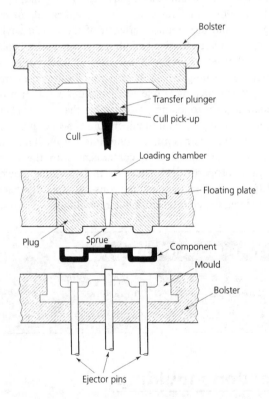

1. Die shown open ready for removal of component and cull.
2. Floating plate closes on mould and moulding powder is loaded into chamber.
3. Transfer plunger descends and forces plasticised moulding powder through sprue into mould.

is offset by the advantages of the process. Since the plasticised material is forced into the closed mould under very high pressure, complete filling of the mould cavity is ensured irrespective of its complexity. Shrinkage is reduced and improved mechanical properties are obtained from the moulding material.

5.7 Compression-moulding conditions

The moulding material may be fed into the mould as a powder, as granules or compacted into a preformed shape. The latter is used to ensure uniform filling of the mould cavity, particularly when the cavity has a complex form. Correct loading of the mould is critical, insufficient material resulting in voids and porosity through the cavity not being properly filled. A slight excess of material is preferable as it ensures complete and uniform filling of the mould with any excess being allowed to form a 'flash'. Excessive overcharging must be avoided as the powder is incompressible and damage could be done to the mould and to the press. Automatic metering and feeding of the moulding material results in more uniform results than hand feeding, as well as being more productive.

The moulding material can be loaded either cold or preheated. Preheating reduces the curing time and also reduces erosion of the mould cavity since the partially plasticised material is in a less abrasive condition. During curing, volatile gases are released and these must be allowed to escape either through the mould clearances, through vents or by momentarily opening the mould part-way through the cure.

To prevent sticking, a release agent (lubricant) must be sprayed into the mould cavity immediately prior to loading. (Remember that plastic resins are also very good adhesives.) The correct curing time and temperature is also critical as overcuring produces a dull and blistered surface with some crazing, internal cracking, and poor mechanical properties. Undercuring may produce a component with the correct appearance but with poor mechanical properties. Moisture in the moulding powder can also cause blisters and porosity. The correct curing conditions are generally determined by trial and error, based upon previous experience with similar mouldings.

SELF-ASSESSMENT TASK 5.2

1. Sketch out the design for a typical compression-moulding die for a component of your choice. Comment on the importance of your design features.

2. Describe the cycle of events that occur during the moulding process.

5.8 Injection moulding

Unlike the compression-moulding processes described in the previous section, which are usually used in conjunction with thermosetting plastic materials, injection moulding is usually used in conjunction with thermoplastic materials. In the injection-moulding process a measured amount of thermoplastic material is heated until it becomes a viscous fluid whereupon it is injected into the mould cavity under high pressure. In this respect it resembles transfer moulding except that, because a thermoplastic material is used, no curing has to take place and the moulds can be opened as soon as the moulding has cooled sufficiently to become rigid and self-supporting. Injection-moulding machines are generally arranged with the mould parting-line vertical and the axis of injection horizontal, as shown in Fig. 5.8(a). As an alternative to the ram feed, large-capacity machines may use a screw-

Fig. 5.8 *Injection moulding: (a) ram feed injection-moulding machine; (b) screw feed injection-moulding machine*

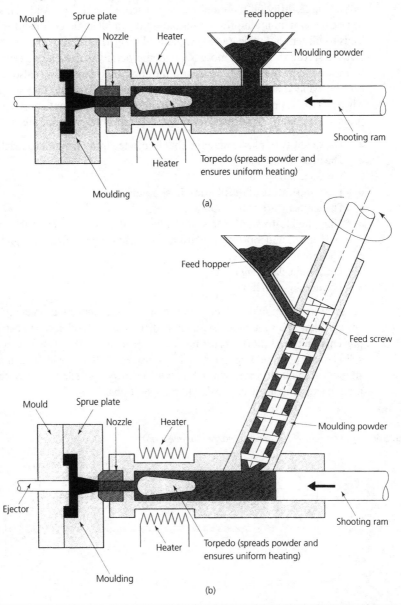

feed mechanism for filling the dies. An additional ram-feed mechanism provides the final pressure in order to ensure complete filling of the mould and compaction of the material. The layout of a typical screw-feed machine is shown in Fig. 5.8(b).

Like so many processes that are simple in principle, the practice of injection moulding is fraught with difficulties. For instance, heating a body of thermoplastic material until it is fluid is not easy. If care is not taken, the surface becomes overheated and degraded before the interior of the plastic mass has reached its moulding temperature. Also the injection temperature of the plastic is extremely critical as this controls its viscosity. Again, if the moulding material is not sufficiently heated, it may chill on contact with the mould and it will not necessarily fill the mould cavity. There may also be voids, sinks and shorts, and some cavities in multi-impression moulds may not fill at all. On the other hand, overheating leads to blistering and degraded mechanical properties. The 'torpedo' shown in Fig. 5.8(a) helps to spread the powder more thinly in the vicinity of the heating elements and this helps to ensure uniform heating. Ejection of the completed moulding is also difficult if distortion is to be avoided and, since plastic materials are also adhesive, they will stick to the mould whenever the opportunity presents itself. A mould lubricant must be used immediately prior to injection. The principle variables that must be controlled are:

- The quantity of plastic material that is injected into the mould at each 'shot'.
- The injection pressure.
- The injection speed.
- The temperature of the plastic whilst moulding.
- The temperature of the mould.
- The plunger-forward (pressure) time – that is, the time the plastic material is maintained under pressure whilst it cools and becomes rigid enough to eject.
- The mould closed time.
- The mould clamping force.
- The mould open time.

As soon as the mould is filled, the pressure is increased and 'packing' commences. This ensures that the moulding has a high density and good mechanical properties. It also prevents sinks and shorts occurring due to shrinkage of the plastic as it cools. The injection cylinder of the moulding machine is connected to the mould by a nozzle. Figure 5.9(a) shows a standard nozzle, whilst Fig. 5.9(b) shows a nozzle with breaker plates which are used to assist mixing in colour-blending operations.

Fig. 5.9 *Injection nozzles (a) standard nozzle; (b) mixer nozzle*

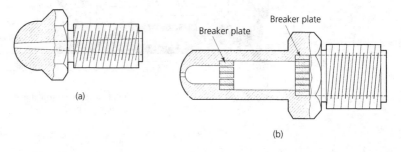

(a)

Breaker plate

Breaker plate

(b)

To ensure economical use of the machine, multiple-cavity moulds are used when moulding small components. This allows the machine to be used to its full capacity at each shot. Figure 5.10 shows a typical multi-impression mould, and it can be seen that the individual cavities are connected by passages, called *runners*, to the central sprue that connects with the injection nozzle. A 'gate' is provided at the cavity end of each runner to produce a local thinning so that the moulding may easily be separated from the runner. The combination of mouldings, runners and sprue produced at each shot is called a *spray*. Unlike thermosetting plastics, thermoplastic materials may be recycled, thus, the sprues and runners may be ground up and recharged into the machine to prevent waste. Since some degradation takes place each time the material is reheated, recycled material is not used when stressed components are being moulded. To ensure uniform filling of the cavities, and to ensure uniform loading on the mould clamping mechanism of the injection-moulding machine, the mould cavities must be balanced about the central sprue as shown in Fig. 5.11.

Fig. 5.10 *Multiple-impression mould*

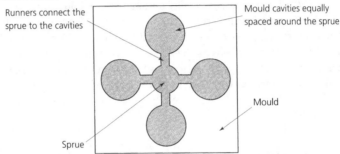

Runners connect the sprue to the cavities

Mould cavities equally spaced around the sprue

Mould

Sprue

Fig. 5.11 *Mould balancing: (a) poor design (unbalanced clamping forces); (b) good design (cavities balanced about sprue)*

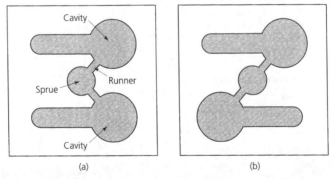

Cavity

Runner

Sprue

Cavity

(a)

(b)

SELF-ASSESSMENT TASK 5.3

1. Compare and contrast the relative advantages and limitations of compression moulding and injection moulding in terms of materials, economics of production, and typical components.

2. Describe the injection-moulding cycle.

5.9 Use of inserts

Inserts are used extensively in plastic products to provide such features as threaded anchorages, bearings and shafts, electrical terminals, and reinforcements. Metal inserts may be moulded into a component or added later by pressing them into cored or drilled holes. Thermosetting plastics are too hard and brittle to allow inserts to be pressed into the finished moulding.

Inserts are most secure when included during the moulding process, but are inconvenient to use where quantity production is required as placing and locating the inserts in the mould slows down the process and increases production costs. Moulded inserts can also interfere with the ejection of mouldings from the mould cavity. Despite careful design, it is difficult to keep the moulding material out of the threads of screwed inserts, and any subsequent correction is, again, an added cost that detracts from the advantages of inserts added during the moulding process.

An ever-increasing use of fully automated moulding processes for both thermosetting and thermoplastic materials has increased the use of inserts that are added after moulding. These inserts are pressed into cored or drilled holes. The slotted end of the insert collapses initially and then expands as the screw is driven home. This expansion locks the insert in place. Inserts are usually coarse knurled to prevent them turning in the hole or being withdrawn. Examples of typical inserts are shown in Fig. 5.12.

Fig. 5.12 *Inserts: (a) 'moulded-in' type inserts; (b) pressed-in self-locking insert*

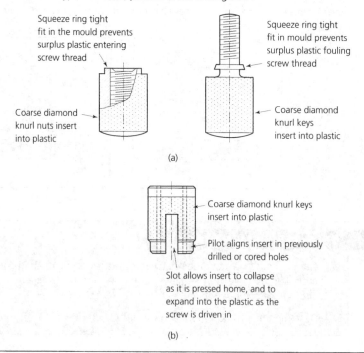

Squeeze ring tight fit in the mould prevents surplus plastic entering screw thread

Coarse diamond knurl nuts insert into plastic

Squeeze ring tight fit in mould prevents surplus plastic fouling screw thread

Coarse diamond knurl keys insert into plastic

(a)

Coarse diamond knurl keys insert into plastic

Pilot aligns insert in previously drilled or cored holes

Slot allows insert to collapse as it is pressed home, and to expand into the plastic as the screw is driven in

(b)

5.10 Extrusion

The extrusion of thermoplastic materials is, in principle, a continuous injection-moulding process. Any thermoplastic material can be extruded to produce lengths of uniform cross-section such as rods, tubes, sections and filaments. To produce a continuous flow of plastic material through the die, a screw conveyor is used in place of the piston and cylinder of the injection-moulding machine. The general arrangement is shown in Fig. 5.13. Unlike the injection-moulding machine, no 'torpedo' is required in the heating chamber when a screw conveyor is used because the screw itself keeps the moulding powder in contact with the hot walls of the chamber and no cold, central core of material can occur.

Fig. 5.13 *Extrusion moulding*

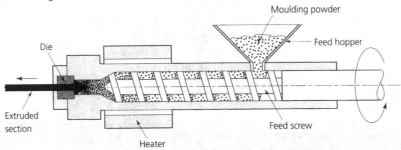

To allow for shrinkage as the extruded material cools, the die profile has to vary from the shape of the finished product, as shown in Fig. 5.14. The plastic is still soft as it leaves the die and must be supported to prevent distortion. Cooling is provided by a water tray, mist spray or air blast, depending upon the specific mass and temperature of the extrusion. A conveyor draws the cooled plastic extrusion from the die ready for coiling or cutting to length.

Fig. 5.14 *Extrusion die (shrinkage allowance)*

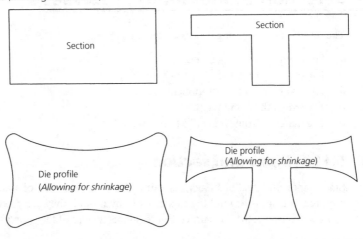

Wire coating is a special application of extrusion, as shown in Fig. 5.15. Since the wire runs at right angles to the axis of extrusion, this device is called a *cross-head die*. After passing through the die the covered wire is water-cooled and tested for its insulation characteristics.

Fig. 5.15   *Cross-head die*

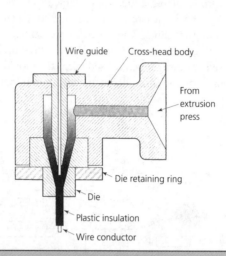

SELF-ASSESSMENT TASK 5.4

1. Discuss the design needs for inserts, production problems that can arise through their use in plastic mouldings, and how these problems can be minimised.

2. Explain why an extrusion die orifice often has to differ from the required section of the product, and describe the precautions that must be taken to prevent distortion of the extruded section as it flows from the die.

5.11 Selection of moulding processes

The selection of moulding processes depends upon:

*   The component being made.
*   The quantity required.
*   The geometry of the component.
*   The size of the component.
*   The material from which it is to be made.

5.11.1 *Pipes and sections*

Since pipes and sections, such as rainwater guttering, are of uniform section and are required in lengths beyond the capacity of any mould, they are produced by the extrusion process (Section 5.10). This process necessitates the use of thermoplastic materials.

5.11.2 *Thermoplastic mouldings*

These are usually made by the injection-moulding process (Section 5.8) which is suitable for the quantity production of both large and small components and is the most widely used moulding process. Small components can be made in multi-impression moulds and left on the sprue until required to prevent loss. Examples of typical components made by injection moulding can range from model kit parts made from polystyrene, and small nylon gears for office machinery, to motor vehicle rear-light clusters made from transparent acrylic plastics and even complete motor vehicle bumpers moulded from impact-resistant plastics.

5.11.3 *Thermosetting plastic mouldings*

Only in very special circumstances can thermosetting plastics be injection moulded. Almost invariably, thermosetting plastics are moulded by compression or transfer techniques (Sections 5.4 and 5.5). Since the plastic resin can be readily blended with a wide variety of filler materials and pigments, mouldings made from thermosetting plastics can be given a wide range of properties and appearances. Compression mouldings are used for components such as: meter cases, electric fan bodies and blades, electrical insulators for switch gear, contactors and distribution equipment, and tableware. In all these examples rigidity and strength are required, coupled with good surface finish and scratch resistance. Only thermosetting plastics have all these properties at the same time.

5.12 Justification

Providing the quantities being manufactured can offset the high cost of the moulds, the use of plastic mouldings can be justified on the following grounds:

- Wide range of materials with a correspondingly wide range of properties available.
- Polymeric materials have a lower density than metals.
- Relatively low cost of raw materials.
- High rates of production – ease of full automation.
- Highly polished finish straight from the moulds, little finishing required beyond trimming.
- Wide colour range available – multi-colour mouldings can be made.
- Little post-moulding machining required – many components can be used straight from the mould.

EXERCISES

5.1 Explain the essential differences between *thermoplastic* and *thermosetting* plastic materials.

5.2 Explain the mechanism of polymerisation by condensation and how this effects the design of the component and the moulds.

5.3 Describe the composition of a typical thermosetting moulding powder and explain the purpose(s) of the various constituents.

5.4 With the aid of diagrams, describe any **two** of the following compression-moulding processes and compare the advantages and limitations of the process chosen:
(a) positive-die moulding
(b) flash-die moulding
(c) transfer-die moulding

5.5 Describe the compression-moulding cycle, paying particular attention to the precautions that must be taken to ensure successful production.

5.6 With the aid of sketches, compare the principles of compression moulding, injection moulding and extrusion moulding. Describe where each would be used in terms of materials and components.

5.7 Explain what is meant by the term *inserts*, and explain where they are of use in plastic moulding. Sketch a typical component showing the position of the inserts and explain their purpose.

5.8 Justify the selection of a plastic material and moulding process for a component of your choice.

6 Cutting processes

The topic areas covered in this chapter are:

- Sheet metal cutting operations including piercing tools, 'follow-on' tools, transfer tools, and combination tools.
- Process selection including surface geometry, accuracy and surface finish, tooling and cutting costs, and cutting tool materials.
- Automatic lathes.
- Screw-thread production.
- Gear-tooth production.
- Broaching.
- Centreless grinding.
- Electrical discharge machining (EDM).
- Electrochemical machining (ECM).
- Chemical machining (CM).
- Laser cutting.

6.1 Sheet metal

You were introduced to the cutting of sheet metal by shearing and blanking in *Manufacturing Technology*, Volume 1. This section examines some further processes for cutting sheet metal.

6.1.1 *Piercing*

This process is used to produce holes in previously blanked out components. The clearance required between the punch and the die, and the force required to cut out the hole, is calculated in the same manner as for blanking. However, when piercing, the punch is made the same diameter as the required hole size whilst, when blanking, the die orifice is made the required blank size. Figure 6.1 shows the principle of piercing and a typical small piercing tool.

Fig. 6.1 *Piercing: (a) piercing action (hole diameter = punch diameter = d; die diameter = D = d + clearance); (b) piercing tool*

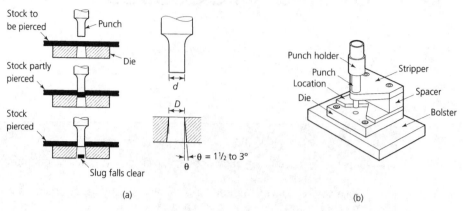

(a)

(b)

6.1.2 'Follow-on' tools

To speed production, piercing and blanking operations are often combined together in one tool. Figure 6.2 shows a simple 'follow-on' tool for making washers. To ensure concentricity of the hole, the blanking punch is provided with a pilot to locate in the previously pierced hole. This type of tool gets its name from the sequential arrangement of the piercing and blanking stations.

Fig. 6.2 *Simple 'follow-on' tool*

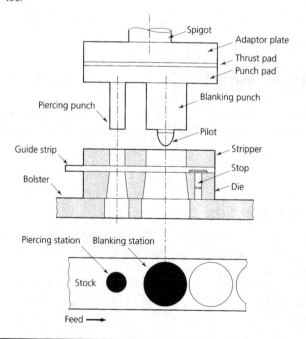

6.1.3 *Transfer tools*

These develop the 'follow-on' design a stage further and incorporate forming operations. The blank and the partially formed and pierced component is moved from stage to stage of the tooling by a transfer feed mechanism which is actuated by, and synchronised with, the crank or eccentric drive of the press. During each up-stroke of the ram, the partially formed components are moved to the next stage of the tool.

6.1.4 *Combination tools*

Like 'follow-on' tools, combination tools also combine operations. In these tools, various combinations of piercing, blanking operations and forming operations are combined together and are performed at the same time and not sequentially as in a 'follow-on' tool. Figure 6.3 shows a simple 'cut and cup' combination tool in which the two operations take place at the same stroke of the press. If required, one or more holes could be pierced in the base of the cup in the same operation.

Combination tools and 'follow-on' tools increase the rate of production of pressed components and reduce handling and inter-operation storage. However, they are more costly than simple tools and only become economical if the number of components warrants their extra cost, also they are less flexible. That is, only one combination of blank and hole piercing pattern is available, whereas with separate tools the same blank can be used with a variety of piercing tools to provide a range of components. The cutting force is also greater when the piercing and blanking operations are combined and care must be taken to ensure that the press is not over-loaded. The decision whether or not to use combination tools may be governed by the availability of suitable press capacity rather than optimum production rates. Remember also that the cutting force when piercing and blanking is only part of the load on the press. The cut metal tends to spring back onto the punch and considerable force is required to 'strip' the metal off the punch. Under some conditions this can be as high as 50 per cent of the cutting force and the stripping force

Fig. 6.3 *Combination 'cut-and-cup' tool*

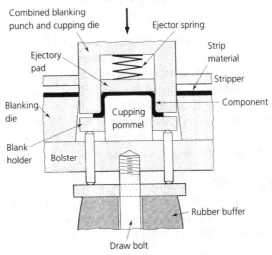

must be allowed for when selecting a press of suitable capacity. Also the mass and drag of large tools (car body panels) can themselves add considerably to the load on the press.

Fig. 6.4 *Self-assessment task (dimensions in millimetres)*

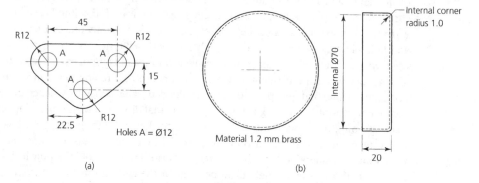

(a)   (b)

6.2 Process selection (introduction)

You were introduced to the principles of metal machining *Manufacturing Technology*, Volume 1. This introduction included the action of the cutting wedge, tool angles, and chip formation, together with the application of these principles to single-point tools, and multipoint tools such as milling cutters and twist drills. The use of lubricants and coolants and the calculation of tool life were also discussed. Some further aspects of metal machining will now be considered.

When selecting a machining process, a compromise often has to be accepted between the ideal process to achieve the specified quality with maximum productivity and minimum unit cost, and the processes that are available within the limitations of the existing plant. It is only when very large quantities of a specific component are required – and the requirement will be repeated – can the purchase of a particular item of plant be justified. If this outlay cannot be justified, it may be necessary to subcontract some processes to specialist companies. The selection process can be broken down into a number of stages, and there are certain important factors that have to be taken into account.

6.2.1 *Material*

The ease with which a metal can be cut (its machineability) will affect the choice of process and machine. Some very hard alloys can only be cast and ground and some metal removal is therefore required. To machine components made from high-duty alloys will require a heavier duty machine with a more powerful drive motor than that required to machine the same type of components from a free-cutting low-carbon steel.

6.2.2 *Process*

Modern machines and cutting tools can remove metal at such a high rate that the overall economics of the process can be easily overlooked. The conversion of stock material into swarf is wasteful, not only of the material, but also of energy to drive the machine. Further, high rates and high volumes of metal removal lead to the need for increased tool refurbishment and replacement costs. Minimum metal removal strategies should be aimed for and machining should be looked upon, wherever possible, as a finishing operation after *preforming* by casting, forging or powder particulate processes, for example.

6.2.3 *Human resources*

The process that keeps human resource costs to a minimum should be selected. To achieve this, the time spent on any operation should be minimal, including the initial tool setting, the actual operation of the equipment, and the loading and unloading of the machine. Physical effort should be kept to a minimum by the use of mechanical aids so that the operator does not tire and work less effectively as the shift progresses. The skill of the operator must also be taken into account, because highly skilled workers invariably command higher wages than workers with lower grade skills. Let us now look at some more specific factors affecting process selection.

6.3 Process selection (surface geometry and process)

Figure 6.5 shows a component that consists of a number of surfaces that are basic geometrical shapes. The production of all these geometrically shaped surfaces can be achieved by selecting the appropriate machining process. For example:

1. The circular plain surfaces on the end of the stem could be produced on a lathe at the same setting as the conical and cylindrical surfaces (1) and (2).
2. The cylindrical surface could be produced on a lathe.
3. The conical surface could also be produced on a lathe.
4. The rectangular plain surface perpendicular to the axis of the cylindrical surface could also be produced on the lathe at the same setting as (2) and (3).
5. The plain surfaces which are mutually parallel or perpendicular could be produced on a milling machine or on a shaping machine.
6. The inclined plain surfaces could also be produced on a milling machine or on a shaping machine.
7. The hole could be produced on a drilling machine.

Should a high level of finish be required, the turned surfaces could be finished on a cylindrical grinding machine and the plain surfaces could be finished on a surface grinding machine. Most engineering components are designed as combinations of geometrical shapes with specific machining processes in mind. Where more complex surfaces are required, copy-machining techniques, as shown in Fig. 6.6, or computer-controlled (CNC) machining techniques (see Chapter 8) can be employed. In Fig. 6.6 The master component is shown located between centres at the rear of the lathe bed. As the saddle traverses along the machine bed, a 'tracer' follows the profile of the master component and actuates a valve that controls the cross-slide movement hydraulically. The cross-slide and, therefore, the cutting tool move towards or away from the axis of the workpiece reproducing the profile of the master component.

Fig. 6.5 *Basic geometric shapes*

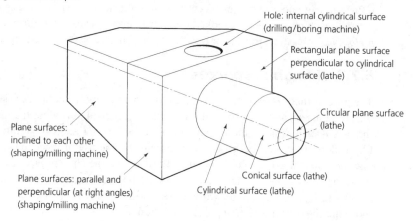

Hole: internal cylindrical surface (drilling/boring machine)

Rectangular plane surface perpendicular to cylindrical surface (lathe)

Circular plane surface (lathe)

Plane surfaces: inclined to each other (shaping/milling machine)

Conical surface (lathe)

Plane surfaces: parallel and perpendicular (at right angles) (shaping/milling machine)

Cylindrical surface (lathe)

Fig. 6.6 *Copy turning*

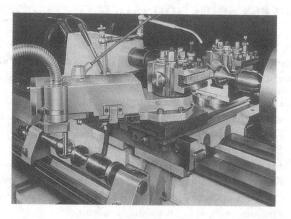

The *generation* and *forming* of geometrical surfaces was considered in *Manufacturing Technology*, Volume 1. Let us now consider the machining processes available for producing various geometrical surfaces.

- *Plain surfaces* Shaping, planing, milling, surface broaching, surface grinding.
- *Cylindrical and conical surfaces (external)* Turning, cylindrical grinding and centreless grinding (Section 6.17).
- *Cylindrical and conical surfaces (internal)* Drilling, parallel and taper reaming, boring, internal grinding, honing and lapping.
- *Screw threads* (Sections 6.7 to 6.10) Tapping, turning, grinding, and use of die heads.
- *Gear teeth* (Section 6.11 and 6.12) Milling, shaping, planing, hobbing, shaving and grinding.
- *Contoured surfaces* Copy milling, internal and external broaching, CNC machining (Section 8.2), electric discharge machining (EDM) (Section 6.19), and electrochemical machining (ECM) (Section 6.20).

6.4 Process selection (accuracy and surface finish)

From the above list it can be seen that there are alternative processes for producing any of the surfaces required. The alternative processes offer varying degrees of accuracy, varying size and mass of the component that can be machined, and varying process costs. Let's now consider the accuracy and surface finish of various manufacturing processes.

In Chapter 1 you were introduced to the concept that the closer the limits (smaller the tolerance) the more difficult and expensive it is to manufacture a component. On the other hand, it is no use choosing a process just because it offers low-cost production if it cannot achieve the tolerance specified. Table 6.1 is based upon BS 4500 and shows the standard tolerances from which tables of limits and fits can be derived (see Volume 1). It also shows how the international tolerance (IT) number is related to standard tolerances for various ranges of linear dimension. The figures given are in microns ($0.001\,\mathrm{mm} = 1\,\mu\mathrm{m}$), so a tolerance grade of IT9 for a dimension of 12 mm would have a standard tolerance of $43\,\mu\mathrm{m}$, that is 0.043 mm. It can be seen from the table that, as the IT number gets larger, the tolerance increases. The recommended relationships between processes and standard tolerances are as follows:

IT16	Sand casting, flame cutting
IT15	Stamping, hot rolling
IT14	Die casting, plastic moulding
IT13	Presswork, extrusion
IT12	Light presswork, extrusion
IT11	Drilling, rough turning, boring
IT10	Milling, slotting, planing, cold rolling
IT9	Low-grade capstan, turret and automatic lathe work
IT8	Centre lathe, capstan, turret and automatic lathe work
IT7	High-quality turning, broaching, honing
IT6	Grinding, fine honing
IT5	Machine-lapping, fine grinding
IT4	Gauge making, precision lapping
IT3	High-quality gap gauges
IT2	High-quality plug gauges
IT1	Slip gauges, reference gauges

Table 6.1 Standard tolerances

Nominal sizes | Tolerance grades

Tolerance unit 0.001 mm

Over (mm)	Up to and incl. (mm)	IT01	IT0	IT1	IT2	IT3	IT4	IT5	IT6†	IT7	IT8	IT9	IT10	IT11	IT12	IT13	IT14*	IT15*	IT16*
—	3	0.3	0.5	0.8	1.2	2	3	4	6	10	14	25	40	60	100	140	250	400	600
3	6	0.4	0.6	1	1.5	2.5	4	5	8	12	18	30	48	75	120	180	300	480	750
6	10	0.4	0.6	1	1.5	2.5	4	6	9	15	22	36	58	90	150	220	360	580	900
10	18	0.5	0.8	1.2	2	3	5	8	11	18	27	43	70	100	180	270	430	700	1100
18	30	0.6	1	1.5	2.5	4	6	9	13	21	33	52	84	130	210	330	520	840	1300
30	50	0.6	1	1.5	2.5	4	7	11	16	25	39	62	100	160	250	390	620	1000	1600
50	80	0.8	1.2	2	3	5	8	13	19	30	46	74	120	190	300	460	740	1200	1900
80	120	1	1.5	2.5	4	6	10	15	22	35	54	87	140	220	350	540	870	1400	2200
120	180	1.2	2	3.5	5	8	12	18	25	40	63	100	160	250	400	630	1000	1600	2500
180	250	2	3	4.5	7	10	14	20	29	46	72	115	185	290	460	720	1150	1850	2900
250	315	2.5	4	6	8	12	16	23	32	52	81	130	210	320	520	810	1300	2100	3200
315	400	3	5	7	9	13	18	25	36	57	89	140	230	360	570	890	1400	2300	3600
400	500	4	6	8	10	15	20	27	40	63	97	155	250	400	630	970	1550	2500	4000
500	630	—	—	—	—	—	—	—	44	70	110	175	280	440	700	1100	1750	2800	4400
630	800	—	—	—	—	—	—	—	50	80	125	200	320	500	800	1250	2000	3200	5000
800	1000	—	—	—	—	—	—	—	56	90	140	230	360	560	900	1400	2300	3600	5600
1000	1250	—	—	—	—	—	—	—	66	105	165	260	420	660	1050	1650	2600	4200	6600
1250	1600	—	—	—	—	—	—	—	78	125	195	310	500	780	1250	1950	3100	5000	7800
1600	2000	—	—	—	—	—	—	—	92	150	230	370	600	920	1500	2300	3700	6000	9200
2000	2500	—	—	—	—	—	—	—	110	175	280	440	700	1100	1750	2800	4400	7000	11000
2500	3150	—	—	—	—	—	—	—	135	210	330	540	860	1350	2100	3300	5400	8600	13500

* Not applicable to sizes below 1 mm.
† Not recommended for fits in sizes above 500 mm.

Surface finish quality is also related to process selection and accuracy. It is no use specifying close dimensional tolerances to a process whose inherent surface roughness lies outside that tolerance. This concept was introduced in Fig. 1.1. It shows how the dimensional tolerance and the surface finish of the manufacturing must be matched. Surface finish and its assessment is discussed in *Manufacturing Technology*, Volume 1, in some detail. Table 6.2 relates the roughness grade (N) number to the R_a value in microns and to various manufacturing processes, where the R_a number is a measure of surface finish.

Table 6.2 *Relationships between surface texture and process*

R_a value (mm)	Roughness grade number (N-value)	Process	Typical N-value
50	N12	Casting, forging, hot-rolling	N11–N12
25	N11	Rough turning	N9
12.5	N10	Shaping and planing	N7
6.3	N9	Milling (HSS cutters)	N6
3.2	N8	Drilling	N6–N10
1.6	N7	Finish turning	N5–N8
0.8	N6	Reaming	N5–N8
0.4	N5	Commercial grinding	N5–N8
0.2	N4	Finish grinding (tool room)	N2–N4
0.1	N3	Honing and lapping	N1–N6
0.05	N2	Diamond turning	N3–N6
0.025	N1		
0.0125	—		

6.5 Process selection (tooling and cutting costs)

The calculation of cutting speeds and feeds, the power required for a given rate of metal removal and the calculation of tool life were all considered in Volume 1. Let's now consider them further in order to determine the overall tooling costs for a particular process. At first sight it would appear that since increasing the cutting speed and feed reduces the process time, this should also reduce the process cost. However, any increase in cutting speed and feed also increases the rate at which the tools wear out so that 'downtime' for tool changing will be more frequent. The cost of tool refurbishment will be increased and there will be the added cost of lost production during tool replacement. Further, there is a greater chance of tool failure and the production of scrap components. Thus, an optimum set of operating conditions has to be arrived at which attempts to achieve a balance between these various conflicting influences.

Experimental work by Takeyama and Murata showed that the relationship between tool temperature and tool life for any given set of cutting conditions is logarithmic. Other experimental work showed that tool temperature is related to cutting speed and feed by an equation of the form:

temperature (K) = some function of (cutting speed × feed)

Increasing the cutting speed increases the temperature more rapidly than increasing the feed, but increasing the feed not only increases the forces acting on the tool but also produces a rougher surface texture.

Since tool life is related to tool temperature and tool temperature is, in turn, related to the cutting speed and feed, it is reasonable to expect that tool life is also related to cutting speed and feed. Taylor showed that an empirical relationship does exist between these variables and his tool life equation

$$Vt^n = C$$

has already been introduced in *Manufacturing Technology*, Volume 1. A typical curve relating V and t is shown in Fig. 6.7(a). Cost is also related to tool life (see Example 6.1), because the longer the tool life, the lower will be the tool refurbishment and resetting costs. Thus, tool cost is proportional to $1/t$ and this relationship is shown in Fig. 6.7(b). By combining these figures, as shown in Fig. 6.7(c), it is possible to arrive at the optimum process conditions for lowest cutting cost and most economical cutting speed for a given set of conditions.

Fig. 6.7 *Effect of cutting speed on tooling cost: (a) relationship between tool life and cutting speed; (b) relationship between tooling costs and cutting speed; (c) variation in machining costs with cutting speed*

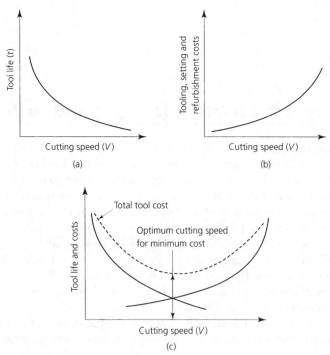

The following data refer to the turning of a batch of 1000 components.

Cutting speed	40 m/min
Value of n for the tool/work combination	0.2
Value of C	80
Time to change tool tip and reset	3 min
Turning time per component	4 min
Machine cost	£0.15/min

$$Vt^n = C \qquad V = 40 \text{ m/min}; \quad n = 0.2; \quad C = 80$$

$$t = \left(\frac{C}{V}\right)^{1/n}$$

$$= \left(\frac{80}{40}\right)^{1/0.2}$$

$$= 32 \text{ min between tool changes}$$

$$\text{Number of components per tool change} = \frac{32}{4} = 8$$

$$\text{Tool change cost 3 min} \times £0.15 = £0.45$$

$$\text{Tool change cost per component} = \frac{£0.45}{8} = £0.056$$

Cost of turning a component = 4 min × £0.15 = £0.60

∴ Total cost per component = £0.60 + £0.056 = £0.656

∴ Total cost for a batch of 1000 components is £656

(*Note*: Cost of tool changing per batch is £56.)

6.6 Process selection (cutting tool materials)

It has already been stated (Section 6.5) that tool life is related to temperature and therefore to cutting speed. Figure 6.8 shows how the hardness of tool materials is related to temperature. It also shows that the hardness at ambient temperatures can be misleading. For example, hardened and tempered high-carbon steel is much harder at room temperature than high-speed steel; however, the carbon steel soon loses its hardness (its temper is 'drawn') as the cutting temperature increases, whilst high-speed steel remains hard up to a dull red heat.

6.6.1 *Carbide-tipped tools*

Cemented carbide- and ceramic-tipped tools can withstand even higher temperatures than high-speed steel without softening and are very much more resistant to abrasion. However, such materials are relatively weak in tension and are also more brittle. Their lower mechanical strength requires greater support of the cutting edge. Therefore, negative rake

Fig. 6.8 *Hardness–temperature curves for cutting-tool materials*

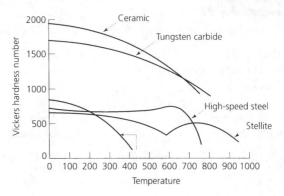

Fig. 6.9 *Negative rake cutting: (a) positive rake; (b) negative rake*

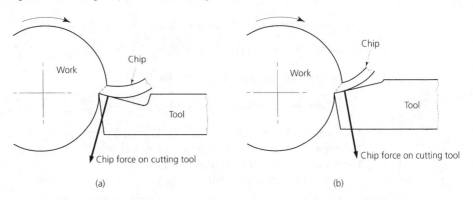

tool geometry is frequently adopted when using carbide and ceramic tool materials as shown in Fig. 6.9. To benefit fully from negative rake geometry the power and rigidity of the machine tool needs to be increased in order to increase the work done on the workpiece material at the cutting zone. This, in turn, increases the temperature of the cutting zone, resulting in increased plasticity of the workpiece material and the chip at the point of cutting. Increasing the plasticity of the workpiece and chip at the cutting zone reduces the mechanical forces acting on the cutting tool even when high-duty alloys are being machined.

The choice of carbide tip must be matched to the material being cut and the cutting conditions. The hard particles of carbide are dispersed through a softer and tougher matrix. This can be compared with the hard granite chippings set in a matrix of tar as used for road surfacing. Too much tar and not enough chippings would result in a surface that would quickly wear away. On the other hand, more chippings and less tar would result in a harder surface that would be long lasting for light traffic conditions but would be too brittle and would break up quickly at the first glimpse of a juggernaut.

Similarly, more particles of carbide and less cobalt in the matrix results in a wear-resistant cutting-tool material, but it would lack toughness. It would be very suitable for taking finishing cuts and for machining low-strength and abrasive materials (such as grey iron castings and laminated plastics). Less carbide and more cobalt in the matrix will produce a less wear-resistant cutting-tool material, but one that is very much tougher and suitable for machining high-tensile materials or where an intermittent cut is being taken.

The grain size of the carbide particles is also important. A tool tip with very fine grains will hold a sharp-cutting keen edge and is suitable for finish machining. Such a tool tip would be highly suitable for machining non-ferrous metals and alloys. Coarse-grained tips will not sharpen to such a keen edge but are more suitable for the rough machining of high-tensile materials that tend to result in heavy wear and cratering of the rake face of the tool tip.

Tool tips made up from a mixture of titanium carbide particles and tungsten carbide particles (mixed carbide tools) are less hard and abrasion resistant than straight tungsten carbide tips, but they are very much stronger and tougher and are used for the machining of high-strength materials. Mixed carbide tools are less porous than straight tungsten carbide tools and are therefore less prone to forming built-up edges. The previous comments concerning particle size and quantity apply equally to mixed carbides. Table 6.3 lists the standard grades of carbide together with some typical applications.

Table 6.3 *Carbide grades for metal-cutting tools*

ISO code	ISO grades	General applications
P (blue)	P01–P50	Ductile materials such as plain carbon and low-alloy steels, stainless steel, long-chipping malleable cast iron, ductile non-ferrous metals and alloys
M (yellow)	M10–M40	Tough and difficult materials such as high-carbon steels and high-duty alloy steels, manganese steels, cast steels, alloy cast irons, austenitic stainless steel castings, malleable cast iron, heat-resistant alloys
K (red)	K01–K30	Materials lacking in ductility and components which cause intermittent cutting. Cast iron, chill-cast iron, short-chipping malleable cast iron, hardened steel, non-ferrous free-cutting alloys, free-cutting steels, plastics, wood, titanium alloys

Cutting properties

P01 ◄─────────────────────────── K30
Increasing hardness and ability to withstand wear
High cutting speeds and fine feeds

P01 ───────────────────────────► K30
Increasing toughness and ability to withstand
interrupted cutting with coarse feeds
Rough machining high-strength materials

Modern minimum metal removal strategies are leading to lighter and faster cutting conditions in which only small volumes of metal are removed in what are little more than finishing operations. This, coupled with developments in carbide tool-tip material, has resulted in a return to positive rake-cutting geometry for many disposable-tip carbide tools in the small and medium sizes.

6.6.2 *Coated tools*

For automatic and computer-controlled machines, *coated carbide* tipped tools are now widely used. The coating is usually *titanium nitride* (TiN) and the coating technique is either physical vapour deposition (PVD) or chemical vapour deposition (CVD). Coating thicknesses are typically between 1 and 6 μ and the hardness of the coating can be as high as Vickers 3000 DPHN (Diamond Pyramid Hardness Number). The coefficient of friction for the coated surface is very much lower than for uncoated surfaces. This not only allows the chip to flow more freely across the rake face of the tool, but it also prevents chip welding and the formation of a built-up edge. Both these factors lead to a better workpiece surface finish. It is this lower coefficient of friction that permits coated tools to be operated at significantly higher speeds and feed with no increase in temperature and with no loss of tool life of the substrate.

Typically, cutting speeds can be increased by up to 50 per cent and the tool life can be extended by up to eight times that of a similar uncoated tool. Tests on drills and turning tool inserts have shown that the coating reduces both flank and crater wear. Such tests have shown that the cutting forces on coated tools remains constant over long periods, whereas the forces on uncoated tools increases progressively due to tool wear and the formation of a built-up edge. The coating can also be applied to high-speed steels.

Where tools are reground, as in the case of twist drills, the performance of the drill is not affected significantly, as long as the coating is still present in the flutes which provide the rake face of the tool; the antifriction properties of TiN enhance chip clearance from the hole. Generally, holes drilled with coated drills have a smoother finish and improved dimensional and geometrical accuracy, often eliminating the need for reaming.

Coating is expensive and the cost of coated tools can be up to two and a half times the cost of uncoated tools. Therefore, such tools can only be justified where their full potential can be fully exploited by the machining equipment available.

6.6.3 *Ceramics*

Ceramic tool tips are even harder than carbide tips, but unfortunately, they are also more brittle. The ceramic material most commonly used is aluminium oxide, either commercially pure or mixed with other metallic oxides such as chromic oxide. As ceramic tool tips are weak in tension and susceptible to edge chipping, they are only used for high-speed finishing cuts with fine feeds where high standards of surface finish are required. Machines used in conjunction with ceramic cutting tools must be rigid and capable of maintaining high cutting speeds without vibration, as vibration and chatter will cause immediate failure of ceramic tooling. Let's now consider the nitride ceramics.

6.6.4 *Sialon ceramics*

Sialons are a family of silicon nitrides containing differing levels of of aluminium and oxygen substitution (Si–Al–O–N). These materials are particularly successful when cutting cast irons and nickel steels. They combine the high-temperature (high cutting speed) performance of conventional solid ceramic tips (Al_2O_3) with the impact resistance of normal coated carbides and are thus suitable for intermittent cutting. Unlike conventional ceramic tool materials, sialons are also resistant to thermal shock and can be used with coolants. This improves the tool life even further and also improves the finish of the machined surface. The position of sialon ceramics in the cutting tool material hirearchy is shown in Fig. 6.10. From the same figure it can be seen that boron nitride heads the list for maximum cutting speed, but only at very fine feeds.

Fig. 6.10 *Position of sialon ceramics in the cutting-tool materials hierarchy*

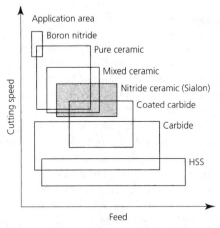

6.6.5 *Boron nitride*

Polycrystalline cubic boron nitride (PCBN) has a hardness comparable with diamond, the hardest known substance. Tool life can be as high as 20 times that expected from coated carbide tooling provided that the same cutting parameters are maintained. By reducing the 'downtime' for tool changing, the productivity of a machine tool using PCBN-tipped tools can be considerably (up to 20 per cent) improved. As for all very hard cutting tool materials, PCBN tips are easily chipped and care must be taken in tool approach and choice of feed rate. Again it must be remembered that sophisticated cutting-tool materials such as the sialons and cubic boron nitride are substantially more expensive than more conventional materials such as coated and uncoated carbides. Therefore they should only be used where their properties can be fully exploited.

6.6.6 *Disposable-tip tools*

Brittle cutting-tool materials, such as *carbides* and *ceramics*, tend to be easily chipped. Therefore care is required in the fast approach of the tool to the workpiece. For this reason the more accurately and consistently controlled cutting conditions of automatic and computer-controlled machine tools results in improved tool life and fewer breakages when using carbide- and ceramic-tipped tools.

When using such tool materials the *cost of replacement and refurbishment* must be considered more carefully. To achieve the maximum performance from such materials the initial tool geometry must be strictly maintained, yet these materials are more difficult to regrind than steel tools and usually need to be honed after grinding. For these reasons, modern practice favours the use of 'throw-away' disposable tool tips clamped into special holders. Such disposable tips are mass-produced relatively cheaply to a high degree of accuracy so that consistency of performance is guaranteed. To extend the life of the tips, they usually have more than one cutting edge and, to present a fresh cutting edge, can be indexed round in the tool holder as each cutting edge becomes dulled. The cost of replacement is usually much less than attempting 'in-house' refurbishment.

6.6.7 *Diamond*

Industrial diamonds are used for finish-machining operations. Although very hard, diamonds have a low shear strength and can only be used with very fine feeds so that the forces acting on the tool are minimised. However, because of the very high measure of hardness possessed by diamonds, which is sustained at high temperatures, very high cutting speeds can be used. Diamond turning tools are used for finish turning such components as aluminium alloy pistons for internal combustion engines, aluminium alloy lens components for cameras, and copper rolls for printing colour illustrations. None of these components lends itself to finishing by grinding because of the material from which they are made, but the finish attainable from diamond turning can be equally as good.

Finally, in selecting a suitable cutting-tool material it must be remembered that a compromise has to be achieved between all the parameters discussed previously in this section. For example, despite the fact that the harder cutting tool materials are more brittle than conventional materials and have to be used with a finer feed, their productivity is greater because of the higher cutting speeds that can be used and the fact that tool changing is less frequent. On the other hand, they cost more and this additional outlay can only be recovered if the machines in which they are to be used have the power and rigidity to exploit the full potential of such materials. Again, the component itself and the material from which it is made must be compatible with the tooling selected. It would not be cost-effective to use a very sophisticated cutting tool on a low-powered and worn machine. Similarly, it would not be cost-effective to use a cutting tool capable of very high rates of material removal if the component is of slender design which limits the clamping forces that can be applied and which, itself, cannot sustain the forces exerted upon it by such a cutter.

SELF-ASSESSMENT TASK 6.2

1. For a component of your own choice discuss the selection of a suitable machining process in terms of workpiece material, surface geometry, accuracy and finish, tooling and cutting costs.

2. Discuss the selection of suitable cutting materials for:
   (a) rough turning grey iron castings
   (b) milling high-tensile steel forgings
   (c) turning small components from free-cutting brass on a CNC lathe
   (d) turning i/c engine pistons from aluminium alloy castings on a CNC lathe

3. With the aid of manufacturers' literature, discuss the economics of using standardised disposable tooling systems.

6.7 Capstan and turret lathes

In Section 6.2.3 we considered the importance of the economic use of human resources. The lathes, milling machines and drilling machines used in jobbing workshops require to be operated by high-cost, skilled labour. This would be uneconomical for medium–high-volume production. Capstan lathes were originally developed for the medium-volume production of turned parts using relatively low-skilled, low-cost labour. Figure 6.11(a) shows an example of a capstan lathe and Fig. 6.11(b) shows typical turret tooling for such a machine. By using preset tooling which can be presented to the work sequentially and fed to preset stops, the skill of the centre lathe operator is not required, whilst the time per part is greatly reduced. Further, the accuracy and repeatability of the process is improved. For larger work a turret lathe is used. Apart from size, the essential difference between these two types of machines lies in the fact that the turret movement of a capstan lathe is limited by the length of the slide on which the turret is mounted. On a turret lathe the turret is free to move along the full length of the bed of the machine.

Fig. 6.11 *Capstan lathe and tooling: (a) capstan lathe; (b) tooling*

(b)

(a)

Both capstan and turret lathes have largely been superseded by CNC turning centres. However they are still in use in industry and, although no longer manufactured, there is an increasing demand for reconditioned capstan lathes by small (start-up) firms both in this country and abroad. In this country, the low capital cost of reconditioned capstan lathes compared with a comparable CNC machine makes it an economic investment despite the need for manual operation. In 'third-world' countries low labour costs makes them even more attractive. In both situations their simple, rugged and reliable technology resulting in ease of setting and maintenance is an added advantage. The continuing popularity of capstan and turret lathes, in the face of modern machine tool technology, is proved by the sustained high level of demand for their tooling equipment in the UK and abroad.

6.8 Automatic lathes

The next logical step in the development of the lathe was to remove the need for the capstan lathe operator and make all the machine movements automatic. This was achieved mechanically by the use of cams that have to be changed for each new component. The manufacture and setting of the cams increased the 'lead time' compared with setting up a capstan lathe and the cams are themselves a considerable extra cost. However, for very long runs this is not such an important factor, and this is especially so where there are regular repeat orders and the cams can be reused – for example, the manufacture of standard bolts and nuts.

Where very long runs of the same or similar components are required, automatic lathes can out-perform CNC lathes. This is particularly true of multi-spindle automatic lathes where many operations can be performed at the same time. However, the demarcation lines are becoming increasingly blurred as CNC technology is being merged into the design of the latest automatic lathes to produce highly versatile and highly productive machines with the flexibility to respond to the demand for the quick turnaround of smaller batches. Traditionally, the three most popular types of automatic lathe are:

- Single-spindle sliding-head automatic lathes.
- Single-spindle turret-type automatic lathes.
- Multi-spindle bar and chucking automatic lathes.

6.8.1 *Sliding-head automatic lathes*

These machines are used for the production of very small components such as watch and instrument pivots from small-diameter rod. Because this type of machine was developed for the Swiss watch and clock industry it is often referred to as a Swiss-type automatic lathe. The general layout of the machine is shown in Fig. 6.12(a). In all other lathes, the headstock position is fixed, as is the length of the workpiece material protruding from the spindle collet. The cutting tools then traverse along the protruding bar to take the required cuts. However, for very small components such as watch, clock and instrument spindles and pivots, the rod from which they are made is not sufficiently strong to resist the cutting forces. In the sliding-head automatic lathe:

- The tool bracket supports four or five tool slides arranged radially around the workpiece material, together with a *bush* that supports the workpiece material right up to the point of cutting, as shown in Figs 6.12(b) and 6.12(c).
- The headstock slides along the bed of the machine and feeds the rotating material through the bush and past the cutting tools to manufacture the component.
- A feed base is provided in place of the conventional turret for bar-end operations such as drilling, screw threading, etc.
- A front camshaft controls the head movement and, therefore, the amount of feed and the feed rate. It also carries cams that control the movements of the radial tool slides on the tool bracket, and cams that control the feed base.

Fig. 6.12 *Single-spindle, sliding head (Swiss-type) automatic lathe: (a) general layout of sliding head automatic lathe; (b) tool slide and bush bracket; (c) principle of support of the bar stock at the point of cutting*

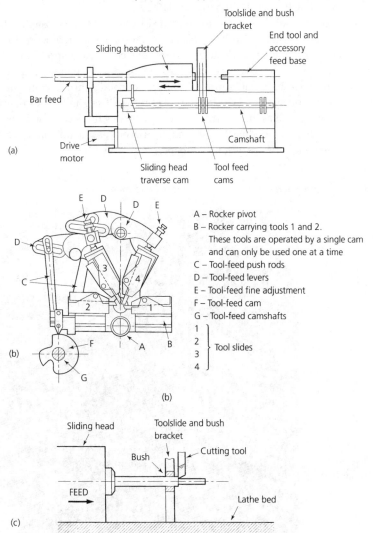

A – Rocker pivot
B – Rocker carrying tools 1 and 2.
   These tools are operated by a single cam
   and can only be used one at a time
C – Tool-feed push rods
D – Tool-feed levers
E – Tool-feed fine adjustment
F – Tool-feed cam
G – Tool-feed camshafts
1 ⎫
2 ⎬ Tool slides
3 ⎪
4 ⎭

The camshaft is driven through its cycle time from a constant-speed backshaft that also drives the headstock spindle. One revolution of the front camshaft produces one component, so various component cycle times can be made available by changing the gears connecting the camshaft to the constant-speed backshaft.

6.8.2 *Single-spindle turret-type automatic lathes*

This type of automatic lathe (also known as an automatic screw machine since they are widely used for nut and bolt manufacture) derives directly from the capstan lathe. The bar stock protrudes from the headstock spindle and the cutting tools are mainly carried on a turret. Unlike a capstan lathe, the turret of an automatic lathe is in the vertical plain (its axis is horizontal), as shown in Fig. 6.13. The tool feed is provided by movement of the tool turret along the bed as for a capstan lathe.

All the tool movements are operated from the front camshaft. These include infeed to the front and rear cross-slide tools and the turret tools. Again, one rotation of the camshaft equals one complete cycle and produces one complete component. The camshaft is driven from a constant-speed backshaft through cycle time change gears. The cams and clutches on the backshaft will cause the turret to rotate through one or two stations, speed changing, opening and closing the collet for bar feed, and forward and reverse drive to the spindle. As well as turning from the bar, these machines can also be used for chucking work with magazine loading for second operation production and the machining of small castings and forgings of suitable shape.

Fig. 6.13   *Single-spindle turret-type automatic lathe (cutting zone)*

6.8.3 *Multi-spindle automatic lathes*

These may have from 4 to 8 spindles and may be used for bar or chucking work. We will consider a typical 6-spindle chucking machine. The spindles are arranged at 60° intervals around the axis of the *spindle carrier*, as shown in Fig. 6.14(a); this figure also shows the Geneva mechanism that is used to index the spindle carrier. Figure 6.14(b) shows the *end toolslide* that carries the various turning tools and attachments. This slide has a common axis with the spindle carrier but does not index. However, it can move axially to and from the spindles to provide tool traverse, as shown in Fig. 6.14(b). Each spindle position also has a side-slide to provide additional tooling facilities.

Fig. 6.14 *Spindle carrier and end toolslide for 6-spindle automatic lathe: (a) spindle carrier for 6-spindle automatic lathe; (b) end toolslide for 6-spindle automatic lathe (reproduced by courtesy of BSA Machine Tools Ltd)*

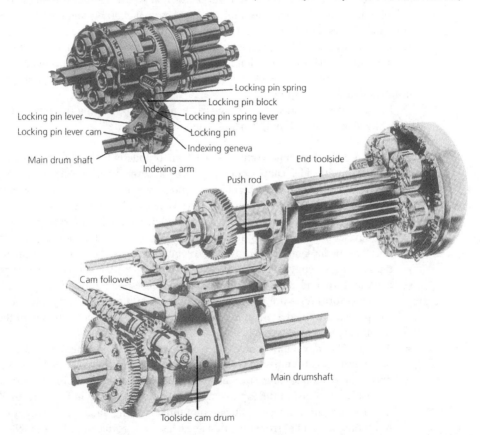

As the carrier indexes the spindles from one position to the next, the work meets each set of tools in turn, so by the time it reaches the final (unload/load) position it has been presented to all the tools. It has, therefore, received all its machining operations, and the component is finished. An example of a component and the machining operations it receives at each station is shown in Fig. 6.15. Since 12–16 tools can be operating at the same time, this enables components to be produced from 3 to 10 times faster than by single-spindle machines. The rate of production is very high indeed, which is where the multi-spindle automatic has advantages over a conventional CNC lathe. However, the CNC lathe has more flexibility and an easier changeover from one job to the next for small and medium-sized batch production. Setting an automatic lathe is a complex, highly skilled and time-consuming job, but providing the batch sizes are large enough the multi-spindle automatic lathe is more productive and more economical despite its higher initial cost. As for single-spindle automatics, multi-spindle machines also have a wide range of attachments that can be used to increase their versatility even further.

Traditionally multi-spindle automatics achieve high reliability using mechanical linkages, cams, and complex gearboxes capable of revolving the spindles and the cams from one motor. This complexity has entailed long changeover times between components, meaning that large batches were required to be economical. Many of these cam-type automatic lathes are still in use and will be for many years as they are highly productive, reliable and robust. However, technology marches on inexorably and the new generation of automatic lathes now being produced by BSA Tools Ltd combine the flexibility of CNC technology with the high productivity associated with multi-spindle automatic lathes. Another advantage is that by eliminating the need for cams the changeover time is much reduced – as is the noise level, which is an important factor in complying with current health and safety legislation.

Let us briefly consider two of the new generation of turning machines produced by BSA Machine Tools Ltd for the manufacture of turned parts.

- The *Speedturn* machine combines the high productivity of the conventional single-spindle, multi-slide, cam-operated automatic lathe with the flexibility of a CNC machine. The cross-slides operate independently and there is X-axis movement to the tool turret, as shown in Fig. 6.16(a). The Windows NT control system developed for this machine removes the need for specialised programming knowledge. This machine can have control of up to 8 axes of movement. Any combination of all slides can be operated at the same time, provided no collision exists. Thus the cycle time is exactly the same as for conventional cam-type automatics. With 8-axis control it is possible to contour turn and cut taper threads using single-point tools. Unlike the cam-type machine, form tools are not required. The setting time for this machine for a typical component is quoted as 45 minutes at the computer screen compared with the 6 hours required to set a cam-type machine.
- The new *Autoflex* multi-spindle series of automatic lathes combines all the advantages of the traditional multi-spindle machine with the flexibility of CNC techniques. This makes the economic batch size smaller and production more flexible. In addition, CNC side-slides are fitted in the top two slide positions, as shown in Fig. 6.16(b). These are used for contour turning and single-point thread cutting. Increased accuracy is achieved by automatic spindle error correction as each spindle corrects itself to its CNC slide. The advantages of this generation of machines over traditional multi-spindle lathes can be summarised as follows. Tool setting and job changeover are flexible since the spindle speeds, feeds, forms and threads are all programmable from the CNC unit. Accuracy is improved since the offset provides a common base position for each spindle as it is presented to the CNC slides. Reduced noise and oil mist result from the use of a separate drive motor for feed rates and spindle speeds in place of complex gear trains.

Fig. 6.15 *Manufacture of a clutch cylinder on a 6-spindle automatic lathe: (a) component to be machined; (b) machining sequence; (c) spindle sequence (reproduced courtesy of BSA Machine Tools Ltd)*

$7\frac{1}{4}$" BRB-6.

BSS 1452 grade 14 iron clutch cylinders machined on one side. The facing slide used was operated by a pusher mounted on the 4th position crossline.

Cycle time: 59 seconds

(a)

2nd position
Counterbore and recess.
Knee turn.
Face.

1st position
Drill
Trepan.
Knee turn.
Face.

6th position
Centre drill.
Knee turn.
Face – 2 tools.

5th position
Unload/load

4th position
Face cup and flange.
Chamfer – 3 tools.

(b)

3rd position
Chamfer – 2 tools.
Knee turn.
Form recess.

Straight magazine mounted
on headstock face

Stripper unit mounted
on tool slide

Pick-up and transfer unit
mounted on 6th position cross slide

(c)

Unload chute mounted
on gearbox face

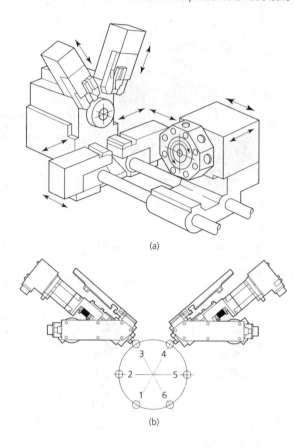

(a)

(b)

6.9 CNC turning centres

Capstan, turret and single-spindle automatic lathes have developed progressively from the basic centre lathe. Many CNC lathes have departed from this configuration for reasons of structural rigidity and for ease of chip disposal, as shown in Fig. 6.17. The change to slant-bed machines allows the chips to fall freely onto a conveyor built into the base of the machine for removal directly into a suitable disposal unit. Workholding, spindle control, tool selection, and tool movement are all controlled from a dedicated computer built into the machine. The programming of CNC machines was introduced in *Manufacturing Technology*, Volume 1, Chapter 5, and is developed still further in Chapter 8 of this book.

Fig. 6.17 *Configuration of CNC lathe to facilitate swarf removal*

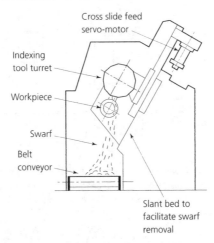

Cross slide feed servo-motor

Indexing tool turret

Workpiece

Swarf

Belt conveyor

Slant bed to facilitate swarf removal

SELF-ASSESSMENT TASK 6.3

1. Compare the advantages and limitations of capstan lathes, automatic lathes and CNC lathes for the production of small turned parts in medium and large batches.

2. Research the manufacturers' literature and write a presentation for your senior management justifying the replacement of a conventional multi-spindle automatic lathe with a modern machine incorporating a CNC system. Pay particular attention to the costs involved.

6.10 Screw-thread production (lathes)

So far, we have only considered the production of plain, cylindrical, and conical surfaces and combinations of these surfaces. Let us now consider some more complex geometrical surfaces. For example, a screw thread is a practical application of a helix, and a helix can be defined as: *the locus (path) of a point travelling around an imaginary cylinder so that its axial and circumferential velocities maintain a constant ratio.* This is most easily understood by considering the configuration for cutting a screw thread on a centre lathe.

You can seen from Fig. 6.18 that the spindle of the lathe provides the rotational movement and that the lead screw provides the axial movement necessary to generate a helix. To maintain a constant velocity ratio between the two movements, the spindle and the lead screw are connected by a gear train. By altering the gear ratio, the *lead* of the helix can be altered. The terms *lead* and *pitch* are often confused and their true meaning is shown in Fig. 6.19. Only single start threads, where lead and pitch are the same, will be considered in this chapter.

Fig. 6.18 *Screw-thread generation*

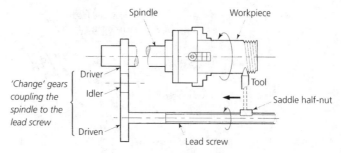

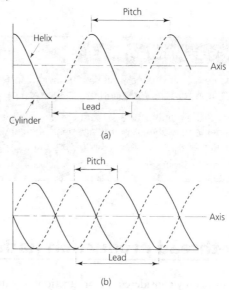

Fig. 6.19 *Pitch and lead: (a) single-start helix ('pitch' is the distance between adjacent corresponding points on the helix measured parallel to the axis; 'lead' is the distance a point moves along the helix in one revolution, i.e. the distance a nut would move along a bolt in one revolution); (b) two-start helix (lead = pitch × number of starts)*

There are two main considerations when screw cutting:

- The lead and pitch of the thread.
- The profile of the thread (see Section 7.8).

The tool is ground to the correct profile, and is set perpendicularly to the axis of the workpiece. It is fed more deeply into the rotating workpiece with each successive pass until the correct depth of thread is obtained. In practice the process is rather more complex and considerable skill is required on the part of the operator. Although widely used for prototype development, screw cutting on the centre lathe is too slow and demanding in craft skill to be used for medium- and high-volume production. We have already considered the flow forming of screw threads by rolling in Section 2.9. Let us now look at some alternative techniques used for quantity production.

6.10.1 *Die heads*

Self-opening die heads are used on capstan lathes for cutting internal and external threads, and two types are shown in Fig. 6.20 together with their thread chasers. Both types of die head can be preset to size and are fitted with mechanisms whereby they can take a roughing cut followed by a finishing cut. When the end of the thread is reached the die head opens automatically and can be withdrawn over the work without having to stop or reverse the spindle.

Fig. 6.20 *Die heads: (a) radial-type die head; (b) tangential-type die head*

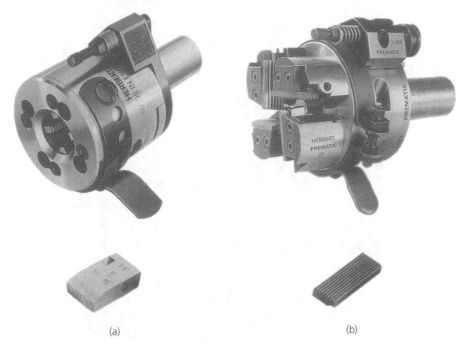

(a)   (b)

Solid die heads are used on turret-type single-spindle automatic lathes as there is no operator to reset the die head between operations. The machine spindle is reversed so that the die head can be withdrawn when the thread has been cut. Sliding-head automatic lathes use single-point tools, and the axial component of the helix is controlled by the traverse cam instead of a lead screw.

6.10.2 *Thread generation on CNC turning centres*

Screw-thread cutting with single-point tools (similar to the centre lathe process) is used on CNC turning centres. To aid the programming a 'canned' cycle is available. Programming for screw cutting is considered in Section 8.3.3.

6.10.3 *Thread chasing*

Figures 6.21(a) and 6.21(b) show the difference between a thread cut with a single-point tool and a thread cut with a chaser. The nose radius of a single-point tool forms the root radius, but as the crest of the thread is left flat it does not have a true form. The crest radius can only be formed by using a multi-tooth cutting tool called a *chaser*. Some typical chasers are shown in Fig. 6.21(c). Thread chasing using die heads has already been described but single chasers can also be set in the tool posts mounted on the cross-slide of a capstan lathe, as shown in Fig. 6.22. To provide the axial feed for the chaser, a short lead screw is fitted. Internal and external thread chasing is used where the diameter of the screw thread is too large for a die head or where the screw thread is obstructed by a shoulder.

Fig. 6.21 *Thread chasing: (a) thread-form left by a single-point tool; (b) thread-form left by a chaser; (c) thread chasers*

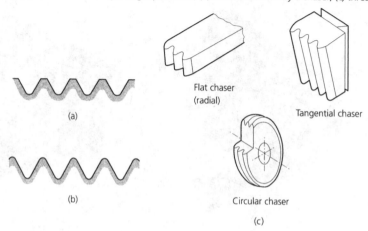

Fig. 6.22 *Thread chasing: internal and external threads*

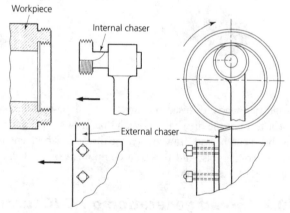

6.11 Screw-thread production (milling)

This is a rapid production process for cutting screw threads that are too large to be produced using die heads. Internal threads can also be cut. The two techniques used are:

- Hobbing.
- Form milling.

6.11.1 *Hobbing*

This process used a multi-tooth cutter and a full thread form is cut with radiused roots and crests. The principal of thread hobbing is shown in Fig. 6.23, where it can be seen that the threads on the hob are not helical but are a series of discrete, annular, parallel threads. Some interference with the threads being cut does occur but this is minimal in large-diameter, fine-pitch threads. The hob is gashed axially to produce cutting edges, and each tooth is form relieved to provide clearance. The hob should be a few threads longer than the thread being cut and this limits the length of the workpiece thread. External hobs are set on a stub arbor, whilst hobs for internal threads have a shank for fitting into the machine spindle.

Fig. 6.23   *Thread hobbing*

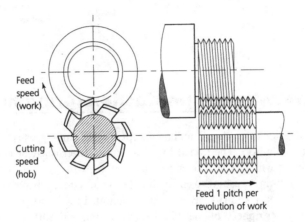

The workpiece and hob are carried on parallel spindles. The hob is rotated at the optimum cutting speed for the workpiece material, and the workpiece is rotated at a speed calculated to give the correct rate of feed. The rotating hob is fed into a stationary workpiece blank until the correct depth of thread has been achieved. To commence thread cutting, the feed is engaged and the workpiece rotates at the correct feed speed. At the same time, the hob moves axially at the rate of one pitch per revolution of the workpiece. The relative directions of rotation and axial movement, as shown in Fig. 6.24, would produce a right-handed thread. By reversing the axial feed of the hob a left-handed thread would be cut. This process allows for threading close up to a shoulder.

6.11.2 *Form milling*

This process is shown in Fig. 6.24. It is used for producing components such as worm gears. These are essentially large-diameter coarse pitch screw threads. A single form-relieved cutter is used and the axis of the cutter arbor and cutter is inclined to the axis of the workpiece spindle. This aligns the cutter with the angle of the helix and prevents interference with the flanks of the thread. Unlike thread hobbing, there is no limit to the length of thread that can be cut. Unfortunately, the thread cannot be cut up to a shoulder. The workpiece is rotated at one revolution per pitch for single-start threads and one revolution per lead for multi-start threads.

Fig. 6.24 *Form milling (θ = helix angle of thread)*

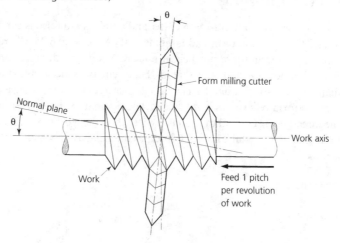

6.12 Screw-thread production (grinding)

This is similar to thread milling except that a grinding wheel is used instead of a milling cutter or a hob. The thread form can either be crushed or cut into the periphery of the grinding wheel. The former technique is used for fine pitch-threads. The wheel rotates slowly during crushing to ensure that (a) there is no slip between the abrasive wheel and the formed roller and (b) the roller is not abraded away. Alternatively, the profile can be cut into the periphery of the wheel using a diamond cutter controlled by a template and pantograph system, as shown in Fig. 6.25. This technique is used for coarse-pitch threads and worm gears.

Fig. 6.25 *Diamond dressing (pantograph) – pantograph ratio = A/B = C/D*

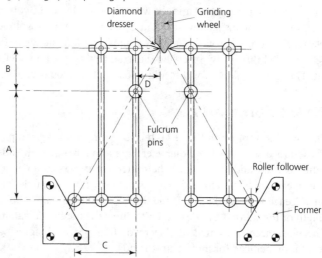

6.12.1 *Traverse grinding*

The principle of screw-thread production by traverse grinding is shown in Fig. 6.26. For fine threads a multi-ribbed wheel is used in which a number of annular thread-forms are arranged side by side as in a thread-milling hob. For coarse threads a single-ribbed wheel is used and the wheel axis is inclined to the helix angle to prevent interference with the thread flanks. Fine threads are often ground from the solid from pretoughened blanks but coarse threads are pre-machined, heat treated, and thread-ground merely as a finishing and sizing operation.

Fig. 6.26 *Traverse grinding: (a) coarse-thread grinding; (b) fine-thread grinding*

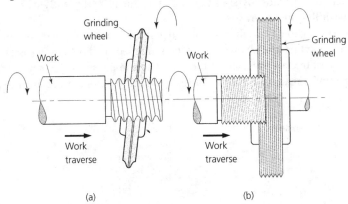

6.12.2 *Plunge-cut grinding*

The principle of plunge-cut grinding is shown in Fig. 6.27. The wheel is plunged into the work to full thread depth with the work stationary. The work then makes one revolution whilst the grinding wheel traverses one pitch for a single-start thread or one lead for a multi-start thread. The thread can be cut up to a shoulder when plunge-cut grinding.

Fig. 6.27 *Plunge-cut grinding*

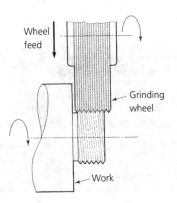

6.13 Production tapping

Tapping machines and tapping attachments are available for use with conventional taps. The machines and attachments provide automatic forward and reverse rotation of the tap depending upon the direction of feed. A slight downward pressure on the tap engages the forward drive, whilst a slight upward pressure engages reverse drive and allows the tap to unscrew itself out of the threaded hole it has just cut. Another type of tapping attachment uses a spring-loaded clutch so that when the tap reaches the bottom of the hole, or should the tap jam in the hole, the clutch slips and the machine spindle can be stopped and reversed. This latter type of tapping attachment is frequently used with large radial arm-type drilling machines.

Having to reverse the tap out of the hole is a waste of time. For nut tapping, where very large quantities are involved, hook taps are used, as shown in Fig. 6.28. The nuts run up the shank of the tap and spin off through a hole in the side of the tap carrier to be collected in an annular manifold. They fall away from the manifold by gravity into the work pallet.

Fig. 6.28 *Production tapping*

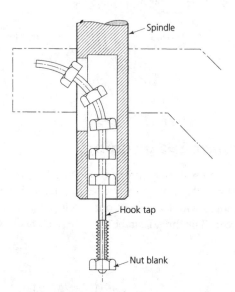

Spindle

Hook tap

Nut blank

SELF-ASSESSMENT TASK 6.4

1. Discuss the advantages and limitations of screw-thread production by single-point turning and the use of die heads in terms of production economics and the quality of the thread form produced.

2. Compare and contrast the process of thread production by hobbing and form milling and give examples where these processes would be used, with reasons for your choice.

6.14 Gear-tooth production (forming)

A second complex surface that has to be considered is the *involute*. Figure 6.29 shows that an involute curve is generated by an imaginary taught cord BTC being unwound from a base circle centre O, commencing at the point A. The curve AC, so generated, has an *involute form*. Modern gear teeth have an involute profile and this has the advantage that correctly meshing gear teeth of this form roll together with no sliding or scuffing to cause wear.

Fig. 6.29   *The involute curve*

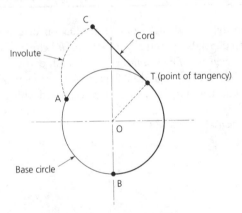

Milling is a form-cutting operation confined to single gears for prototype purposes, or very small batches of gears. Compared with the gear-generation process to be described in Section 6.15, gear forming using an involute milling cutter is a very slow and uneconomical method of production. Figure 6.30(a) shows a typical form milling gear cutter, and Fig. 6.30(b) shows the set up for producing the gear on a horizontal milling machine. The cutter is the profile of the space between the gears, and it is passed through the blank at a depth of cut dependent upon the strength of the cutter, the rigidity of the set up and the machineability of the workpiece material. After each pass the gear blank is rotated (indexed) to the next position by a device called a *dividing head* upon which the gear blank is mounted. When all the spaces have been cut, the cutter is set deeper into the blank and the process is repeated until the required tooth height is attained. Dividing heads and their associated indexing calculations were considered in *Manufacturing Technology*, Volume 1, Section 3.5.4.

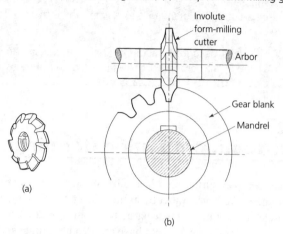

(a)

(b)

6.15 Gear-tooth production (generation)

For the medium- and high-volume production of gears, generating processes are used. Generation depends upon the fact that a *rack*, with easily machined straight-sided teeth, is an *involute gear of infinite radius* and will mesh correctly with any corresponding involute gear, as shown in Fig. 6.31. All gear generation processes are superior in respect of accuracy, finish, and rate of production compared with gear forming on a milling machine.

Fig. 6.31 *Rack and pinion*

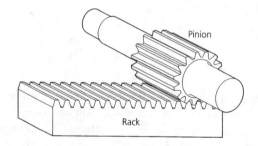

6.15.1 *Gear planing*

Figure 6.32 shows a gear blank, devoid of teeth, mounted on a mandrel that is geared to a slide carrying the rack cutter. The slide carrying the cutter can move in the direction XY and is placed so that the distance L between the pitch line of the cutter teeth and the axis of the blank will give the correct depth of cut. The gearing between the cutter slide and the mandrel carrying the blank is arranged so that as the blank rotates, the cutter slide moves along the path XY exactly as if the teeth being cut engaged the teeth of the rack cutter without slip.

Fig. 6.32 *Use of rack cutter (gear planing)*

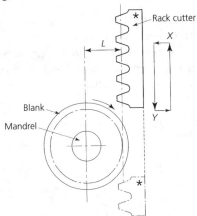

In order to cut the teeth, the rack edges are made into cutting edges by providing them with rake and clearance angles. The cutter slide and cutter are arranged so that they reciprocate in a plane parallel to the blank axis. Compared with this reciprocating motion the rotation of the blank and the movement of the cutter along XY is relatively slow since this movement provides the feed of the cutter. The feed takes place intermittently during the return (non-cutting) stroke of the cutter. Figure 6.33 shows the generating action of a rack cutter. For the sake of clarity the blank is shown fixed and the rack is shown rolling around it. This provides the same relative motion between the blank and the rack. When the rack reaches the end of its XY travel it is automatically withdrawn and returns to its starting position.

Fig. 6.33 *Involute generation*

Gear planing is a relatively simple process using relatively cheap and easily made cutters. These cutters can be made to a high degree of accuracy and are easy to regrind. It is the only process for cutting large-diameter gears and can be used for the production of single- and double-helical gears.

6.15.2 *Gear shaping*

This is similar to gear planing except that the workpiece blank is set with its axis vertical and the cutter also reciprocates in a vertical plane. Further, a circular pinion-shaped cutter is used in place of a rack cutter. Figure 6.34(a) shows a diagram of a pinion cutter and indicates the rake and clearance angles. Figure 6.34(b) shows the configuration of the set-up for cutting involute teeth in the blank.

Fig. 6.34 *Gear shaping: (a) pinion cutter; (b) cutting action*

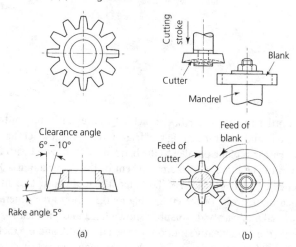

Whilst cutting takes place the cutter and blank rotate together slowly as though they are two gears in mesh. Gear shaping is quicker than gear planing because the cutting process is continuous and the cutter does not have to be constantly stepped back. Gear shaping is the most versatile of all gear-cutting processes and is capable of producing internal and external gears. It can also be used for manufacturing close-coupled cluster gears and splines. The pinion cutter is itself produced by planing since the rack cutter is the fundamental tooth form.

6.15.3 *Gear hobbing*

Hobbing is the most productive of all gear-cutting processes since cutting takes place continuously. Compare this continuous cutting action with planing and shaping where cutting only takes place on the forward stroke of a reciprocating cutter that is retracted clear of the work on the return stroke. Hobbing cutters for gear cutting have a tooth form which is cut helically, unlike threading hobs where the tooth form is annular. For this reason the hob and its spindle are set at an angle to the work. This angle is equivalent to the helix angle of the cutter, as shown in Fig. 6.35. The set-up for generating a gear by use of a hobbing cutter is shown in Fig. 6.36. Hobbing can only be used for producing spur gears and worm wheels. It cannot be used to cut internal gears and it cannot work up to a shoulder.

Fig. 6.35 *Effect of hob helix*

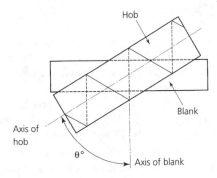

Fig. 6.36 *Principle of gear hobbing*

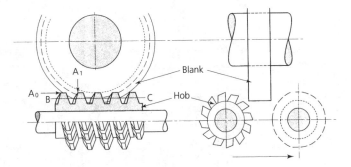

6.15.4 *Gear finishing*

Gears can be finished by shaving or by grinding where high accuracy and surface finish are required to reduce noise in high-speed applications. Gears are usually deburred using electrochemical machining, as described in Section 6.19. Before moving on to non-conventional machining processes, let's consider two further machining processes associated with medium- and high-volume production.

SELF-ASSESSMENT TASK 6.5

1. Discuss the advantages and limitations of gear tooth production by *generation* and *forming*.

2. Compare and contrast the processes of gear planing and gear shaping and explain where the use of each process would be appropriate.

3. Discuss the advantages and limitations of gear hobbing compared to planing and shaping.

6.16 Broaching

This process uses a cutting tool called a *broach*, and was originally developed as a means of machining non-circular holes. More recently the broaching process has been adapted for the machining of external surfaces. Some typical broached surfaces are shown in Fig. 6.37. The principle of broaching is shown in Fig. 6.38. This shows the simple broaching operation of cutting a keyway. The hole is first bored to the correct diameter and the broach is then pushed or pulled through it. Since the teeth of the broach stand progressively further out from the broach body, until the end is reached, each tooth successively deepens the keyway until the finished depth is reached. The broaching process can produce accurate internal and external surfaces and, for external surfaces, it competes with milling for large volume production where the quantity warrants the higher cost of the cutter. Figure 6.39 shows the elements of a typical broach and its teeth. There are three basic broaching techniques: *pull*, *push* and *surface*.

Fig. 6.37 *Typical broached surfaces: (a) internally broached surfaces; (b) externally broached surfaces (heavy outline denotes broached surface)*

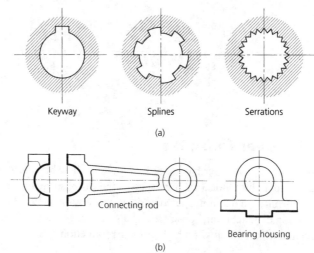

Fig. 6.38 *Principle of broaching*

Fig. 6.39 *Broach details: (a) internal broach layout; (b) tooth details – (i) roughing teeth (α = rake angle (to suit workpiece material), β = primary clearance $\frac{1}{2}°$ to 2°, θ = chip clearance 20° to 40°), (ii) finishing teeth*

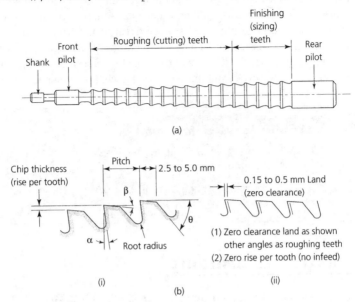

6.16.1 *Pull broaching*

As the name implies, the broaching machine pulls the broach through the workpiece that must always be pierced with a pilot hole equal to the root diameter of the finished profile. Because of the inherent stability of drawing the broach through the workpiece, long broaches can be used and the work finished in one pass. Since the broach needs to be threaded through the component it must be fitted with a quick-release shank for attachment to the machine.

6.16.2 *Push broaching*

This uses short broaches to prevent deflection and buckling. However, a short broach cannot remove as much metal in one pass as a long broach. Where large amounts of metal have to be removed a sequence of separate broaches are pushed through the same component. Push broaches are easier and cheaper to make than the long pull broaches and they are also less likely to distort or crack during hardening. Because of the shorter stroke required, the press used with push broaching is simpler and less costly. Since push broaching is usually done in the vertical plane, workholding is easier and machine loading is quicker as the broach does not have to be coupled to the machine ram.

6.16.3 *Surface broaching*

The push- and pull-broaching techniques discussed so far have been concerned only with internal broaching. However, external- or surface-broaching techniques have been developed as an alternative to the milling process. Surface broaches are very expensive and are usually built up in short sections, not only for ease of manufacture but also that, in the event of uneven wear or accidental damage, only part of the broach has to be repaired or replaced. Compared with milling, broaching is more expensive in tooling costs, but gives greater accuracy and improved surface finish to the work, coupled with higher rates of production, because the broaching teeth that take the roughing cuts never have to take the finishing cuts. Also, the slide that carries the broach is the only moving part of the machine during the cutting cycle, and the machine as a whole can therefore be made more rigid.

SELF-ASSESSMENT TASK 6.6

1. Describe, with the aid of sketches, the types of hole profiles for which broaching would be an appropriate process to use.

2. Compare and contrast the processes of broaching and milling a profiled surface of your choice.

6.17 Centreless grinding

Cylindrical grinding was introduced in *Manufacturing Technology*, Volume 1. However, for many applications, the cost of mounting components in chucks or between centres is prohibitive. In other instances, the components may not lend themselves to such conventional methods of workholding.

Centreless grinding is an alternative manufacturing process which lends itself to the accuracy, production rates and cost control demanded by the high-volume industries. Centreless grinding can provide metal removal rates of 0.25 mm per pass when roughing out, yet maintain a dimensional tolerance of 0.005 mm on finishing cuts. The process readily adapts to automatic sizing techniques and also to automatic component feed systems.

As the name suggests, the work is not held between centres but is supported on a workrest blade and held up to the face of the grinding wheel by a second abrasive wheel called the control or regulating wheel. This control wheel not only rotates the workpiece but also provides the through feed in the case of long parallel work. Figure 6.40(a) shows the layout of a typical centreless grinding machine, and Fig. 6.40(b) shows the relationship between the grinding wheel, the control wheel and the workpiece. Figure 6.40(c) shows the work guide and end stop in plan view. The end stop is not always required. The angularity of the workrest blade is necessary to:

- Assist in 'rounding-up' the workpiece and prevent lobing.
- Keep the workpiece from being drawn into the grinding wheel and to keep the workpiece in contact with the control wheel.
- Provide the control to ensure that the workpiece is brought gradually into contact with the grinding wheel when plunge grinding.

The workrest is set so that the axis of the workpiece is slightly below the centre line of the grinding wheel and the control wheel to help prevent a form of out-of-roundness called *lobing*.

Fig. 6.40 *External centreless grinding: (a) main features of a typical centreless grinding machine; (b) relationship between work and grinding/control wheels; (c) work guides and end stop*

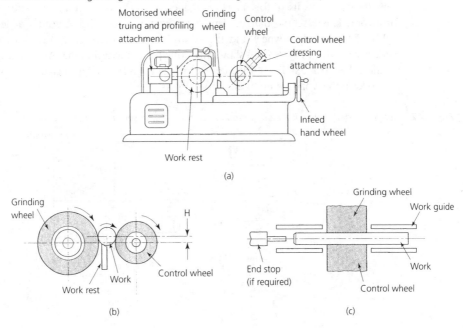

(a)

(b)

(c)

6.17.1 *Through feed (traverse) grinding*

This technique is used for parallel work of any length such as rods of silver-steel or the rollers of roller bearings. These components are of constant diameter and have no shoulders to prevent them from passing between the grinding wheel and the control wheel. In order to provide axial movement to the workpiece, the control wheel is inclined, as shown in Fig. 6.41. This inclination produces an axial component velocity and, as is evident from the velocity diagram, the greater the inclination the greater will be the feed rate (F).

Fig. 6.41 *Control wheel inclination for 'through-feed' grinding: (a) control wheel inclination; (b) velocity triangle*

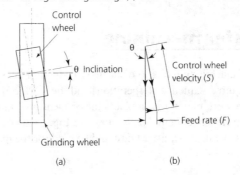

(a)   (b)

6.17.2 *Plunge grinding*

This technique is used for short, shouldered workpieces, multi-diameter work, and formwork. It is essential that the length of the component being ground is less than the width of the grinding wheel and the control wheel. Figure 6.42(a) shows some typical workpieces, whilst Fig. 6.42(b) shows the principle of the technique. An end-stop is used to position the work axially and the control wheel is given a slight inclination not exceeding 0.5° to keep the work fed up to the stop.

Fig. 6.42 *Plunge grinding: (a) typical plunge-ground components; (b) set-up for plunge grinding*

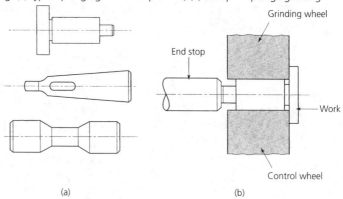

(a)   (b)

After the workpiece has been positioned against the stop, the control wheel is fed forward and advances the rotating workpiece up to the grinding wheel. When the control wheel slide is arrested by a positive stop, the control wheel is allowed to dwell whilst the grinding wheel 'sparks-out' to leave the workpiece the correct diameter. The slide and the control wheel are withdrawn and the work is automatically ejected by the end stop. As previously mentioned, the angularity of the workrest blade ensures that the workpiece falls back against the control wheel and prevents it from being drawn into the grinding wheel.

6.17.3 *End-feed grinding*

This is a hybrid technique embodying the principles of both through-feeding and plunge-grinding techniques. It is used for work that is too long for plunge grinding but which cannot be fed through the machine because of a shoulder or other obstruction.

6.18 Transfer machining

Transfer machining refers to dedicated production lines, where the machines are arranged in the required sequence of operations and the work is transferred from one machine to the next on a work platen that replaces the conventional machine worktable. An example of a linear transfer machining line is shown in Fig. 6.43(a) and typical unit heads are shown in Fig. 6.43(b). Such a transfer line could be used for the mass production of motor car engine cylinder blocks.

Fig. 6.43 *Dedicated transfer machine (linear): (a) typical transfer machining line (unit heads omitted for clarity); (b) typical unit heads (reproduced courtesy of Staveley Machine Tools Ltd)*

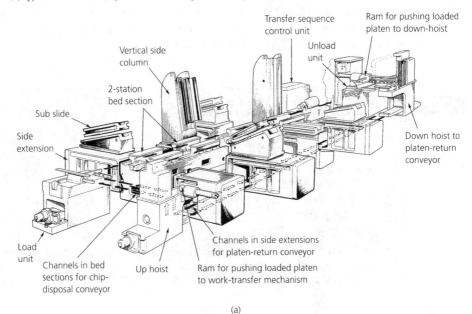

Transfer sequence control unit

Ram for pushing loaded platen to down-hoist

Vertical side column

Unload unit

2-station bed section

Sub slide

Side extension

Down hoist to platen-return conveyor

Load unit

Channels in bed sections for chip-disposal conveyor

Up hoist

Channels in side extensions for platen-return conveyor

Ram for pushing loaded platen to work-transfer mechanism

(a)

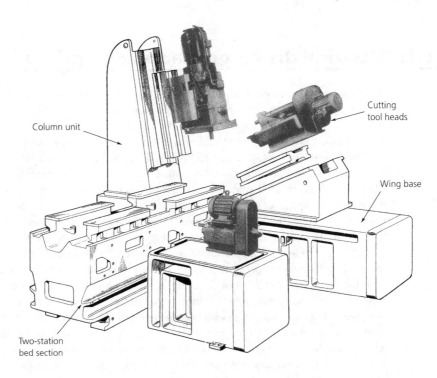

Column unit

Cutting tool heads

Wing base

Two-station bed section

(b)

For smaller components a rotary transfer machine is used. The work is mounted on a rotary table and is indexed under each workhead in turn. One workstation is left clear for loading and unloading. In the present age of rapid-response manufacturing philosophy there is a tendency to move away from dedicated, single-product transfer lines to *flexible manufacturing systems* (FMS) as discussed in Chapters 8 and 9.

SELF-ASSESSMENT TASK 6.7

1. Discuss the advantages and limitations of centreless grinding compared to conventional cylindrical grinding.

2. Compare and contrast the following centreless grinding operations and explain where each would be used giving reasons for your choice:
   (a) through feed (traverse) grinding
   (b) plunge grinding
   (c) end-feed grinding

3. Sketch out a schematic design for a transfer production line to manufacture a component of your choice.

6.19 Electrical discharge machining – principles

In addition to the conventional metal-cutting processes previously described there are several other processes that are increasingly being used. The first of these that we will consider is electrical discharge machining (EDM), which is also called 'spark erosion' since it uses the thermal energy of electric sparks to remove workpiece material.

The earliest practical system of EDM was developed during the 1940s using the circuit shown in Fig. 6.44. The authors of this system were the Lazarenko brothers in Russia. This EDM system is included since its simplicity is useful in describing the principle of the process. The capacitor is charged from a direct-current source through the resistor. Charging continues until the potential difference (voltage) across the capacitor exceeds the breakdown potential of the spark gap between the tool electrode and the workpiece. The spark then jumps across the shortest distance between the workpiece and the electrode. The extremely high spark temperature (circa 20 000 °C) causes local melting and vaporisation of both the workpiece and the tool electrode. However, by connecting the workpiece to the positive side of the supply and the tool electrode to the negative side of the supply, the rate of erosion of the workpiece is made very much greater than that of the tool electrode. After each discharge, the capacitor is recharged through the resistor and the process is repeated. The resistor has two purposes: firstly, to control the time-constant of the circuit, i.e. the frequency of the discharges; secondly, to limit the current flow from the supply to a safe value during the charge and discharge cycle.

Fig. 6.44 *Lazarenko RC (resistance–capacitance) circuit*

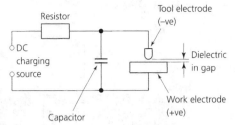

Both the tool electrode and the workpiece are immersed in a dielectric fluid, such as paraffin, or deionised water. The functions of the dielectric fluid are to cool the tool electrode and to flush away the debris caused by the discharge. Although the dielectric is normally an insulator, the very high potential across the spark gap immediately preceding the discharge causes local ionisation of the dielectric. This allows the spark to jump across the gap.

6.19.1 *Controlled pulse generation*

This simple circuit is not satisfactory for actual production requirements as too much time is spent in charging the capacitor and too little time in effective machining. This problem has been overcome by the development of controlled pulse generators that allow precise control of spark frequency and duration, thus allowing much higher metal removal rates.

6.19.2 *Control of gap width*

To attain maximum material removal an optimum gap width has to used and maintained at a constant value. This is achieved by means of a servo system that controls the movement of the tool electrode slide. The electronic control system constantly monitors the potential across the spark gap and compares it with a reference potential. If the potential across the spark gap rises relative to the reference potential, the servo closes the gap slightly. If the potential across the gap falls, the servo widens the gap slightly. A typical gap lies between 0.02 and 0.08 mm.

6.19.3 *Dielectric flushing*

Continuous flushing of the spark gap by the dielectric fluid is necessary in order to remove the debris of the erosion process. If the debris is allowed to build up in the spark gap a short circuit will occur. It is essential that the dielectric is filtered before it is recirculated. Further, continual flushing of the machining area improves the cooling of the tool electrode. Figure 6.45(a) shows a conventional flushing arrangement. Other techniques that are used are to:

- Pump the dielectric through a hole or holes in the electrode, as shown in Fig. 6.45(b).
- Draw the dielectric through the component by suction, if it already has holes drilled in it, as shown in Fig. 6.45(c).

Fig. 6.45 *Dielectric flushing: (a) conventional dielectric flushing; (b) dielectric flushed through tool electrode; (c) dielectric exhausted through workpiece*

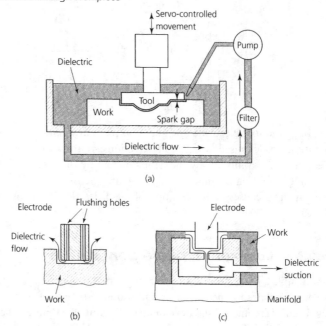

(a)

(b)

(c)

6.19.4 *Electrode design*

Electrodes range from simple rods to complex three-dimensional shapes for producing cavities in dies and moulds. A typical machining sequence for a complex cavity would require several replica electrodes. All but one of the electrodes would be required for roughing out, using relatively high current densities resulting in rapid metal removal. The final electrode would be used to obtain the finished size and required surface finish at a lower current density and lower metal-removal rate.

Electrode materials must be electrically conductive (as must the workpiece material). Preferably, they must have a low erosion wear rate. The basic cost of the electrode material, the cost of shaping the material to the required size and shape, and the working life of the material must all be taken into account when selecting an electrode material. Typical electrode materials include: copper alloys, zinc-based alloys, aluminium alloys, tungsten carbide, and graphite. Graphite is the most widely used material as it has the best all-round capability. Its main drawbacks are its susceptibility to damage whilst being handled and set in the machine, and the fact that the dust produced whilst machining it to shape represents a severe health hazard if inhaled. Comprehensive dust-extraction facilities must be provided.

6.19.5 *Application of CNC to EDM*

Conventional CNC systems can be applied to the X and Y axes of the EDM machine table and also to the Z motion of the electrode slide. This greatly increases the flexibility of the process and allows complex contoured shapes to be machined by the use of very simple circular or rectangular rod electrodes.

6.19.6 *Wire cutting*

The EDM wire-cutting technique is an important variant of the process. The erosion principles remain the same as those previously described, but the electrode is a continuously fed wire as shown in Fig. 6.46. The process is particularly suitable for cutting complex profiles in prehardened press tool die blanks using CNC profiling. The use of prehardened blanks avoids distortion and cracking problems that can occur when heat treatment is carried out after conventional machining. Although a relatively slow process, it does not require attention and can continue 'lights out' if required. Further, the final pass of the electrode not only provides an accuracy and surface finish which is sufficiently good to avoid the necessity for further machining, but can also provide the taper or 'draught' necessary in blanking dies.

Fig. 6.46 *EDM wire cutting*

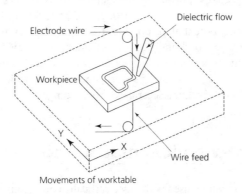

6.19.7 *Electrical discharge machining – applications*

The EDM process can be used to cut any electrically conducting material regardless of its hardness. Most of its applications are associated with the 'one-off' production of tools and dies. One advantage of the use of this process for producing tools and dies is the fact that the blanks can be hardened and tempered before machining. This prevents the cracking and distortion that can occur when heat treatment is carried out after machining owing to variations in cross-section.

There are some instances where EDM is used for low-volume production, particularly in the aerospace industry, for machining exotic materials that are not easily cut by other means. The process can also be used to cut very fine holes (0.05–1.0 mm diameter), particularly fine holes with a high depth/diameter ratio, without the usual drill breakage problems. The advantages and limitations of the process can be summarised as follows.

Advantages

- Any electrically conductive material can be cut regardless of hardness and without causing distortion.
- Neither the workpiece nor the cutting electrode have to resist any mechanical cutting forces.
- The process is readily adaptable to automatic operation.

- The process is compatible with CNC systems.
- May be the only process available for machining some of the more 'difficult' aerospace alloys.
- The process leaves a non-directional surface finish. The surface consists of tiny craters with no definite pattern or lay.
- The surface texture can be varied as required, usually in the range 1.5–5.0 mm R_a. However, mirror finishes of 0.05 mm R_a are possible, as are deliberately created coarse-patterned surfaces.

Disadvantages

- The metal-removal rate is slow compared with conventional machining.
- EDM can only be used with electrically conducting workpiece materials.
- The repeated vaporisation and melting forms an 'as cast' surface layer. This may later cause problems of surface cracking, leading to fatigue failure.
- There are potential health hazards from contact with the dielectric fluid and the inhalation of spark-induced fumes from the dielectric fluid.
- Electrode wear can vary from almost zero with some workpiece material/electrode combinations, to situations where the tool electrode wears faster than the workpiece material necessitating several electrode changes.

6.20 Electrochemical machining

This process can be used with any electrically conducting workpiece material and dissolves away the material by electrolysis. The principles involved are the same as for electroplating, except that electroplating takes place at the cathode (negative electrode) of the cell, whereas electrochemical machining (ECM) takes place at the anode (positive electrode) of the cell. Figure 6.47 shows the essentials of the process. The tool electrode is fed into the workpiece at a controlled rate. Note that the shank of the tool is insulated except for a small land at the end. This is to confine the reaction to the end of the tool only, otherwise it would continue along the full length of the tool causing a tapered hole to be machined in the workpiece. In the following description of the reactions that take place, it is assumed that a *steel* workpiece is being electrochemically machined.

Fig. 6.47 *Electrochemical machining (ECM)*

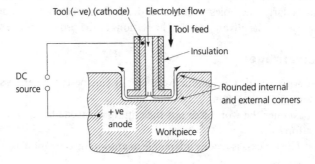

When a potential difference is applied across two electrodes, the reaction that takes place at the anode is called *anodic dissolution* of the workpiece – that is, metallic ions are released from the workpiece surface as shown in equation (6.1).

$$Fe \rightarrow Fe^{++} + 2e \tag{6.1}$$

Note that the metallic ion has lost two electrons and, therefore, carries a positive charge. Also two free electrons have been released.

At the cathode the reaction is the generation of hydrogen gas – that bubbles off – and the production of negative hydroxyl ions. The electrolyte is a water-based solution (e.g. sodium chloride in water). Equation (6.2) shows what happens at the cathode:

$$2H_2O + 2e \rightarrow H_2 + 2(OH^-) \tag{6.2}$$

The positively charged metallic ions released at the anode combine with the negative hydroxyl ions released at the cathode to form ferrous hydroxide which precipitates out and is flushed away by the electrolytic flow. Thus the overall reaction is as shown in equation (6.3):

$$Fe + 2H_2O \rightarrow Fe(OH)_2 + H_2 \tag{6.3}$$

The ferrous hydroxide [$Fe(OH)_2$] precipitate initially forms a dark-green sludge, but this later oxidises in air to form the typical reddish-brown sludge of ferric hydroxide [$Fe(OH)_3$].

6.20.1 *ECM tools and applications*

ECM tools appear in a variety of configurations but, typically, consist of a table for the workpiece and a vertical ram with controlled feed on which can be mounted the tool electrode. ECM systems pose many problems including electrical insulation, electrolyte containment and handling, and fume extraction since hydrogen is an explosive gas. A typical ECM configuration is shown in Fig. 6.48.

Fig. 6.48 *Typical ECM configuration*

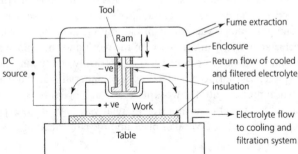

6.20.2 *Control of the process*

Compared with EDM, there is no significant electrode wear of the cutting electrode during ECM, so there is no need for a closed loop servo system to maintain a controlled gap width. It is only necessary to feed the tool electrode into the work at a constant speed dependent upon the metal removal rate. The tool cuts a mirror image of itself into the workpiece. The rate of metal-removal is proportional to the electrical current flow between the electrodes. Too high a current flow can cause overheating and boiling of the electrolyte, excess generation of gases, and a poor surface finish. Further, the increased potential difference between the electrodes, required to cause such a heavy current flow, could result in arcing between the electrodes.

6.20.3 *Electrolytes*

The electrolyte completes the circuit between the anode and the cathode and allows the electrochemical reaction to occur. It also transports away the machined particles and the heat of the reaction. The electrolyte has to be filtered and cooled before being recirculated. Water-based solutions of sodium chloride and sodium nitrate are the most commonly used electrolytes.

6.20.4 *ECM applications*

As an alternative to EDM, electrochemical machining (ECM) can be used to cut any electrically conducting material regardless of hardness. The main uses for the electrochemical-machining process are as follows:

- The 'one-off' production of complex internal cavity shapes in dies and moulds.
- The low-volume production of components made from materials that are difficult to machine by conventional methods.
- The distortion-free machining of components that have already been hardened.
- The deburring of components produced by conventional machining.

A variation on the process is electrolytic grinding. Special grinding wheels are used which are porous and electrically conductive. These can be similar to conventional wheels but contain graphite powder, or diamond impregnated copper wheels for the finest work. The wheel is kept lightly in contact with the work and the abrasive acts mainly as an insulator controlling the distance between the electrically conducting elements of the wheel and the workpiece. The surface finish and rate of material removal depend largely upon how hard the wheel presses onto the workpiece. At high pressures some EDM as well as ECM takes place so that metal-removal rates are relatively high at the expense of the surface finish. At low pressures only ECM takes place and the equivalent of precision-lapped surfaces can be produced.

Advantages

- ECM can cut any electrically conducting materials regardless of hardness and without distortion.
- There are no cutting forces acting on the workpiece.

- ECM may be the only economically viable process for materials which are difficult to machine or that have been prehardened.
- The surface produced is a very close copy of the electrode surface so, for example, a highly polished electrode will produce a highly polished workpiece surface.
- A high order of accuracy is obtainable due to the small gap between work and tool electrodes and the absence of wear on the tool electrode (typical gap 0.25 mm).
- The process is readily adaptable to automatic deburring. If a tool electrode is placed close to an area needing deburring, the current densities are highest at the peaks of the surface irregularities (the burrs) and these are rapidly removed.

Disadvantages

- Metal-removal rates are slow compared with conventional machining.
- ECM can only be applied to electrically conducting workpiece materials.
- There are difficulties with handling and containing the electrolyte.
- There are difficulties in safely removing and disposing of the explosive hydrogen gas generated during the process.
- It is necessary to clean and oil the workpiece immediately after machining because of the corrosive effects of the electrolyte residue.
- The process cannot produce sharp internal or external corners and allowance also has to be made for overcutting, as shown in Fig. 6.49. For deep holes the tool electrode is often fitted with an insulating sleeve so that cutting only takes place at the end of the electrode.
- The pumping of high-pressure electrolyte into the narrow gap between the workpiece and the tool electrode can give rise to large forces acting on the work and on the electrode.

Fig. 6.49 *Overcutting (ECM)*

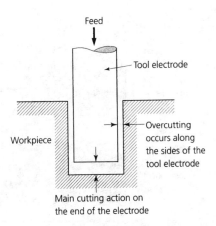

Feed

Tool electrode

Overcutting
occurs along
the sides of the
tool electrode

Workpiece

Main cutting action on
the end of the electrode

6.21 Chemical machining

The term *chemical machining* (CM) covers a variety of processes that use chemical reactions to remove material. The machining action is localised by protecting areas that do not need machining with a chemically resistant film called a *resist*. Usually the resist is applied all over the component and then removed in selected areas, either by mechanical cutting and, peeling or by photographic processing. Alternatively, the resist can be applied only to the required areas by a screen-printing technique. The chemical solutions used to remove material are called *etchants* and can be either acid or alkaline depending upon the material being machined. When the workpiece is to be chemically machined only to a certain depth the process is called *chemical milling* or *chemical etching*, but when the material is cut right through leaving a profile or a hole behind, the process is called *chemical blanking* or *chemical piercing*. In the former case the piece of metal removed from the sheet is the required component, whilst in the latter case it is the hole that is required. Figure 6.50(a) illustrates the general principles of chemical milling (etching) whilst Fig. 6.50(b) illustrates the general principles of chemical blanking.

Fig. 6.50 *Chemical machining: (a) chemical milling; (b) chemical blanking*

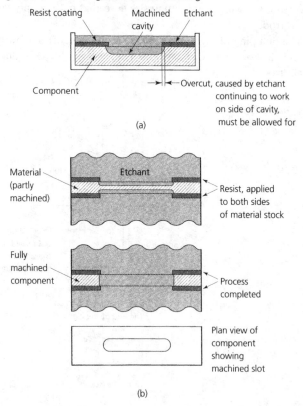

6.21.1 *Chemical milling*

In this process the resist (also called a *maskant*) is applied to the component material surface by dipping or spraying. When the resist has dried, the required pattern is cut using a hand knife guided by a template and the unwanted areas of maskant are peeled off. Alternatively, where greater accuracy is required, a CNC low-power laser may be used to cut the required shape in the maskant. A photographic technique similar to that described in the next (chemical blanking) section may also be used.

The workpiece material is then immersed in the etchant for a predetermined time to obtain the depth of machining required. The etchant only acts on the unmasked areas, although some overcut occurs in the sides of the machined cavities, as previously shown in Fig. 6.50(a), and this must be allowed for. Finally, the component is rinsed to remove the etchant, after which the remaining maskant is either peeled off or dissolved by immersion in a suitable solvent.

A typical application for this process is the production of aerospace components, where the objective is to remove metal selectively to give a lighter component but leave behind a lattice of integral strengthening ribs. The related process of chemical engraving uses exactly the same principles but the workpiece is only given a very brief exposure to the etchant to produce decorative artwork or labelling that is lightly etched into the workpiece surface.

6.21.2 *Chemical blanking*

This process almost always uses a photographic resist. This is a material that polymerises after exposure to ultraviolet light and, thereafter, becomes resistant to the action of the etchant. The process has the following stages.

1. The required product shape is drawn out by hand or by CAD at a known magnification onto a plastic sheet, as shown in Fig. 6.51(a). The magnification is necessary to capture the fine detail of the components since this will, typically, be small in size and complex in shape.
2. This artwork is then photographically reduced to actual size to produce a photographic negative of the component. This is called a *phototool*. Where large quantities of a small component are required, a single phototool might have hundreds of economically spaced components reproduced over its entire area, as shown in Fig. 6.51(b).
3. The sheet of material to be processed is cleaned and coated on both sides with the photopolymer. Each side is then covered with a matched pair of phototool negatives.
4. Both sides of the workpiece sheet are then exposed to ultraviolet light, as shown in Fig. 6.51(c). The exposed sheet is then dipped into a developer that causes the coating to polymerise in the exposed areas and become resistant to the etchant, as shown in Fig. 6.51(d).
5. The unexposed, unwanted, photoresist is then removed by means of a solvent so as to expose the surface of unwanted areas of the metal sheet.
6. As shown in Fig. 6.51(e), the sheet of material and exposed resist is dipped into the etchant for the time required to dissolve away all the material not protected by the photoresist. All that remains are the required component shapes, which are then removed and rinsed to halt the reaction.

Fig. 6.51 *Stages in chemical blanking: (a) enlarged drawing of component produced by hand or CAD; (b) negative phototool produced with components reduced to actual size; (c) one of a matched pair of phototools is applied to each side of the sheet which has been pre-coated on both sides with photopolymer; (d) metal sheet dipped in developer – this causes the photopolymer to develop its etchant-resistant qualities in the areas exposed to ultraviolet light; (e) metal sheet dipped in etchant – this dissolves away all the unprotected metal leaving the components behind*

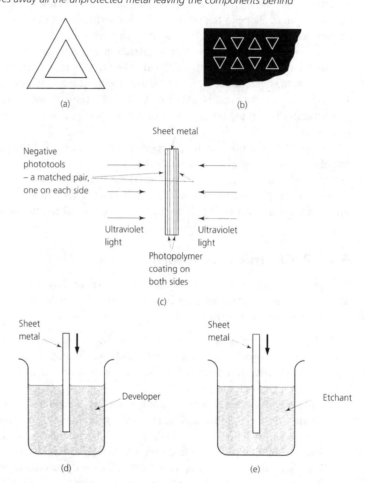

6.21.3 Applications

For blanked components from thin foils and sheets, the process is an alternative to fine blanking and laser cutting. When the process is used to cut a cavity or form into the workpiece, it is an alternative to conventional milling, EDM or ECM. The main advantage of chemical machining is that it does not require extensive investment in expensive capital equipment and tooling where 'one-off' and low-volume production are required. Further, when using chemical machining the lead times are short.

Where high-volume production is required, automated plant is available and this can be very costly depending upon the degree of automation and accuracy required. Chemical machining is widely used in the electronics industry where it is used to produce printed circuit boards and solid-state devices.

Advantages

- Chemical machining can be used with almost any metal or alloy.
- It can be used for hardened materials.
- No cutting force is exerted on the workpiece.
- Chemical machining does not leave any residual stress or surface phenomena.
- When the complexity requirements are too great for fine blanking or the accuracy requirements are too great for laser cutting, chemical machining may be the most economical process even for high-volume production.
- The components are free from burrs or fraze.
- For low-volume production only simple equipment is required.

Disadvantages

- Production rates are very slow compared with conventional machining or blanking.
- The process is limited to metals.
- Safety hazards – as with all corrosive chemicals care must be taken in the handling and disposal of the etchants. Fumes produced during etching must also be disposed of safely.
- Only rounded internal and external corners can be produced.

SELF-ASSESSMENT TASK 6.8

1. Compare the relative advantages and limitations of electric discharge machining (EDM), electrochemical machining (ECM), and chemical machining (CM).

2. Explain the purpose of, and the essential differences between, the dielectric fluid used in EDM and the electrolyte used in ECM.

3. For a component of your choice, describe how you would manufacture it by chemical machining, and state the safety precautions that must be taken.

6.22 Laser cutting

When transistors first replaced thermionic valves in radio receivers, all small receivers containing these devices were loosely referred to as 'transistors' or just 'trannies' and were looked upon as something of a desirable novelty. Similarly, the term *laser* also tends to be used loosely for all sorts of devices using laser light. Laser light can be used for a wide range of applications including the cutting and reading of compact discs, precision measurement and, in high power applications, the cutting of metals. Let's see what we mean by the term *laser*.

LASER is an acronym for Light Amplification by the Stimulated Emission of Radiation.

- Laser light is *monochromatic*, i.e. of one wavelength.
- Laser light is *coherent*, i.e. all the light waves vibrate at the same frequency, in the same direction and in phase with each other so that their peaks and troughs match, as shown in Fig. 6.52(a).

These features mean that laser light can be transmitted in an almost perfectly parallel beam, as shown in Fig. 6.52(b), and unlike a point source of ordinary light which is highly divergent, as shown in Fig. 6.52(c). Laser light can also be focused onto a small area so that its energy is concentrated. With some types of laser this gives rise to localised high temperatures causing melting and vaporisation of the material.

Fig. 6.52 *Characteristics of laser light: (a) monochromatic, coherent laser light; (b) parallel beam of laser light; (c) divergence of ordinary light*

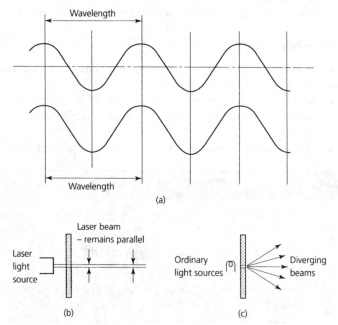

6.22.1 *Principles of laser operation*

The types of laser suitable for metal cutting use either a transparent solid material or a clear gas mixture as the lasing material. An example of the solid-state type is Nd–YAG that consists of a small amount of neodymium in a matrix of yttrium, aluminium and garnet. A typical metal-cutting CO_2 gas laser uses a mixture of carbon dioxide, nitrogen and helium in the ratio $6:66:40$ as the lasing material.

Regardless of whether the lasing material is a solid or a gas, the basic laser configuration is as shown in Fig. 6.53. An intermittent external energy source is needed to initiate and maintain the lasing action. For solid-state lasers, the energy is derived from light flash sources similar to those used for flash photography and stroboscopes. The light energy passes through the transparent walls of the lasing material. This is called *optical pumping*.

In a gas laser, the lasing medium is contained within an electrical discharge tube and pulses of energy are passed into the gas via electrical discharges between the anode and the cathode. This is called *discharge pumping*.

Fig. 6.53 *Basic laser configuration. An intermittent external energy source is needed to initiate and maintain the lasing action. For solid-state lasers, the energy is derived from flash lamps. The light energy passes through the transparent walls of the lasing material. This is called optical pumping*

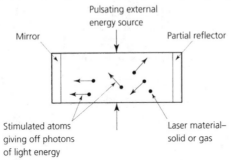

No matter which system is used, atoms within the lasing material are stimulated by the external energy source into giving off photons of light energy in all directions. Some of this stimulated light emission occurs along the axis of the laser and is reflected back and forth between the mirror and the partial reflector, as shown in Fig. 6.54. This causes a 'cascade effect' where more and more atoms are stimulated into giving up their photon energy so that they are all in phase with the pulsing of the light source. Some of the energy is emitted as a continuous stream of laser light through the partial reflector and can then be focused by an optical system. Only optically pumped solid-state lasers and discharge-pumped CO_2 gas lasers have sufficient power and reliability for cutting operations.

Fig. 6.54 *Configuration of a laser with external optical system*

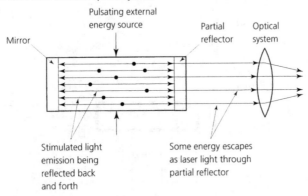

6.22.2 *Optically pumped solid-state laser*

The Nd–YAG solid-state laser-cutting configuration is shown in Fig. 6.55. It is relatively compact, robust and reliable. It is also easy to maintain in a harsh workshop environment. The main limitation is the difficulty in keeping the solid lasing material cool enough to prevent degradation, limiting the power to a maximum of 1 kW, with 0.5 kW being a more common rating. A laser of this type is ideal for drilling and cutting profiles in sheet metal up to 5mm thick. It can also be used for spot and seam welding.

Fig. 6.55 *Optically pumped solid-state laser*

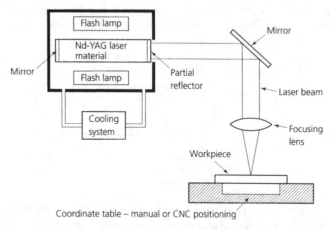

6.22.3 *Discharge-pumped CO₂ gas laser*

The CO_2 laser is not as compact as the Nd–YAG type, but is much more powerful with up to 15 kW available as output. Figure 6.56 shows the configuration of a gas-discharge laser. The greater output power is obtainable because of the inherently greater efficiency of the lasing medium and the relative ease of cooling. The gas itself can be passed at high velocity through a heat exchanger, giving very effective heat control. The applications are as for the solid-state laser except that the greater power allows profile cutting of metal plate up to 20 mm thickness.

Fig. 6.56 *Discharge-pumped CO₂ gas laser*

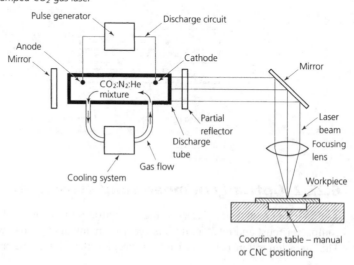

6.22.4 *Laser-cutting applications*

For cutting operations the laser light is focused on a small spot. The concentrated energy causes local melting, evaporation and ablation. Most practical cutting applications of lasers use an assisting gas to increase the efficiency of the process. The functions of the assisting gas are cooling and cleaning the cut area of molten and evaporated material. With materials such as steel a reactive gas (oxygen or air) is used to cause an *exothermic* reaction. The additional heat energy generated increases the cutting speed. With highly reactive materials such as aluminium and its alloys, a protective gas atmosphere (nitrogen or argon) is used to prevent excessive oxidation. The assisting gas is usually delivered to the cutting zone through a co-axial nozzle, as shown in Fig. 6.57. The process can be used to mark, drill, and cut a wide range of materials, including metals such as the nimonic alloys that are difficult to cut with conventional tools, and non-metals such as ceramics, plastics, glass, wood, leather and cloth.

Fig. 6.57 *Gas-assisted laser cutting*

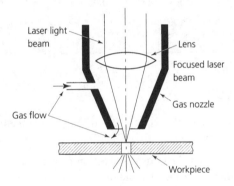

Advantages

- High cutting speed.
- Small heat-affected zone.
- Good profiling accuracy due to small cutting area.
- Smooth finish along the cut edge.
- Compatibility with CNC systems.
- Readily adaptable to robotic operation and guidance.
- The workpiece is not subjected to any cutting forces.
- The cutting beam can be directed and focused over long distances and into inaccessible positions.

Disadvantages

- Relatively high capital cost of the equipment and the relatively high operating cost of the consumable gases.
- Hazards resulting from the laser beam itself and from the vaporised workpiece materials.

6.23 Justification

Despite the fact that all cutting processes are wasteful of materials and energy, their use can be easily justified. No other techniques for shaping materials can give the consistent high-dimensional accuracy and quality of surface finish that can be achieved by machining and grinding, lapping and honing. Further, by adopting minimum material-removal strategies, the volume of chip formation and energy used can be kept to a minimum. Machining is often the only means of producing 'one-off' prototypes and small-quantity batches where the volume of production precludes the pattern and tooling costs of casting, flow forming and sintered particulate technique moulding.

SELF-ASSESSMENT TASK 6.9

1. With the aid of sketches, explain the meaning of the term 'laser', state how a beam of laser light differs from a beam of ordinary light, and give the basic principle of operation of a simple laser device.

2. Describe the essential differences between optically pumped lasers and discharge-pumped lasers.

EXERCISES

6.1 With the aid of sketches, explain the difference between *combination and follow-on* press tools.

6.2 Justify the cutting process selection for a component of your choice in terms of cost, accuracy, repeatability and finish.

6.3 Compare and contrast the advantages and limitations of the following cutting tool materials:
    (a) high-speed steel (HSS)
    (b) sintered carbide
    (c) 'Sialon'

6.4 Compare the advantages and limitations of 'coated' carbide tool tips with conventional carbide tool tips.

6.5 Research manufacturers' literature and draw up a tooling layout for producing a component of your choice on a multi-spindle automatic lathe.

6.6 For a component of your choice, and paying particular attention to the costs involved, economic batch size, and lead times, compare the advantages and limitations of producing it on:
    (a) a capstan lathe
    (b) a single-spindle automatic lathe
    (c) a multi-spindle automatic lathe
    (d) a CNC lathe

6.7   Compare and contrast the production of screw threads by turning, milling, hobbing and grinding, in terms of cost and quality.

6.8   Describe, with the aid of sketches, the difference between gear-tooth production by *forming* and by *generation*.

6.9   (a)   Describe the process of:
        (i)   gear planing
        (ii)   gear shaping
        (iii)   gear hobbing
    (b)   Compare the advantages and limitations of each of the above processes in terms of their applications, accuracy, finish and process costs.

6.10   With the aid of sketches, describe the basic principles of the broaching process.

6.11   (a)   With the aid of sketches describe the principles of the centreless grinding process.
    (b)   Explain how 'out of roundness' can be avoided when centreless grinding.
    (c)   Compare and contrast the advantages and limitations of centreless grinding with plain cylindrical grinding.

6.12   With the aid of sketches, describe the basic principles of:
    (a)   electric discharge machining (EDM)
    (b)   electrochemical machining (ECM)
    (c)   chemical machining (CM)

6.13   For a component of your choice, justify its manufacture by an appropriate process from those listed in Exercise 6.12, giving reasons for your choice.

6.14   (a)   With the aid of sketches, describe the principle of the laser-cutting process.
    (b)   Compare and contrast laser-cutting processes with other thermal-cutting processes (e.g. oxy-fuel gas cutting).

7 Measurement and inspection

The topic areas covered in this chapter are:

- The kinematics of measuring equipment and instrument mechanisms.
- Linear measurement.
- Angular measurement.
- Laser measurement.
- Touch-trigger and analogue probing.
- Three-dimensional (3D) measurement and coordinate-measuring machines (CMMs).
- Screw-thread measurement.
- Gear-tooth measurement.
- Geometric tolerancing.

7.1 Kinematics of measuring equipment

The measurement and inspection of engineering components was introduced in *Manufacturing Technology*, Volume 1, Chapter 10. In this chapter the equipment and techniques for linear and angular measurement will be developed further, together with the inspection of screw threads and gear teeth, the manufacture of which was discussed in Chapter 6. First, let us consider the *kinematics* of measuring instruments. The word 'kinematics' is derived from the Greek word *kinema*, meaning movement. Hence kinematics, when applied to mechanisms, relates to movement and position without reference to mass or force. Machine tools and measuring instruments are dependent upon kinematic principles for the location and controlled movement of their various components and assemblies. Thus the aim of kinematic design is to allow a component to move in a specified manner with the utmost freedom, whilst restraining it in the remaining degrees of freedom with the utmost rigidity (stiffness).

You may wonder in the age of electronics why it is necessary to consider kinematics and the mechanisms of conventional, mechanical measuring instruments. There are several reasons for doing this.

- Conventional, mechanical measuring devices such as the Sigma comparator and the Johansson Microkator were ruggedly built with a minimum of moving parts to wear out, and many are still in use in standards rooms and metrology laboratories (*metrology* means 'the science of fine measurement').

- Dial test indicators (DTI) of various types are still widely used for machine setting and measurement.
- Conventional measuring devices do not require batteries, nor do they require any external supply of electricity. They are unaffected by electromagnetic interference.
- Conventional measuring equipment is robust and reliable and, properly used and looked after, will maintain their accuracy over long periods of time.
- The probes used with electronic measuring equipment are mechanical in the sense that they convert the displacement of a stylus into electrical signals. The stylus and the electromechanical transducer to which it is attached still need *mechanical support* so that they can move with little restriction in the desired plane but have maximum resistance to deflection (stiffness) in all other planes.

So why bother with the complexities of electronic measuring equipment? The main advantage of electronic measuring is that a range of applications and read-outs are available and it is comparatively easy to switch between them. For example, it is easier to read a digital display accurately than it is to read an analogue display. Not only can the measuring head be situated in a position remote from the display (connected by only an electric cable) but the output from a measuring head can be fed directly into a computer. This not only enables the read-out of each individual measurement to be instantly displayed but it also allows the cumulative results of all the measurements of a sample batch to be displayed as a graph after automatic statistical analysis by the software. Further, these results can also be retained on disk or as a hard-copy printout for future quality assurance purposes. The range of electronic applications to measuring instruments is increasing constantly both in diversity of use and accuracy.

For the time being, let us return to the basics of the kinematics of measuring equipment and consider some of the mechanisms and their applications.

7.1.1 *Six degrees of freedom*

As a reminder, Fig. 7.1 shows the six degrees of freedom of a body in space. The body is free to slide along or to rotate about any of the three axes. Figure 7.2 shows *Kelvin's coupling* – the classic solution to prevent movement in all six degrees of freedom without duplicated or redundant restraints. It can be seen that the top plate is supported on a tripod of balls and is dependent only upon gravity to keep it in place. Providing that the gravitational force, and any extraneous forces, acts within a triangle joining the centres of the balls, the plate will remain firmly in position. The balls rest in a trihedral hole, a V-groove, and a plain recess. The coupling satisfies the following conditions:

- The plates are mutually restrained without the application of external forces other than the force of gravity.
- The plates are free to expand or contract without distortion and without affecting the integrity of the coupling.
- The top plate is supported without distortion.

Fig. 7.1 *Six degrees of freedom*

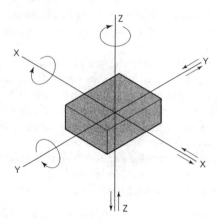

Fig. 7.2 *Kelvin's coupling: (a) underside of platform showing spherical feet; (b) top of stand; (c) platform and stand assembled*

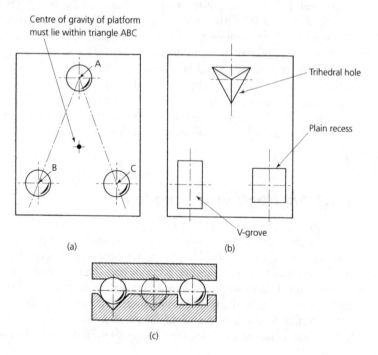

For most instrument applications only five degrees of freedom need constraint, whilst total freedom of movement is required in a single specified direction. An adaptation of Kelvin's coupling to satisfy these criteria is shown in Fig. 7.3. No redundant or duplicated constraints are present and the minimum number of constraints has been applied. This type of slideway arrangement is used in the floating carriage diameter-measuring machine (see Section 7.9).

Fig. 7.3 *Adaptation of Kelvin's coupling*

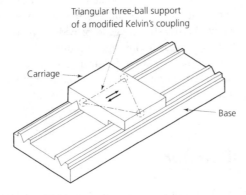

Triangular three-ball support
of a modified Kelvin's coupling

Carriage

Base

7.2 Instrument mechanisms

Since only relatively small movements and deflections are required in instrument mechanisms, conventional pivots, hinges and slides can frequently be dispensed with and replaced with simple, robust mechanisms that are free from wear and backlash. For example, Fig. 7.4(a) shows how two reinforced spring steel lamina can provide parallel

Fig. 7.4 *Instrument mechanisms (principles): (a) strip support – parallel movement; (b) strut and strip support – parallel movement; (c) split diaphragm – parallel movement; (d) cross-strip hinge – rotary movement*

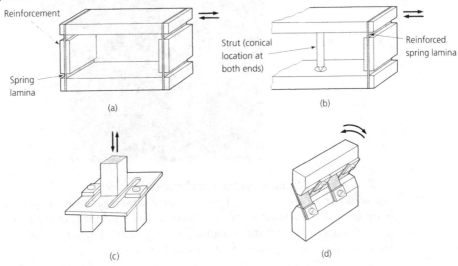

Reinforcement

Spring
lamina

(a)

Strut (conical
location at
both ends)

Reinforced
spring lamina

(b)

(c)

(d)

movement free from sliding and wear. Although the distance between the members varies, this is of little practical importance over the small deflections encountered. Figure 7.4(b) is a variation on this mechanism and is to be found in the screw pitch-measuring machine described later in the chapter. Figure 7.4(c) shows how a split diaphragm can be used to provide parallel motion. A pair of such devices would be used to support and guide the stylus and armature (core) of an electronic probe for linear measurement. Finally, Fig. 7.4(d) shows a cross-strip hinge to provide angular or rotary motion. Since there is no pivot pin, there is no wear in this type of hinge. All the above devices use spring steel strips and lamina, therefore they are self-restoring when used in measuring instruments and can provide the necessary measuring pressure that is required to keep the stylus in contact with the workpiece.

7.3 Magnification

In the dial test indicators discussed in *Manufacturing Technology*, Volume 1, magnification was provided by gear trains or by lever and scroll mechanisms. Such mechanisms are subject to wear and develop inaccuracies, and electronic test indicators with digital read-outs, as shown in Fig. 7.5, are now widely used. The instrument shown has a direct RS232 output for interfacing with computers – a facility that no traditional measuring device can provide.

Fig. 7.5 *Digital electronic dial gauge and comparator stand*

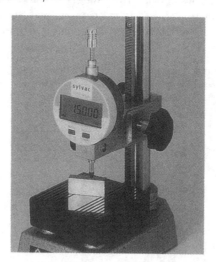

Prior to the advent of electrical and electronic measuring devices, a wide variety of ingenious mechanisms were developed to provide the magnification required and these can be categorised as mechanical, optical and pneumatic. Since many of these are still in current use, although no longer manufactured, we will look at some of the more interesting examples before moving on to more modern devices.

7.3.1 *Mechanical magnification*

Figure 7.6 shows a simple, but effective, mechanism using a cross-strip hinge and slit diaphragms to provide high magnification, high accuracy and repeatability, with negligible wear and wear-related errors. Figure 7.7 shows how tension on a twisted strip can be used to provide magnification. The pull of the flexure spring on the twisted strip tends to unwind the strip and this, in turn, moves the pointer across the scale. A slit diaphragm is used to locate the plunger and provide the measuring pressure. Again, there are no pivots or gears that may wear and introduce errors.

Fig. 7.6 *Mechanism of the Sigma comparator: magnification = A/B × R/r. (Note: Frame members omitted for clarity)*

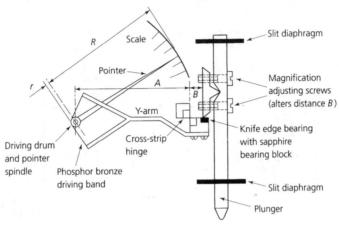

Fig. 7.7 *Mechanism of the Johannson Mikrokator*

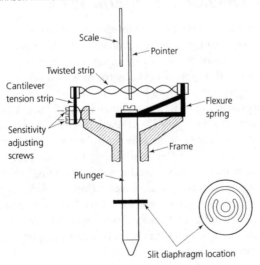

7.3.2 Optical magnification

Figure 7.8 shows one method of using an optical system for magnification. The object is to minimise the use of mechanical moving parts and replace them with an inertialess light beam. Unfortunately such an instrument has to be viewed in subdued light and also tends to be bulky because of the optical systems.

Fig. 7.8 *Eden-Rolt optical comparator: (a) mechanical amplification (×400); (b) optical magnification (×50). (Overall magnification 400 × 50 = 20 000)*

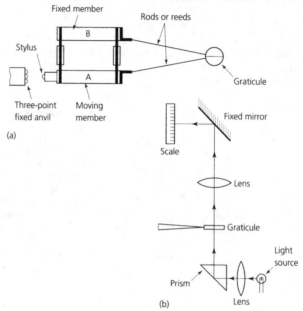

7.3.3 Pneumatic magnification

Pneumatic comparators can be divided into *back-pressure types* and *flow types*. The back-pressure types can be further subdivided into high-pressure and low-pressure types. Like electrical comparators, their measuring heads can be remote from the display. The main advantage of this type of device is the lack of mechanical contact with the surface being measured (*proximity gauging*).

The principle of operation is shown in Fig. 7.9(a). Air is passed into the measuring head at a controlled pressure P_c. It then passes through the control orifice O_1, into the intermediate chamber. Whilst the control orifice has a constant size, the effective size of the measuring orifice O_2 will vary according to its proximity to the work surface distance (D). Variation in the distance D produces corresponding variations in the back pressure P_b. An indicating device such as manometer (low-pressure) or a bourdon-tube pressure gauge (high-pressure) is used to measure the back pressure and is calibrated to read directly the distance D in units of linear measurement. Figure 7.9(b) shows a typical low-pressure pneumatic comparator with an oil dashpot pressure regulator. Figure 7.10 shows some typical measuring heads.

Fig. 7.9 *Solex-type low-pressure pneumatic comparator with oil dashpot pressure regulator: (a) principle of the pneumatic transducer; (b) low-pressure pneumatic comparator*

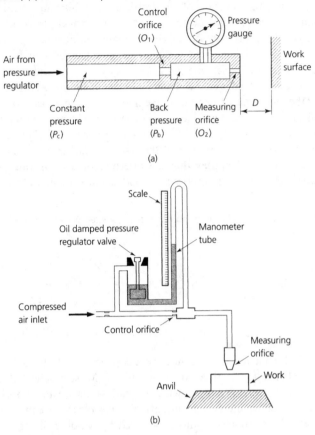

(a)

(b)

Fig. 7.10 *Pneumatic comparator measuring heads: (a) indirect measuring head for shallow and blind holes; (b) indirect ball contact measuring head for rough surfaces*

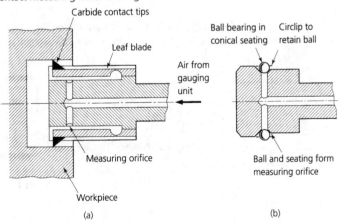

(a)

(b)

7.3.4 *Electrical/electronic magnification*

Let us now turn to some of the electrical/electronic magnification systems used in modern metrology equipment. Electrical and electronic measuring devices have largely displaced the devices previously described for many production applications. The use of digital electronics and integrated circuits allows instruments to be built more cheaply, yet be accurate and reliable. One simple approach for electrical comparators can be based upon some form of bridge circuit. Figure 7.11 shows the circuit for a simple bridge using resistors for the ratio arms. When the bridge is in a state of balance, no current will flow through the galvanometer and it will have zero deflection. The relationship between the resistors in the ratio arms is $R_1/R_2 = R_3/R_4$. The galvanometer will show a reading either side of zero should the value of any one of the resistors change to unbalance the current flow in the circuit. It would appear that the insertion of a measuring head containing a variable resistor into one of the ratio arms would produce a practical comparator.

Fig. 7.11 *Resistance bridge. When in balance the galvanometer, G, will read zero and $R_1/R_2 = R_3/R_4$*

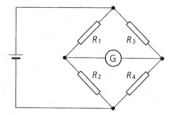

Unfortunately, this is not the case because of the difficulty of producing reliable variable resistors with adequate sensitivity. Most practical electrical comparators use an *alternating current bridge* in which the ratio arms are inductors (coils of wire which impede the flow of an alternating current). The *inductance* (ability to impede alternating current flow) of the inductors can be increased by winding the coil of wire around a soft iron (ferrite) core.

Figure 7.12(a) shows the circuit of a practical comparator based upon an inductance bridge, and a suitable measuring head (transducer) is shown in Fig. 712(b). When the

Fig. 7.12 *Electronic comparator (AC bridge): (a) AC bridge comparator circuit; (b) twin coil transducer for use with bridge. Bridge is balanced when $L_1/L_2 = L_3/L_4$ (core is central between L_3 and L_4)*

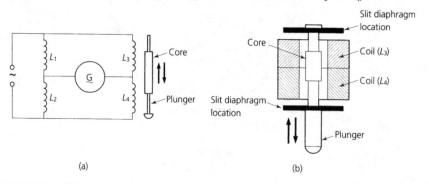

plunger and core assembly rises, more of the core is inserted into L_3 and its inductance increases. Correspondingly, less of the core is inserted into L_4 and its inductance decreases.

This unbalances the bridge and a reading is indicated by the alternating-current galvanometer, G. Unlike a variable resistor, there are no sliding contacts and this is where the alternating current bridge gains its reliability and sensitivity.

The differential inductance transducer shown in Fig. 7.12(b) contains only two external coils to form the ratio arms of the bridge. With the introduction of reliable high-gain amplifiers in the form of integrated solid-state circuits, it has become a practical proposition to use the arrangement shown in Fig. 7.13(a). The measuring head is shown in Fig. 7.13(b). You can see that it has three coils and is called a *linear variable differential transformer* (LVDT). This type of transducer does not require a bridge circuit. In operation a steady, high-frequency alternating current from the frequency stabilised oscillator is applied to the centre (primary) winding of the LVDT and induces equal and opposite alternating currents in the two secondary windings. As the core move from the point of equilibrium it changes the magnetic flux pattern. This results in the induced current in one secondary winding being increased at the expense of the current in the other winding, which is decreased. This provides a signal for the amplifier that its magnitude and direction have changed when compared with the steady-state signal from the oscillator. By using high-frequency alternating current the sensitivity and size of the instrument is increased whilst, at the same time, the size of the transducer is reduced by the use of smaller coils and a core of lower mass.

Fig. 7.13 *Electronic comparator (LVDT): (a) general arrangement of comparator circuit; (b) general arrangement of transformer-type transducer*

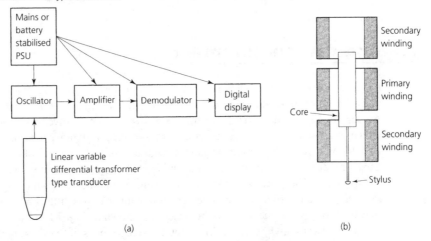

Electrical and electronic measuring devices are becoming ever more widely used as they become lower in cost, increasingly more reliable, and more compact. Some of the advantages of electronic measuring devices are:

- They do not contain mechanisms which are liable to introduce friction, backlash, and inertia effects.
- The magnification can be changed by simply switching the electrical circuit.

- Zero-adjustment can be effected electrically, thus avoiding the need for fine mechanical adjustments to the position of the measuring head.
- The system is energised externally and the movement of the plunger/stylus does not have to drive the magnification mechanism. This allows extremely low measuring pressures to be used, so that there is negligible distortion of the workpiece. This is important when measuring soft and thin-walled components.
- The output of the device can be used in conjunction with analogue or digital displays. It can also be used to operate machines, recording instruments, sorting gates, light signals, computers, etc.
- Remote readings with the measuring head up to 30 m from the rest of the equipment can be taken. This is particularly true with electronic instruments using LVDT-type transducers as they are less susceptible to the capacitance effects of long interconnecting leads.

SELF-ASSESSMENT TASK 7.1

1. Discuss the advantages and limitations of linear-measuring devices using digital electronic magnification compared with those using mechanical magnification.

2. Discuss the kinematics of measuring instruments and explain how backlash and 'lost movement' can be avoided.

3. Research manufacturers' literature and describe the construction and operation of a digital electronic linear-measuring device of your choice.

7.4 Angular measurement

Angular measurement by such devices as plain and vernier protractors and, where greater precision is required, sine bars were discussed in *Manufacturing Technology*, Volume 1. Angles can also be measured by comparison. Just as comparative measurements of linear dimensions can be made using slip gauges, so too can comparative measurements of angles be made using combination angle gauges. Figure 7.14(a) shows a set of combination angle gauges made from hardened, stabilised, and lapped blocks of alloy steel in the same manner as slip gauges. The blocks are made to precise angles and the surface finish of the measuring faces enables them to be wrung together. The manner in which they may be combined so that their angles may be added or subtracted is shown in Fig. 7.14(b) and sizes of typical sets of blocks is shown in Table 7.1. A set of twenty blocks of the sizes shown enables any angle to be built up in steps of 3 seconds of arc. An additional 9° block extends the range to 90°, and the use of the precision square block enables the complete range of 360° to be obtained. Alternatively, the combination angle gauges are also available in decimal notation as shown. These gauge blocks are normally used in conjunction with optical measuring devices such as the Angle Dekkor (Section 7.4.2). Alternatively, they may be used in the workshop for direct measurement and setting in a similar manner to the sine bar.

Fig. 7.14 *Combination angle gauges: (a) a set of angle gauges; (b) use of combination angle gauges*

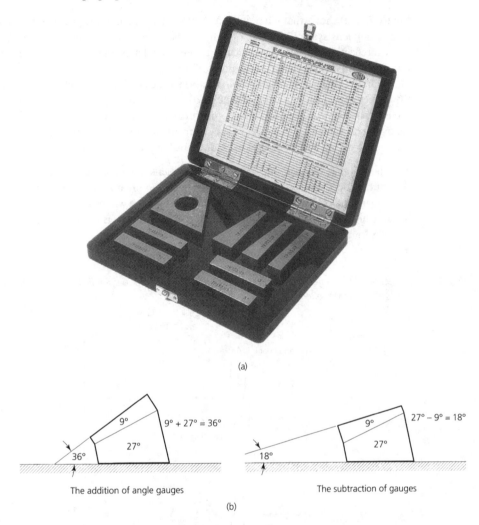

(a)

9°

9° + 27° = 36°

27°

36°

The addition of angle gauges

9°

27° – 9° = 18°

27°

18°

The subtraction of gauges

(b)

Table 7.1 **Combination angle gauges (range of sizes)**			
Degrees	Minutes	Seconds	Decimal minutes
1	1	3	0.05
3	3	9	0.1
9	9	27	0.3
27	27		0.5
41			
A square block is also provided			

7.4.1 Collimation

Figure 7.15(a) shows that if a point source of light is placed at the principal focus of a collimating lens system, the rays of light will be projected as a parallel beam. If these rays strike a reflecting surface perpendicular to the optical axis, they will be reflected back along the same path and refocused at the point of origin.

Figure 7.15(b) shows that if the reflective surface is tilted through some small angle $\theta°$, the reflected, parallel rays will be inclined at $2\theta°$ to the optical axis. These rays will be focused at a point in the focal plane of their origin but displaced by a small distance x. The following characteristics of such an arrangement should be noted since they are of supreme importance in the optical devices to be described in the next two sections.

- The distance between the reflector and the lens has no effect on the separation of the source and the image (distance x). However, too great a distance will result in the rays missing the collimating lens altogether and no image will be formed.
- For high sensitivity a long focal length collimating lens is required to give a large value of x for a small value of $\theta°$.

-

Fig. 7.15 *Reflection of collimated light: (a) collimated light reflected from perpendicular surface; (b) collimated light reflected from tilted surface*

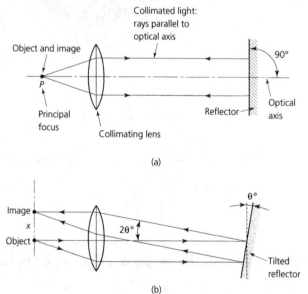

(a)

(b)

7.4.2 The Angle Dekkor

The *Angle Dekkor* does not measure angles directly, but by comparison with a known standard such as a combination angle gauge. Thus it is a comparator with a scale of limited range capable only of giving a difference reading.

The optical system and the scale as seen in the eyepiece is shown in Fig. 7.16. The magnifying eyepiece views both the fixed datum scale and the reflected image of the illuminated scale at right angles to each other as shown. The image does not fall across a simple datum line, but across a similar fixed scale at a right angle to the image. Thus the reading on the reflected scale measures angular deviations in one plane at 90° to the optical axis, and the reading on the fixed scale gives the deviation in a plane which is perpendicular to the former plane. This feature enables angular errors in two planes to be dealt with at the same time and, more importantly, to ensure that the reading on the setting gauge and on the workpiece lie in the same plane, whilst the error is read in the other plane. If there is no error the two scales should cross at the same reading as would have been shown when the instrument was set up against the setting gauge. Normally it would be set with the scales appearing to cross in the middle of the eyepiece. Figure 7.17 shows the displacement of the scales for various errors in the reflected surface of the workpiece.

Fig. 7.16 *Angle Dekkor optical system*

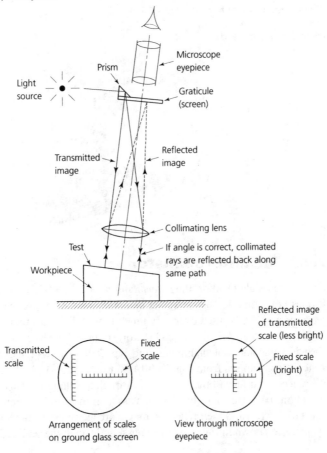

Fig. 7.17 *Angle Dekkor scales (compound angles): (a) shows a horizontal displacement; (b) shows a vertical displacement; (c) shows a vertical and horizontal displacement*

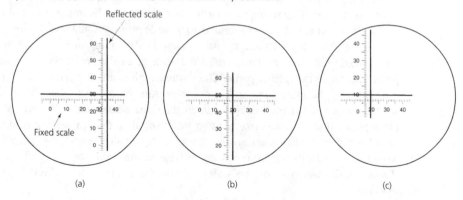

(a)　　　　　　　　　(b)　　　　　　　　　(c)

Although the Angle Dekkor requires considerable skill to set and read accurately, it is a very useful instrument for a wide range of angular measurements at short distances. Direct readings can be obtained down to 1 minute of arc over a range of 50 minutes and, by estimations, readings down to 0.2 minutes of arc are possible with practice. Figure 7.18 shows an Angle Dekkor and its stand.

Fig. 7.18 *Angle Dekkor and stand*

7.4.3 *The autocollimator*

Unlike the Angle Dekkor, the *autocollimator* is a direct-reading angular-measuring device. It does not require to be used in conjunction with a setting master; it is very much more sensitive and can be used over very much longer distances than the Angle Dekkor. Figure 7.19(a) shows a typical autocollimator, Figure 7.19(b) shows a reflector suitable for checking machine tool alignments, and Fig. 7.19(c) shows the light path through its optical system. You can see from Fig. 7.19(b) that the image of the target wires is projected onto the reflective surface of the workpiece through a collimating lens system and returned, after reflection, back through the same lens system in a similar manner to the Angle Dekkor. The target wires and their reflected images are viewed simultaneously through a magnifying eyepiece, as shown in Fig. 7.20, after passing through the semi-reflector.

Fig. 7.19 *Autocollimator: (a) a typical device; (b) a reflector; (c) schematic diagram of Hilger & Watts visual microptic autocollimator*

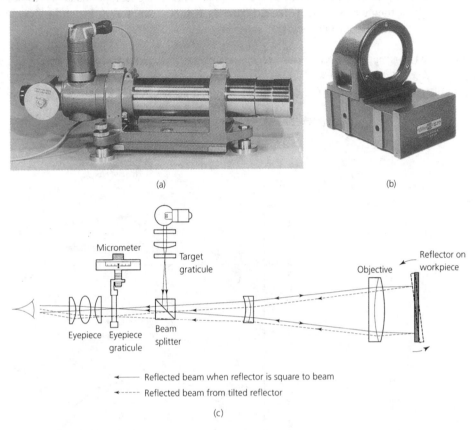

(c)

Fig. 7.20 *Autocollimator field of view*

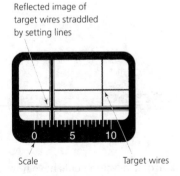

The eyepiece contains an adjustable graticule engraved with setting lines and a scale divided into minutes and half minutes of arc (Fig. 7.20). The setting lines are adjusted by the micrometer control until they straddle the reflected image of the target wires. The scale is then read to the nearest minute of arc plus the micrometer reading. The divisions of the micrometer scale each represent a 0.5 second of arc. Instruments are available with the scales in decimal notation. The autocollimator has two sets of scales and two micrometer-adjustments perpendicular to each other. This enables compound angles to be measured or setting errors to be reduced. Figure 7.21 shows two uses of the autocollimator for testing machine tool alignments.

Fig. 7.21 *Testing machine alignments with the autocollimator: (a) testing a horizontal slideway; (b) testing slideway squareness*

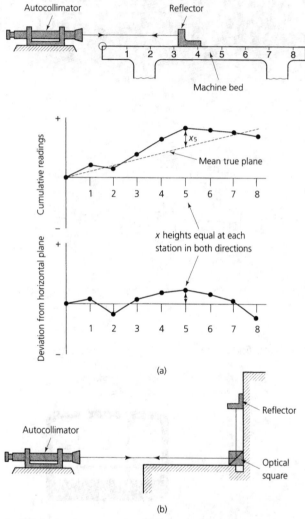

(a)

(b)

Inevitably electronic developments have now influenced the design of instruments such as the autocollimator. In addition to the *visual* instrument just described, *photoelectronic* autocollimators providing dual axis measurement with digital display are now widely used. A typical instrument and its digital display are shown in Fig. 7.22(a), and the light path of a photoelectronic autocollimator is shown in Fig. 7.22(b). You can see that photoelectronic autocollimators still have an eyepiece with an individual visual channel to assist the initial setting-up procedures. The electronic digital display unit is readable from several metres and it also incorporates a signal strength display to ensure correct set up and results. An RS232 interface is provided for connection to various types of accessory computer. Among other things, this allows the computer processing of straightness, flatness, angle and polygon (rotary devices) measurement. A remote control switch is available, enabling the autocollimator and display unit to be positioned remotely from the operator.

Fig. 7.22 *The photoelectronic autocollimator: (a) photoelectronic autocollimator and read-out; (b) light path for photoelectronic autocollimator*

(a)

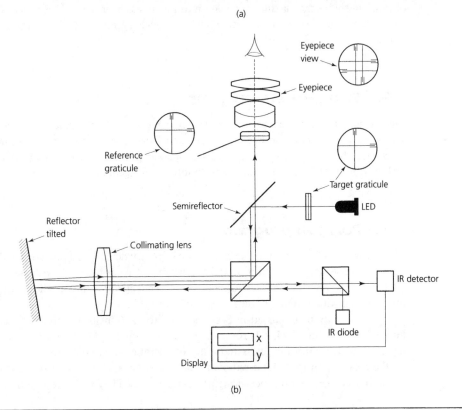

(b)

Microcomputer control of data processing permits automatic calculation using the selected program, with the results shown on the screen display and printout of the measurement results also available. By speeding up the measuring process and minimising the risk of human error when taking readings, direct microcomputer data processing results in increased measuring efficiency and improved quality control. Software packages (compatible with MS Windows™) are available to support the photoeletronic autocollimator and we will now look at some of the facilities that are available.

7.4.4 *Flatness program*

The rectangle and diagonal ('Union Jack' or 'Moody') method of flatness measurement offers simple, interactive, menu-driven software. This displays or prints out an initial diagram of the surface to be measured together with the generator lines and instructions on the method of entering the surface data. Multiple measuring steps can be taken along each generator line. After the measurements have been entered, the program calculates and displays the shape of each generator line and the flatness of the surface. Once the computer has accepted the autocollimator readings, these values are displayed and/or printed out as seconds of arc and then converted to micrometres (μm) or millionths of an inch (0.000 000 1 in.). Measurement results of flatness can be displayed and/or printed out as an ISO diagram or certificate or as a certificate of conformance. In addition, the display and/or printout shows the maximum deviation from flatness over the entire surface. To comply with international standards, a minimum zone calculation is used to generate flatness errors.

7.4.5 *Straightness program*

This is used to measure the straightness of component features such as machine tool slideways. The method and procedures are similar to those used when measuring flatness. Twist and squareness measurement is also available in this software package. Analysis is to LSL or ENDS ZERO, with appropriate graphical representation of the results.

7.4.6 *Polygon program*

This software package permits the calibration of rotary devices and polygons with up to 72 faces. It allows single or bidirectional calibration of rotary devices with results in both angular index accuracy and pyramidal error. The operator has a choice of single or multiple runs with automatic computation of mean values.

The applications discussed in Sections 7.4.4 to 7.4.6 inclusive have been included to indicate the way in which electronics enables a computer, linked with typical measuring device, to increase the usefulness and range of operations of that device. Also, by speeding up the measuring process and removing the chance of human error from calculating the results, the efficiency of the measuring process is increased beyond all recognition.

7.5 Laser measurement

In Chapter 6 we introduced laser light and discussed how it can be used for metal cutting and welding. Let us now see how it can be used for measurement. Laser measurement is accomplished through *interferometry*, which is a technique that uses the wavelength of light as a unit of measurement. Nowadays a laser is used because its light is *coherent* – this means that all the rays have the same wavelength and are exactly in phase (all the peaks and valleys of the component light waves are in perfect synchronisation). The wavelength of laser light from a helium–neon (HeNe) laser is $0.633\,\mu m$ 0.00025 inch) and by further subdividing the wavelength it is possible to achieve a resolution of 1.25 nanometres. (Typically, the Renishaw ML10 laser achieves this resolution with a long-term wavelength stability, in vacuum, of better than 0.1 ppm).

An interferometer system measures the change in distance by counting the number of wavelengths of light seen by detector optics. As the wavelength is known to great accuracy, the overall distance can be counted to great accuracy. An interferometer system (e.g. *Michelson interferometer*) comprises three optics: a polarising beam splitter and two retro-reflectors, or corner cubes. The following is a brief description of the optical interference principles of the Renishaw ML10 system as used to measure relative changes in optical path length between the measurement and reference arms of the interferometer. Figure 7.23 shows a schematic diagram of the ML10 laser head, linear measurement optics, and the interference detectors, together with the laser beam paths.

Fig. 7.23 *Principle of laser measurement (reproduced courtesy of Renishaw plc)*

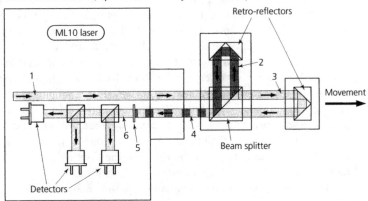

The laser light (1) emerging from the laser head is a circular polarised laser beam of single frequency light. When this beam reaches the polarising beam splitter it is split into two components. The reflected beam (2) is vertically polarised light, and the transmitted beam (3) is horizontally polarised light. As these beams travel to their retro-reflectors they have exactly the same frequency. These two retro-reflectors reflect beams (2) and (3) back towards the beam splitter.

In linear distance (positional) measurement, one retro-reflector is rigidly attached to the beam splitter and the other, which is attached to the machine tool element to be checked (e.g. worktable), moves relative to the beam splitter and forms a variable-length measurement arm. The laser system tracks any change in the separation between the measurement arm retro-reflector and the beam splitter as the measurement arm retro-reflector moves. Whilst it is moving, the frequency of the reflected laser beam from the moving retro-reflector is *Doppler shifted*, so that during movement the frequencies of the two returned beams are not the same. (*Note*: Doppler shift is the effect that causes the apparent rise in pitch of a sound source as it moves towards you (increasing frequency) and the apparent lowering in pitch (decreasing frequency) of a sound source as it moves away from you.) The difference in frequency is directly proportional to the speed of movement. When the reflected laser beams reach the beam splitter they are reflected or transmitted (as before), so that both beams are recombined and returned to the laser head.

The recombined beam (4) consists of two superimposed beams of different polarisations. One beam is vertically polarised and has travelled around the reference arm. The other component is horizontally polarised and has travelled around the measurement arm. At this stage the two beams do not interfere with one another because they have different polarisations. When the beams enter the laser head, they pass through a special optic (5) that causes the two beams to interfere with one another to produce a single beam (6) of plane polarised light. The angular orientation of the plane of this polarised light depends upon the phase difference between the light in the two returned beams. So, if the measurement arm retro-reflector is moving away from the beam splitter, then the plane of polarisation spins in one direction, and if the measurement arm retro-reflector is moving towards the beam-splitter, the plane of polarisation spins in the opposite direction.

Beam (6) is then split into three and focused onto three polarisation-sensitive detectors. As the plane of polarised light spins, each detector produces a sinusoidal output waveform. The polarisation sensitivity of the detectors are adjusted so that their outputs are in phase quadrature. (Three detectors allow the system to eliminate background light, to distinguish the direction of movement, and to measure the distance moved.) These signals are very similar to the feedback signals obtained from many linear and rotary encoder systems, and are processed in a similar way using very high-speed direction discrimination, counting and interpolation electronics. This allows the laser system to measure changes in distance as the measurement arm retro-reflector moves. Because the laser wavelength is very short (633 nm), and can be determined very accurately (to one part per million in air), the laser system just described can provide measurements with very high accuracy and resolution.

The above description assumes a basic understanding of light and optics. Unfortunately constraints of space prohibit further description of the terms used. The reader is recommended to access the internet:

http://www.renishaw.com/calibration/technical.html

for further information concerning the properties of light, the principles of the Michelson interferometer, the retro-reflector, the polarising beam splitter, polarisation, birefringent material, laser cavity, Doppler effect, and deadpath error.

The main application of this equipment is for the calibration of positional accuracy and repeatability of machine tools and coordinate measuring machines. Velocity measurements of machine elements can be carried out with the same equipment. As described above, the basic equipment is a linear interferometer with a computer interface, retro-reflector and beam splitter. Measurements are made by monitoring the change in separation distance (as a function of the wavelength of the laser beam) between the linear interferometer and the second linear reflector. This occurs when either is moved along the axis. The general set up is shown in Fig. 7.24(a).

Fig. 7.24 *Use of a laser for positional calibration: (a) optical set-up for X-axis linear positioning measurement with fixed interferometer and moving retroreflector; (b) screenshots of the system in operation*

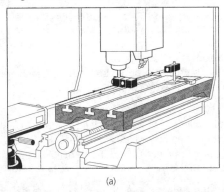

(a)

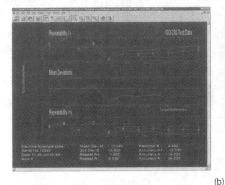

(b)

Using linear analysis software, positional accuracy and repeatability can be displayed, printed, plotted and statistically analysed to most national and international standards. Backlash and cyclic errors can all be displayed with each measurement having a 'live' data plot during data acquisition. Typical printouts are shown in Fig. 7.24(b). The specification of this equipment is typically, range 0–40 m, accuracy $\pm 0.025\,\mu$m, resolution $0.001\,\mu$m, maximum velocity 60 m/min, and velocity measurement accuracy ± 0.05 per cent. Where the greatest accuracy is required, and when measurements are being taken under workshop conditions in which the environmental conditions are constantly changing, continual and automatic environmental compensation should be adopted. This is an additional unit that monitors and compensates for air temperature, air pressure, relative humidity and material temperature.

In addition to linear measurement, the same basic equipment can be used for a range of measurements with the use of further accessories and software packages. Deviation from straight-line motion, parallelism and squareness can all be determined. Pitch and yaw (angular errors) are amongst the largest contributory errors in positional accuracy error for CMM and part-cutting error of machine tools. It is important, therefore, when evaluating a machine tool or a measuring machine to determine the error contributions of pitch and yaw along each axis of movement.

SELF-ASSESSMENT TASK 7.2

1. (a) Compare and contrast the advantages and limitations of the 'Angle Dekkor' and the 'autocollimator' as devices for measuring angles.
    (b) Describe typical applications where these instruments would be used, giving reasons for your choice.
    (c) Describe the selection and assembly of a stack of combination angle gauges to give an angle of 30° 12′ 24″.

2. (a) Discuss and compare the relative advantages and limitations of the visual autocollimator, the photoelectronic autocollimator, and the laser interferometer for measuring the straightness of a machine-bed guideways.
    (b) Explain the meanings of the terms LSL and ENDS ZERO in connection with straightness measurement.

3. Discuss the advantages and limitations of using laser measuring techniques with visual optical techniques for measuring flatness and squareness.

4. Explain the basic principle of using a laser interferometer for checking the positional accuracy.

5. Research the web site given in the text and explain what is meant by:
    (a) Constructive and destructive interference.
    (b) Doppler effect.

7.6 Probing

Coordinate measuring machines have evolved from origins founded in simple layout machines and manually operated systems, to highly accurate automated inspection centres. A major factor in this evolution has been the touch trigger and other forms of inspection probes and subsequent innovations such as the *motorised probe head* and the *automatic probe exchange system* for unmanned flexible inspection. The Renishaw touch-trigger probe was originally developed in conjunction with Rolls-Royce plc, when a unique solution was required for the accurate pipe measurement for the engines for the Anglo-French Concorde aircraft. The result was the development of the first *touch-trigger probe*: a 3D sensor capable of rapid, accurate inspection with low trigger force.

Touch-trigger probes are widely used for toolsetting and for in-cycle inspection. The advantages of using probes for toolsetting can be summarised as:

- Accurate tool length and diameter offset measurement and correction.
- Complete station of tools measured automatically, in minutes.

- Elimination of manual setting errors.
- No presetting of tools required.
- Faster set up times with reduced machine downtimes.
- Precise machining achieved on first-off component from accurate tool offset measurement.
- In-cycle tool breakage detection prevents scrapped components and detects damaged tools.
- Improved confidence in unmanned machining.

The advantages of in-cycle inspection can be summarised as:

- Component position and alignment measured accurately with each datum calculated and corrected automatically.
- Fixture and axis alignment measured simply and quickly, eliminating the need for manual set up.
- First-off component measurement cycles reduces the need to remove components for inspection.
- In-cycle measurement to monitor component size and position, with automatic offset correction.
- Increased productivity and batch size flexibility from automated set up.
- Guaranteed confidence in unmanned process machining from in-process component size monitoring and correction.

The use of probes will be considered further in Sections 7.7 and 8.3.7.

7.6.1 *Scanning*

Scanning is the term used to describe the process of gathering information about an undefined three-dimensional surface. It is used in such diverse fields as tool and die making, mould making, press tool making, aerospace, jewellery, medical appliances and confectionery moulds. It is used when there is a need to reproduce a complex free-form shape. During the scanning process an *analogue scanning probe* (**not** a touch-trigger probe) is commanded to contact and move back and forth across the unknown surface. During this process the system records information about the surface in the form of numerical data. This data is then used to create a CNC program to machine a replica or geometric variant of the shape. Alternatively, the data can be exported in various formats to a CAD/CAM system for further processing. This is shown in Fig. 7.25.

Let's look briefly at the difference between digitising and scanning. Digitising refers to the process of taking discrete points from a surface with a touch-trigger probe. This time-consuming process has since been greatly improved by the introduction of scanning The scanning process uses analogue probes that produce a stream of data at up to 1000 points per second.

In addition to the contact probes discussed so far there are also:

- Video probes for checking two dimensional forms such as circuit boards, holes, and other features.
- Optical probes using visible laser light to scan 3D objects made from pliable or delicate materials. The scanning takes place without deflection of the material.

Fig. 7.25 *The scanning process: (a) scan – a range of devices capture data from an initial part/model; (b) manipulate data – create variant of original; (c) create output – CNC program or export data to CAD system; (d) replica; (e) rough machine mould (CNC mill); (f) finish mould cavity by CNC-type EDM; (g) CAD display if required*

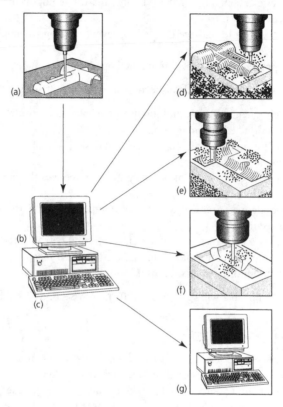

Although surfaces can be scanned using video and optical laser probes, the use of contact probe systems has several fundamental advantages when the object is made from a rigid material:

- Treatment of surfaces to prevent reflections is not required.
- Vertical faces can be accurately scanned.
- Data density is not fixed and is automatically controlled by the shape of the component.
- Time-consuming manual editing of data to remove stray points is not required.

Figure 7.26(a) shows a three-dimensional shape being scanned. The data gathered by the probe is processed by the computer software package, as shown in Fig. 7.26(b). The processed data can then be used to program a CNC machine to make a replica of the original object, as shown in Fig. 7.26(c). Alternatively, the processed data can be used to make a mould for the mass production of replicas of the original object.

Fig. 7.26 *Scanning with an analogue probe: (a) scanning; (b) CAD/CAM computer display; (c) CNC machining of replica*

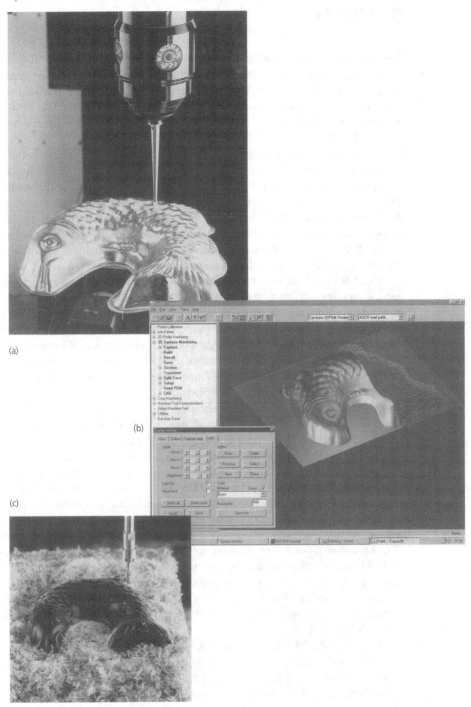

(a)

(b)

(c)

7.7 Three-dimensional coordinate measuring machines

The three-dimensional coordinate measuring machine (CMM) is now used extensively in the manufacturing industries as a powerful and versatile piece of equipment. It is an indispensable tool used in quality control departments that pursue increased efficiency and accuracy in dimensional, geometric and contour measurements. The CNC coordinate measuring machines are playing an increasingly important role in flexible manufacturing systems (FMS).

The coordinate measuring machine may be defined as a 'machine that employs three movable components that travel along mutually perpendicular guideways to measure a workpiece by determining the X, Y and Z coordinates of points on the workpiece with a contact or a non-contact probe. The position of the probe at the moment of contact being determined by displacement measuring systems (scales), associated with the three mutually perpendicular axes.' Since measurements are represented in a three-dimensional coordinate system, a CMM can make many different types of measurements such as dimensional, positional, geometrical, deviation and contour measurements.

Measurement and inspection work using conventional measuring equipment and gauges requires expertise and experience to ensure reliable measurements with high repeatability. Not only is availability of this expertise and experience becoming increasingly limited and costly these days, but the increased use of CNC machining has resulted in the economic manufacture of increasingly complex components. Conventional inspection techniques can no longer be used to check complex-shaped components involving curved surfaces and imaginary origin points. These problems have been eased to a large extent by the increasing availability and use of CMMs.

Three-dimensional coordinate measuring machines may be either CNC or manually controlled machines. Figure 7.27(a) shows a typical manually operated coordinate measuring machine. Such machines only became possible with the introduction of the touch-trigger probes introduced in the previous section. The system configuration is shown in Fig. 7.27(b). Being a manually operated machine, it is driven by the operator by means of a joystick. The resolution of this machine is 0.0005 mm. To ensure smooth and accurate movement, air bearings are used on all three axes and the Z axis is mass balanced.

CNC coordinate measuring machines (CMMs) are used to maximise the efficiency of automated manufacturing processes that demand fast, efficient and accurate measurement. The digital servo-controller automatically seeks the most efficient, and fastest, probing path. An example of a typical CNC coordinate measuring machine is shown in Fig. 7.28(a). This machine can be CNC operated or driven manually by means of a joystick. The system configuration is shown in Fig. 7.28(b). The resolution of this machine is 0.0005 mm. To ensure smooth and accurate movement, air bearings are used on all three axes and the Z axis is air balanced.

Let's consider the basic principles of using a manually operated machine. The component to be measured is mounted on the table of the machine. No clamps are used so as to avoid distortion. The movement of the probe along the X, Y and Z axes is achieved by a simple joystick. When the probe touches the workpiece surface the machine stops no

Fig. 7.27 *Manual coordinate measurement: (a) manual coordinate measuring machine; (b) configuration of the manual system*

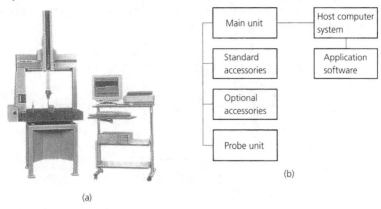

(a)

(b)

Fig. 7.28 *CNC coordinate measurement: (a) CNC coordinate measuring machine (CMM); (b) configuration for CNC system*

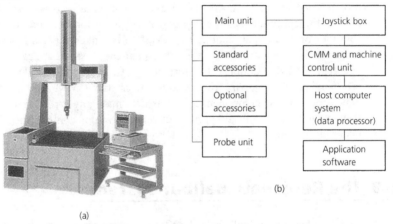

(a)

(b)

matter what the position of the joystick in order to avoid damage to the probe (touch-stop function). The position of the probe is fed automatically into the computer. The system also prevents further movement in the previous direction (restricted area) and the probe must be withdrawn before it can be moved to its next destination. To establish the position of the workpiece on the machine table, the probe touches on at least two points on each of three mutually perpendicular faces, as shown in Fig. 7.29(a). These positions are entered directly into the computer, which automatically assesses the position and alignment of the component on the worktable and establishes the included corner as a point datum, as shown in Fig. 7.29(b). This system avoids the need for elaborate workholding fixtures to position all the components in a batch at exactly the same position on the worktable as each one is measured. The probe is now moved to the next measuring point, P (Fig. 7.29(c)). The computer will display the position of this point relative to the origin. By moving the probe to each of the preplanned, key inspection points on the component in turn, all the dimensions can be checked.

Fig. 7.29 *Principle of use of a coordinate measuring machine: (a) touch probe on position T₁–T₅ inclusive; (b) computer identifies point of origin and alignment of component from the probe signals received in (a) at the common corner for the three faces A, B and C; (c) touching the probe on point P will determine the distance x from the face C*

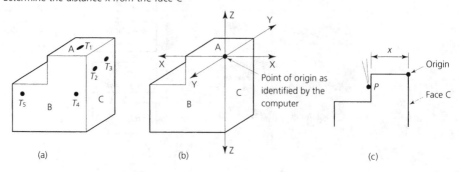

(a)　　　　　　　　(b)　　　　　　　　(c)

Where a large number of components of the same type are to be inspected, a CNC machine can be used. This works in exactly the same way except that the controller is programmed to carry out all the movements necessary to set the point of origin and inspect the components. The components can be loaded onto the machine table by conveyor or robot. This enables the CMM to be linked into an FMS cell for lights-out, unmanned production. Automatic probe-changing systems are also available with CNC operated CMMs, adding still further to their versatility. The computer can be programmed to compare the readings from the CMM and probe with the ideal dimensions for the component and thus determine any errors that may be present and their magnitude. It will also print out a certificate of conformance for the component under inspection.

7.8 The Renishaw Ballbar system

The Renishaw QC10 Ballbar machine tool checking gauge is simple in concept and elegant in design. The test is a method prescribed in international standards for machine testing. It provides an easy way to compare one machine with another, revealing errors and establishing performance benchmarks. The Ballbar consists of a high-accuracy displacement sensor housed inside a telescopic bar. The sensor works on similar electromagnetic principles to the LVDT described in Section 7.3.4. As the Ballbar changes length, the core of the LVDT moves inside the coils and causes an inductance change. This change in inductance is detected by laser-calibrated circuitry that converts the change in inductance into a position read-out with a resolution of $0.1\,\mu m$ and an accuracy of $\pm0.5\,\mu m$ (20 microinches) at $20\,°C$.

The sensor is coupled directly to a precision ball on one end, and to a magnetic cup on the other end. Another precision ball is held in a magnetic base that is set on the machine bed. This ball connects with the cup in the end of the Ballbar body and is retained against a kinematic mount by a powerful magnet. This kinematic mount allows the measuring bar to rotate in any attitude. The ball on the opposite end of the sensor is attached to another magnetic cup held in the machine spindle. In this way, the Ballbar is held magnetically between the machine table and the machine spindle, as shown in Fig. 7.30.

Fig. 7.30 *Use of a Ballbar to measure radial deviation*

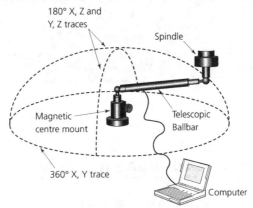

The machine spindle is driven round the centre point by programming the machine to follow a circular path, with a radius equal to the length of the Ballbar. As the machine moves through its programme, the sensor measures any variation in path radius. From this, the associated PC-based Renishaw software can derive volumetric accuracy, squareness values, repeatability, backlash, contouring ability, linearity XYZ, and servo performance. Driving the spindle through one flat circle, for example in the XY plane, will give X and Y lengths, XY squareness, backlash, and servo performance, the entire process taking approximately two minutes.

The conditioning electronics are contained in a small battery-powered unit that can be mounted on the machine magnetically in any convenient position. One cable is connected to the sensor and the other to the COM port of a PC or laptop computer. Data is sent to the computer using an RS232 protocol. The maximum sample rate is 250 values per second and the maximum data points are 3600 per run. The menu-driven software requires no programming knowledge. The results of the test, as displayed on the PC screen, can be retained on disk or printed out as hard copy. These results can be shown numerically and graphically.

The Renishaw QC10 Ballbar system is not only a checking gauge but is also part of standards for machine tool calibration and pass-off. Ballbar testing is included in a number of standards for machine tool accuracies, i.e. ISO 230, ASME B5.54 and BS 3800.

SELF-ASSESSMENT TASK 7.3

1. Research the manufacturers' literature in order to describe, with the aid of diagrams, the constructional and operation principles of:
   (a) touch-trigger probes
   (b) analogue-scanning probes

7.9 Screw-thread measurement

Figure 7.31 shows the elements of a screw thread. In order to inspect a screw thread completely it is necessary to measure the major, minor and effective (pitch) diameters. It is also necessary to measure the pitch and check the thread profile – particularly the flank angle.

Fig. 7.31 *ISO screw-thread form*

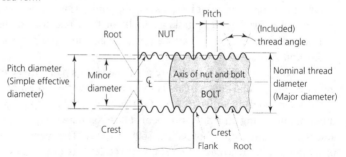

Figure 7.32 shows a floating carriage diameter-measuring machine. The workpiece to be measured is supported between centres, and the micrometer and fiducial indicator are mounted on a carriage. This carriage maintains the measuring axis perpendicular to the workpiece axis and ensures that the measurement is across the diameter of the workpiece. The carriage is free to move parallel to the axis of the workpiece and the top slide, which carries the micrometer and fiducial indicator, is free to move perpendicularly to the axis of the workpiece. The slides follow the principles described in Section 7.1.1 to give maximum freedom (minimum friction) in the required direction and maximum stiffness in all other directions. The micrometer has a large-diameter barrel and thimble that has a reading accuracy of 0.001 mm over a range of 25 mm. As its spindle does not rotate, this reduces the wear on the contact face. The fiducial indicator ensures a constant measuring pressure each and every time the instrument is used.

Adjustable arms are provided for suspending prisms and cylinders, and the importance of these will be described later.

Fig. 7.32 *Floating-carriage diameter measuring machine (reproduced courtesy of J.E. Nanson Gauges Ltd)*

7.9.1 *Measurement of the major diameter*

Measurements taken with a diameter-measuring machine are comparative. The initial reading is taken over a known standard as near to the major diameter as possible, as shown in Fig. 7.33(a). The thread to be measured is then substituted for the setting standard and a reading is taken over the major diameter of the workpiece (see Example 7.1), as shown in Fig. 7.33(b). The major diameter is then calculated as follows:

$$M = D_s \pm (\text{difference between } R_1 \text{ and } R_2)$$

where M = the major diameter of the screw thread
D_s = the actual diameter of the setting standard
R_1 = the reading over the setting standard
R_2 = the reading over the screw thread

Fig. 7.33 *Measuring with a diameter-measuring machine: (a) calibrating the diameter-measuring machine; (b) measuring the major diameter*

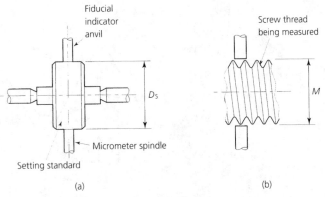

Calculate the major diameter of a screw thread with a nominal diameter of 28 mm, given the following details:

    calibrated diameter of setting standard = 25.000 mm
    reading over setting standard   = 17.908 mm
    reading over screw thread   = 20.898 mm

Since the major diameter is larger than the setting standard, use:

$$M = D_s + (R_2 - R_1) \quad \text{where } D_s = 25.000 \text{ mm}; \quad R_1 = 17.908; \quad R_2 = 20.898$$
$$= 25.00 + (20.898 - 17.908)$$
$$= \textbf{27.990 mm}$$

7.9.2 *Measurement of the effective diameter*

To measure the effective or pitch diameter of a screw thread, contact must be made with the straight flanks of the thread. To do this, precision *cylinders* or 'wires' are introduced between the micrometer anvils and the screw thread, as shown in Fig. 7.34(a). If a floating carriage measuring machine is not available, then three wires have to be used, as shown in Fig. 7.34(b) to align the micrometer and prevent it from twisting out of a plane perpendicular to the axis of the screw thread. Figure 7.34(c) shows the geometry of a cylinder in contact with a thread. You can see that:

$$E = T + 2x$$

where E = effective diameter;
          T = the measured size over the cylinders minus twice the cylinder diameter
       $2x$ = a constant (P) whose value depends upon the cylinder diameter, the pitch of
            the thread and the semi-angle of the thread form.

Fig. 7.34 *Use of measuring cylinders (wires): (a) measurement over cylinders diameter measuring machine; (b) use of three cylinders micrometer caliper; (c) geometry of cylinder contact with thread flanks*

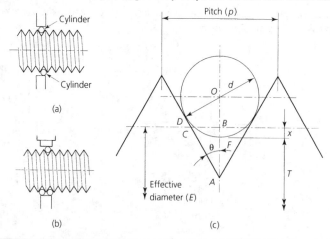

For all practical purposes, when measuring ISO threads where the included flank angle of the thread form is 60°, the following expression may be used to calculate P:

$$P = 0.86603\,p - d$$

where p = pitch
$\quad\quad$ d = cylinder diameter

Dimensions are in inches when unified threads are being measured, and in millimetres when metric threads are being measured.

Again, the measurements taken are comparative and the procedure is as follows. An initial reading is taken over the setting standard with the cylinders or 'wires' in place, as shown in Fig. 7.35(a). A second reading is then taken over the screw thread with the wires in place, as shown in Fig. 7.35(b), and the dimension T can be calculated as follows:

$$T = D_s \pm (\text{difference between } R_3 \text{ and } R_4)$$

where D_s = actual diameter of setting standard
$\quad\quad$ R_3 = measurement over setting standard with cylinders in place
$\quad\quad$ R_4 = measurement over the screw thread with cylinders in place.

If the effective diameter is larger than the diameter of the setting standard, use $D_s + \cdots$; if the effective diameter is smaller than the diameter of the setting standard use $D_s - \cdots$ (see Example 7.2).

Fig. 7.35 *Measuring the effective diameter: (a) calibration of diameter-measuring machine with cylinders in place; (b) measurement of screw thread*

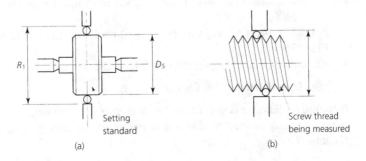

ISD metric screw threads

Maximum wire	Minimum wire	Best wire
1.01p	0.505p	0.577p

p = Pitch

(a) Setting standard

(b) Screw thread being measured

Calculate the effective diameter for an external screw thread given the following data:

pitch	$= 3.500$ mm
diameter of the calibrated standard	$= 30.000$ mm
measurement over the standard	$= 15.377$ mm
measurement over the screw thread	$= 14.343$ mm
cylinder diameter	$= 2.000$ mm

Since the effective diameter is smaller than the setting standard, use:

$$T = D_s - (R_3 - R_4)$$

where $D_s = 30.000$ mm; $R_3 = 35.377$ mm; $R_4 = 34.343$ mm

$$= 30 - (35.377 - 34.343)$$
$$= \mathbf{28.966 \; mm}$$
$$P = 0.86603\,p - d \qquad \text{where } p = 3.5 \text{ mm}; \quad d = 2.0 \text{ mm}$$
$$= 0.86603 \times 3.5 - 2.0$$
$$= \mathbf{1.0311 \; mm}$$

Since:

the effective diameter $E = T + P$

$$E = 28.966 + 1.0311$$
$$= \mathbf{29.971 \; mm}$$

The above calculations hold good provided that (a) the cylinders touch the threads somewhere on the straight flanks and (b) the flank angle is correct. The maximum and minimum wire sizes for a given thread are shown in Fig. 7.35 for metric and unified threads. The best cylinder size is also shown. This is used if there is a possibility of the flank angle being incorrect.

7.9.3 *Measurement of the minor diameter*

The minor or root diameter is measured in a similar way to the effective diameter, except that *prisms* are used instead of cylinders. These are inserted between the micrometer anvils and the screw thread so that contact can be made with the minor diameter of the thread. Again the measurement is comparative and the procedure is as follows:

- A reading R_5 is taken over the setting standard with the prisms in place, as shown in Fig. 7.36(a).
- A reading R_6 is taken over the screw thread with the prisms in place, as shown in Fig. 7.36(b).
- The minor diameter is then calculated from the expression:

$$m = D_s \pm (\text{difference between } R_5 \text{ and } R_6)$$

If the minor diameter is larger than the diameter of the setting standard use $D_s + \cdots$; if the minor diameter is smaller than the diameter of the setting standard use $D_s - \cdots$ (see Example 7.3).

Note that the size of the prisms does not affect the calculation. The only proviso being that the prisms will fit between the minor diameter of the thread and the micrometer anvils without touching any other part of the thread.

Fig. 7.36 *Measuring the minor diameter: (a) calibration of diameter-measuring machine prisms in position; (b) measurement of screw thread*

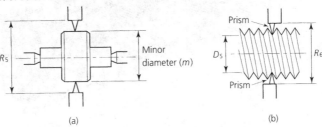

(a)   (b)

EXAMPLE 7.3

Calculate the minor diameter of a screw thread given the following data:

$D_s = 30.000$ mm diameter
$R_5 = 12.377$ mm diameter
$R_6 = 14.513$ mm diameter

Since the minor diameter is larger than the setting standard diameter use:

$m = D_s + (R_6 - R_5)$ where $D_s = 30.000$ mm
$= 30.000 + (14.513 - 12.377)$
$= \mathbf{32.136}$ **mm diameter**

7.9.4 *Screw-thread pitch measurement*

Figure 7.37 shows a typical screw-thread pitch-measuring machine. The carriage is moved parallel to the axis of the thread being measured by a lead screw and micrometer. A fiducial indicator mounted on the carriage indicates when corresponding points have been reached on adjacent threads. The difference between the readings of the micrometer dial indicates the pitch of the screw being measured.

Figure 7.38(a) shows a typical fiducial indicator suitable for use on a screw thread pitch-measuring machine. A round-nosed stylus engages the thread approximately on the pitch line. As the carriage is moved, the stylus rides up the thread and causes the reinforced diaphragm spring and strut to deflect in a radial direction. This does not register on the pointer. However, should the side forces on the stylus (F_1 and F_2) be unequal, the diaphragm spring will twist, causing the pointer to be deflected.

Fig. 7.38 *Use of a fiducial indicator: (a) mechanism (principle); (b) indicator scales*

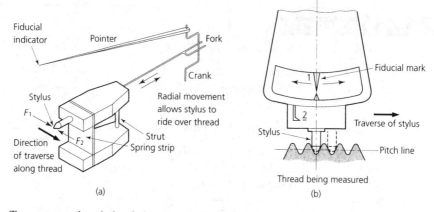

(a)

(b)

To measure the pitch of the thread, the stylus is engaged in the thread groove with sufficient force for the pointer (2) in Fig. 7.38(b) to lie between its datum lines. The micrometer is adjusted until the pointer (1) in the same figure lies opposite its datum mark and the initial reading of the micrometer is noted. Continuing to rotate the micrometer in the same direction, so that no backlash is introduced, the carriage is moved along by one thread until the pointer (1) is again opposite its datum mark and the micrometer is read for the second time. The difference between the readings is the measured pitch of the thread.

7.9.5 *Screw-thread form*

Finally it is necessary to check the flank angles, the root and crest radii, and the profile in general. This is usually done by means of an optical projector. The projection of screw threads presents a special problem due to the helix of the thread that interferes with the passage of the light rays, as shown in Fig. 7.39(a). There are two ways of overcoming this problem.

One method, as shown in Fig. 7.39(b), is to incline the axis of the thread so that the collimated light beam grazes the thread without appreciable interference. Since the thread is inclined to the focal plane of the projection lens, the image formed at the screen is that of a plane cutting the thread normal to the helix and, therefore, lying at an angle to the axis. Although a sharply focused image will result, this image will be foreshortened as shown.

Alternatively the thread axis can be left parallel to the screen and the collimated beam of light can be inclined to the helix angle, as shown in Fig. 7.39(c). Although the collimated beam of light is inclined, the axis of the screw thread lies in the focal plane of the projection lens and, therefore, the projected image will be that of a section plane passing through the axis of the screw thread. This results in a sharp image with a minimum deviation from the true form. The projection lens has to be specially designed for this application, otherwise only very small helix angles can be projected without considerable loss of field and definition. This second method is the one normally used in practice. Figure 7.40(a) shows a typical optical projector suitable for checking screw threads, and Fig. 7.40(b) shows its optical light path.

Fig. 7.39 *Projecting a screw thread: (a) light rays perpendicular to thread axis; (b) thread inclined at helix angle; (c) light rays inclined parallel to helix angle*

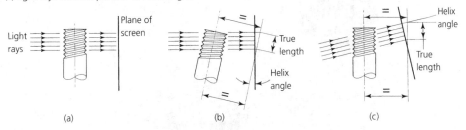

(a)   (b)   (c)

When checking screw-thread forms by projection, two elements are usually measured simultaneously, one is the thread profile and the other is the flank angle. The angle is, of course, part of the profile and is treated as such for production screw threads. However, for gauge measurement it is usual to measure the thread angle directly by a protractor fitted to the rear screen of the projector. A typical protractor for use with a projector is shown in Fig. 7.41(a). The scale is calibrated in intervals of 0.5° and the micrometer scales on the adjusting screws are calibrated in intervals of 1 minute of arc. Protractors with decimal notation are also available.

Root and crest radii together with pitch errors can be checked against templates or overlays, as can the flank angle of production screw threads. The templates or overlays are made to a precise magnification and are used as shown in Fig. 7.41(b). Pitch errors can be measured by means of the micrometer slide on which the screw thread is mounted.

7.10 Screw-thread gauging

As you have seen in the preceding sections, the measurement of screw threads is a complex process involving skilled personnel and costly precision instruments. Except for gauges and special high-precision threaded components made on a 'one-off' basis, it would be too costly to use measurement techniques for checking production screw threads. For threaded components produced on a batch or continuous production basis, thread gauging is used.

Fig. 7.40 *Optical profile projector: (a) profile projector; (b) light path of profile projector*

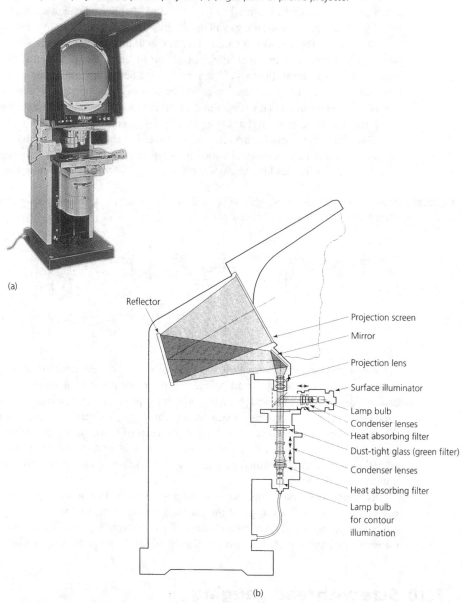

(a)

Reflector
Projection screen
Mirror
Projection lens
Surface illuminator
Lamp bulb
Condenser lenses
Heat absorbing filter
Dust-tight glass (green filter)
Condenser lenses
Heat absorbing filter
Lamp bulb for contour illumination

(b)

As for any other component, or features of components, screw threads cannot be made to an exact size. BS 3643 provides a system of tolerancing for male and female screw threads in a similar manner to the system for plain shafts and holes. In the ISO system, the fundamental deviation is designated by capital letters for nuts and lowercase letters for bolts. A number is used to designate the tolerance grade. Thus the tolerance class is a combination of fundamental deviation letter and the tolerance grade number, as shown in Fig. 7.42(a). It can be seen that just as a hole-based system is the most practical for plain

Fig. 7.41 *Measuring the flank angle: (a) flank-angle protractor; (b) view through microscope ocular; (c) projected image*

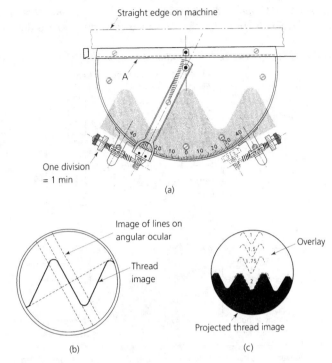

shafts and holes, a nut-based system is the most practical for threaded components because the internal thread is invariably cut with a fixed size tap, whilst the external thread can be varied in size by adjusting the thread-cutting and thread-forming tools. Figure 7.42(b) shows how a metric screw thread should be specified. For the standard metric threads used for nuts, bolts and set-screws, the following standards provide all the data and tolerances necessary to comply with the international requirements, for the manufacture of threaded fasteners:

BS EN 24014: *Hexagon head bolts – product grades A and B*
BS EN 24016: *Hexagon head bolts – product grade C*
BS EN 24017: *Hexagon head screws – product grade A and B*
BS EN 24018: *Hexagon head screws – product grade C*
BS EN 24032: *Hexagon nuts (style 1) – product grade A and B*
BS EN 24033: *Hexagon nuts (style 2) – product grade A and B*
BS EN 24034: *Hexagon nuts (style 1) – product grade C*

The above are just a sample of the wide range of standards available specifying the dimensions and tolerance for standard metric threaded fasteners.

Providing the designer has applied correct limits to the dimensions in accordance with BS 3643, the screw will function satisfactorily if its manufactured size lies within those limits. Therefore, when checking a screw thread, it is not necessary to measure its size exactly but only to determine if its elements lie within the prescribed limits. Providing the number of threaded components being manufactured warrants the outlay, the most economical method of checking screw threads is to use limit gauges. Taylor's Principles of Gauging were discussed in *Manufacturing Technology*, Volume 1, and screw-thread gauges follow these principles closely since they check geometrical form as well as dimensional accuracy.

Fig. 7.42 *Screw-thread specification: (a) screw-thread tolerances; (b) specification of screw threads*

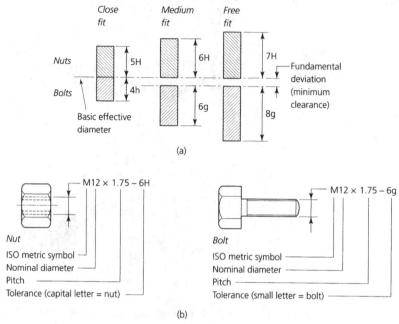

7.10.1 *External screw threads*

Because of the complexity of a screw-thread form, more than one gauge is required to check it thoroughly. In the case of an external thread, current practice favours the use of the following gauges:

- *A screw-thread caliper gauge* has full form GO anvils and form-relieved NOT GO anvils in accordance with Taylor's Principles of Gauging. This enables the gauge to check the effective diameter and, over the length of the anvils, the pitch of the thread.
- *A plain caliper gauge* is used to check the major diameter over the crests of the threads. Only the NOT GO element is required in this instance. It is not usual to check the minor diameter separately, as this feature is checked for interference by the screw-thread caliper gauge at the same time that it checks the effective diameter.

7.10.2 *Internal screw threads*

As for the external screw threads discussed above, internal screw threads also require more than one type of gauge to be used for their inspection and to check the thread form thoroughly. Current practice favours the use of the following gauges.

- *A screw-thread plug gauge*, as shown in Fig. 7.43(a). In compliance with Taylor's Principles, it has a full-form, full-length GO element and a shortened, form-relieved NOT GO element. The NOT GO element has a truncated thread form with the crest removed, and a heavily relieved minor diameter, as shown in Fig. 7.43(b). This permits flank contact only so that it can check the effective diameter.
- *A plain single-ended plug gauge* is used to check the minor diameter. Only the NOT GO element is required to check that the core diameter is not *over size*, thus reducing the strength of the threads. If the core diameter was *under size*, then the screw-thread gauge would not enter, hence there is no need for a GO element on the plain plug gauge. It is not usual to check the major diameter of an internal thread, since the screw-thread plug gauge checks this for interference whilst checking the effective diameter.

Fig. 7.43 *Plug gauge for an internal thread: (a) screw-thread plug gauge; (b) form relief of plug gauge*

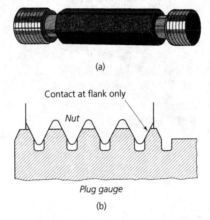

(a)

Contact at flank only

Nut

Plug gauge

(b)

7.10.3 *Screw-thread gauge applications*

The quality and type of screw-thread gauge used depends upon the class of work being checked. Inspection can vary from 'just trying a nut on the screw' to complete gauging and even individual measurement for critical components.

For most screw-thread production of commercial quality fasteners, it is not necessary to use a full set of gauges as set out in Section 7.10.2. Thread-cutting equipment is manufactured to high standards of accuracy and, normally, only the effective diameter has to be strictly controlled. Hence the inspector usually relies solely upon the use of a screw-thread caliper gauge for external threads or a screw-thread plug gauge for internal threads. Screw-thread caliper gauges are usually adjustable so that their anvils can be set to a master gauge. This allows for adjustment when wear takes place.

For the production of commercial quality fasteners, which generally have internal threads, once the tap being employed has been fully inspected, it is not necessary to use a double-ended screw-thread gauge. Only a gauge with a NOT GO element is required, and this speeds up the inspection process with a corresponding reduction in costs.

7.11 Gear-tooth inspection

The inspection of spur gears falls into two distinct areas of investigation:

1. The *measurement* of individual gear-tooth elements to determine whether or not they have been manufactured to the specified design tolerances. The measurement of gears is a very complex subject, therefore, within the confines of this chapter, only the following basic measurements will be considered:
   (a) tooth thickness at the pitch line
   (b) constant chord
   (c) pitch measurement
   (d) base tangent
   (e) involute form testing.

2. The dynamic testing of gears to ensure that they are concentric and that they run quietly and smoothly. Such tests include the use of:
   (a) rolling gear-testing machines
   (b) noise- and vibration-analysing machines.

7.11.1 *Measurement of tooth thickness at the pitch line*

Figure 7.44 shows the elements of involute gear teeth and their relationships when meshing together. One way to measure the tooth thickness at the pitch line is to use a gear-tooth caliper, as shown in Fig. 7.45. It can be seen that this instrument has two sets of scales perpendicular to each other. This enables the tooth thickness to be measured at any given point from the root to the top of the tooth by adjusting the auxiliary slide controlling the height dimension (*h*). In practice, the auxiliary slide is set so that the main scale shows the thickness of the tooth at the pitch circle. This is called the *chordal width* (t_c).

Fig. 7.44 *Elements of involute gear teeth (reproduced courtesy of Butterworth–Heinemann)*

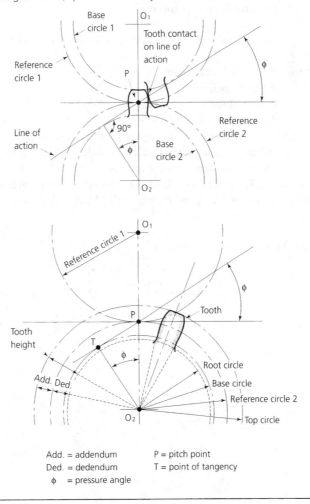

Add. = addendum   P = pitch point
Ded. = dedendum   T = point of tangency
 ϕ = pressure angle

Measurement and inspection **223**

Fig. 7.45 *Gear-tooth vernier caliper*

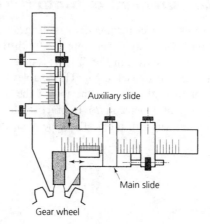

The theoretical value of the chordal width and the setting height of the auxiliary slide can be calculated from the expressions:

$$t_c = m \times N \times \sin(90°/N) \quad \text{and} \quad h = m + \left\{ m \times N/2 \left[1 - \cos\left(\frac{90°}{N}\right) \right] \right\}$$

where t_c = chordal thickness
  h = setting height
  m = modular pitch (as shown in Fig. 7.46)
  N = number of teeth

Thus the auxiliary slide of the gear-tooth vernier caliper is set to the calculated value of the setting height (h), and the actual chordal thickness (t_c) is read from the main scale of the caliper. This is compared with the calculated value of (t_c) and the difference is the error. In practice (see Example 7.4) allowance must be made for variations in tooth form (e.g. stub teeth) and also for designed clearance (backlash) which will reduce the chordal thickness (t_c)

Fig. 7.46 *Tooth thickness at the pitch line*

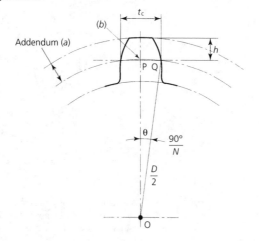

EXAMPLE 7.4

A spur gear has 25 teeth and a modular pitch of 3 mm. Calculate the setting height (h) and the chordal thickness (t_c) of an ideal involute form tooth.

$$t_c = m \times N \times \sin(90°/N) \qquad \text{where} \quad m = 3\,\text{mm and } N = 25 \text{ teeth}$$
$$= 3 \times 25 \times \sin(90°/25)$$
$$= 4.71\,\text{mm}$$
$$h = m + \{m \times N/2[1 - \cos(90°/N)]\}$$
$$= 3 + \{3 \times 25/2[1 - \cos(90°/25)]\}$$
$$= 3.074\,\text{mm}$$

7.11.2 *The constant chord*

In the previous section the chordal thickness (t_c) and the setting height (h) were dependent upon the number of teeth on the gear as well as the modular pitch. The *constant chord* is *independent* of the number of teeth and this is a useful measure where a large number of gears have the same modular pitch and pressure angle. Figure 7.47 shows an involute gear tooth in mesh with an involute rack. The tooth thickness (t) is measured at the points of contact at a dimension (H) from the tip of the tooth (see Example 7.5), and the theoretical values can be calculated from the following expressions:

$$t = m\pi/2(\cos^2\phi) \quad \text{and} \quad H = m(1 - \pi/4\sin\phi\cos\phi)$$

where m = modular pitch and ϕ = pressure angle.

Fig. 7.47 *The constant chord*

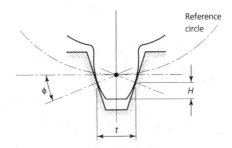

Reference circle

ϕ

H

t

EXAMPLE 7.5

Calculate the constant chord-tooth thickness (t) and its height (H) for a gear with a modular pitch of 3 mm and a pressure angle of 20°.

$$t = m.\pi/2(\cos^2\phi) \qquad \text{where} \quad m = 3\,\text{mm}; \quad \phi = 20°$$
$$= 3 \times \pi/2(\cos^2 20°)$$
$$= 4.161\,\text{mm}$$

$$H = m(1 - \pi/4.\sin\phi\cos\phi)$$
$$= m(1 - \pi/4.\sin 20° \cos 20°)$$
$$= 2.243\,\text{mm}$$

7.11.3 *Pitch measurement*

The circular pitch of the teeth of a spur gear when measured on the base circle is called the *base pitch* (P_B). The circular distance EF in Fig. 7.48(a) is the base pitch for the involute tooth form shown. The geometrical properties of an involute curve ensure that any tangents to the base circle, drawn as shown in Fig. 7.48(b), give the following relationship:

base pitch$(P_B) = \text{EF} = \text{CD} \dots$ etc.,

also:

number of teeth $(N) \times P_B = $ base circle circumference

thus $\quad N \times P_B = \pi D_B$

and $\quad P_B = \pi D_B / N$ $\qquad\qquad\qquad$ (7.1)

The following relationship can be deduced from Fig. 7.48(b):

$(D_B/2) \div (D/2) = \cos\phi$

thus $\quad D_B = D\cos\phi$ $\qquad\qquad\qquad$ (7.2)

Combining equations (7.1) and (7.2)

$$P_B = \pi(D/N)\cos\phi \qquad\qquad\qquad (7.3)$$

where modular pitch $(m) = D/N$

thus $\quad P_B = m.\pi.\cos\phi$ $\qquad\qquad\qquad$ (7.4)

Fig. 7.48 *Base pitch: (a) base-pitch geometry; (b) relationship between D, D_B and ϕ*

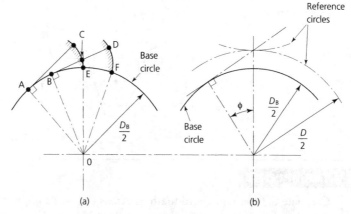

(a) $\qquad\qquad\qquad\qquad$ (b)

Thus the base pitch (P_B) can be calculated from the known data of modular pitch and the pressure angle (ϕ), since DCB is also tangential to the base circle. Although difficult to gauge, base pitch can easily be measured, as shown in Fig. 7.49. The DTI is used as a fiducial indicator and the vernier height gauge, on which it is mounted, is adjusted so that the DTI gives the same reading for each height setting. Thus

$P_B = H_2 - H_1$

For greater accuracy the DTI could be set over slip gauges.

Fig. 7.49 *Measurement of the base pitch*

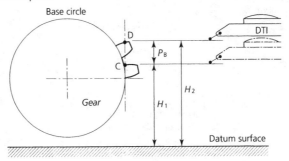

Calculate the base pitch for a spur gear having a modular pitch of 3.5 mm and a pressure angle of 20°.

$$P_B = m.\pi.\cos\phi \qquad \text{where} \quad m = 3.5\,\text{mm}; \quad \phi = 20°$$
$$= 3.5 \times \pi\cos 20°$$
$$P_B = \mathbf{10.332\,mm}$$

7.11.4 Base tangent

This is a useful measure of gear-tooth spacing since it can be made with an ordinary vernier caliper, as shown in Fig. 7.50(a). Since this measurement is taken over several teeth, it also indicates cyclical errors as the caliper is stepped around the gear.

The base tangent measurement can be taken at various positions AB, CD, etc., as shown in Fig. 7.50(b). It should be apparent, from previous discussion in this chapter, that any base tangents measured across opposed involute flanks are equal. Thus:

base tangent (t_B) = AB = CD = arc EF

where the base tangent is measured with a vernier caliper.

The base tangent can also be calculated using the following empirical expression (see also Example 7.7), providing the standard pressure angle of 20° is used:

$$t_B = m\cos 20°[(N/\pi)\,\text{inv}\,20° + k - 0.5]$$

where   inv 20° = involute function of 20°

           k = number of teeth spanned by t_B

           m = modular pitch

           N = number of teeth on the gear wheel

N	10–18	9–27	28–36	37–45	46–54	55–63	64–72	73–81	82–90
k	2	3	4	5	6	7	8	9	10

Fig. 7.50 *Measurement of the base tangent (vernier): (a) measurement with vernier caliper; (b) geometry of the base tangent*

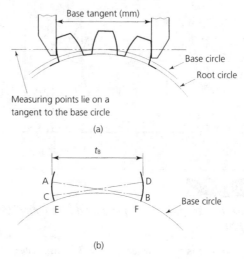

Base tangent (mm)

Base circle

Root circle

Measuring points lie on a
tangent to the base circle

(a)

t_B

A

D

C

B

E

F

Base circle

(b)

EXAMPLE 7.7

Calculate the base tangent for a 50-tooth gear with a modular pitch of 3 mm. The pressure angle is 20°.

$$t_B = m\cos 20°[(N/\pi)\,\mathrm{inv}\,20° + k - 0.5] \quad \text{where} \quad m = 3\,\text{mm}; \, k = 6 \text{ (from table)}$$
$$= 3 \times 0.9397 \times [(50/\pi) \times 0.014\,904 + 5.5]$$
$$= 16.13\,\text{mm}$$

7.11.5 *Measurement over rollers*

Rollers can be placed between the jaws of a measuring instrument and a surface when it is necessary to obtain line contact only. For this reason measurement over rollers can be employed to check gear teeth. The geometrical relationship between a roller and an involute rack tooth is shown in Fig. 7.51. When taking measurements over rollers, it is essential to choose a suitable roller diameter so that the roller contacts the flanks of the tooth at the constant chord position (t_c). The roller diameter (d) can then be calculated by use of the expression:

$$d = (\pi/2)m\cos\phi$$

To check a straight-tooth spur gear, two rollers of equal diameter (d) have to be used, and the distance across them (M_1) has to be calculated and measured. Any difference between the measured size and the calculated size represents a manufacturing error. For a small gear that can be spanned by a vernier caliper or micrometer caliper, the rollers are spaced so that they are as diametrically opposed as possible. Figure 7.52(a) shows the

Fig. 7.51 *Roller and rack-tooth geometry*

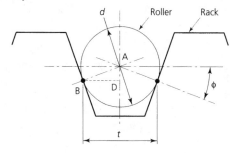

position of the rollers for gears with an even number of teeth so that the gears are diametrically opposite. In this case:

$$M_1 = D + d$$

where D is the base circle diameter and d is the roller diameter.

For gears with an odd number of teeth, it is not possible to arrange the rollers diametrically opposite to each other and allowance has to be made for their angularity, as shown in Fig. 7.52(b). The previous expression is now modified to:

$$M_2 = D(\cos 90/N) + d$$

Fig. 7.52 *Measurement over rollers (small gears); (a) gear with even number of teeth; (b) gear with odd number of teeth*

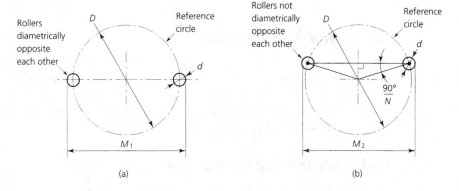

When checking large gears it is not always possible or convenient to span the gear diametrically with the measuring instrument. In these instances, it is more usual to check the gear across a selected number of teeth, as shown in Fig. 7.53. The angle $\theta°$ subtended by a selected number of teeth (N_s) can be determined from the expression:

$$\theta° = (N_s/N) \times 360°$$

thus $M_3 = (D \sin \frac{1}{2}\theta°) + d$ (see Example 7.8)

Fig. 7.53 *Measurement over rollers (large gears)*

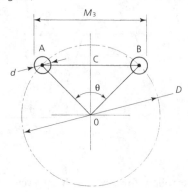

Calculate the diameter of the rollers and the dimension M_3 for a 180-tooth gear having a modular pitch of 8 mm and a pressure angle of $20°$. Selected number of teeth $(N_s) = 8$.

$$d = (\pi/2)m \cos \phi \qquad \text{where} \quad m = 8 \text{ mm}; \quad \phi = 20°$$
$$= (\pi/2) \times 8 \times \cos 20°$$
$$= 11.81 \text{ mm}$$

$$M_3 = (D \sin \tfrac{1}{2}\, \theta°) + d \qquad \text{where} \quad D = mN; \quad \theta° = (N_s/N) \times 360° = 16°$$
$$= (8 \times 180 \times \sin 8°) + 11.81$$
$$= 212.22 \text{ mm}$$

7.11.6 *Involute form testing*

The form of involute gear teeth can be checked either by optical projection for comparison with a template or overlay as in screw-thread inspection or, more usually nowadays, by a 3D coordinate-measuring machine or an electronic involute testing machine. Such machines use an electronic probe to trace the profile of each tooth and, after computer analysis, the signal is used to print out an enlarged profile.

7.11.7 *Dynamic gear testing*

The calculations and tests described previously in this chapter are intended to determine whether or not the gear is dimensionally correct. The dynamic tests are used to determine whether or not the gear will run smoothly and quietly. The two most common tests are:

Rolling test

The gear under test is mounted on a slide and meshed with a hardened and ground master gear. The two gears slowly rotate together and any errors due to eccentric running or irregular pitch will cause the gear under test and its slide to move to or from the master gear. This movement is measured and recorded. Spring tension keeps the gears in mesh.

Vibration and noise test

In a typical gear-tooth noise- and vibration-measuring machine, the finished gears are run in mesh at controlled speeds in both directions with and without brake loads. This enables the character of the vibration and the volume and frequency of the sound to be ascertained and analysed under a variety of working conditions.

SELF-ASSESSMENT TASK 7.5

1. Calculate the constant chord tooth thickness (t) and its height (H) for a gear with a modular pitch of 5 mm and a pressure angle of 20°.

2. Calculate the height (h) and the chordal thickness (t_c) for an involute spur gear with 50 teeth and a modular pitch of 5 mm.

3. Calculate the base pitch for a spur gear having a modular pitch of 5 mm and a pressure angle of 20°.

4. Calculate the base tangent for a 75-tooth gear with a modular pitch of 3.5 mm and a pressure angle of 20°. Take k as 9.

5. For a 170-tooth gear, calculate the modular pitch (m) and the distance over the rollers (M_3) if the diameter of both rollers is 12 mm and the pressure angle is 20° (selected number of teeth (N_s) = 17).

7.12 Geometric tolerance (introduction)

Traditionally, inspection has mainly concerned the measurement and gauging of linear dimensions and angles. However, in those sections of this chapter dealing with screw threads and gear teeth the importance of the component profile has also been emphasised. The importance of correct geometric accuracy as well as correct dimensional accuracy is being increasingly recognised.

Figure 7.58(a) shows a shaft whose dimensions have been given *limits of size*. The difference between the given limits is called the *tolerance*. The reason for providing tolerance dimensions is that it is not possible to manufacture a workpiece exactly to size and, even if it were possible, there is no means of measuring such a workpiece exactly. If the process used for many applications can achieve the dimensional tolerance, it will also produce a geometrically correct form. For example, the shaft shown in Fig. 7.54(a) would most likely be cylindrically ground to achieve the dimensional accuracy specified. Cylindrical grinding between centres should give a satisfactory degree of roundness. However, the shaft could also be mass-produced by centreless grinding and some *out-of-roundness* may occur such as *ovality* and *lobing*. Figure 7.54(b) shows how the shaft can be out-of-round yet still remain within its limits of size.

If cylindricity is important for the correct functioning of the component, then additional information must be added to the drawing to emphasise this point. This additional information is a *geometric tolerance* which, in this case, would be added as shown in Fig. 7.55(a). The symbol in the box indicates cylindricity, and the number indicates the geometric tolerance. Figure 7.55(b) shows the interpretation of this tolerance.

Fig. 7.54 *Need for geometric tolerance: (a) shaft; (b) geometrical error. Dimensions in millimetres*

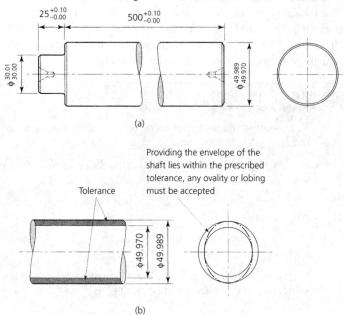

(a)

Tolerance

Providing the envelope of the shaft lies within the prescribed tolerance, any ovality or lobing must be accepted

(b)

Fig. 7.55 *Geometric tolerance: (a) addition of geometric tolerance for cylindricity; (b) interpretation of the geometric tolerance. The curved surface of the shaft is required to lie within two cylinders whose surfaces are coaxial with each other, a RADIAL distance 0.001 mm apart. The diameter of the outer coaxial cylinder lies between 49.989 and 49.970 mm as specified by the linear tolerance. Dimensions in millimetres*

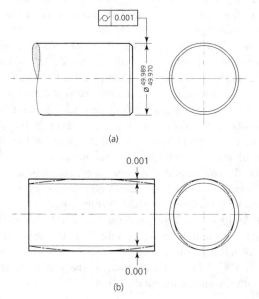

(a)

0.001

0.001

(b)

7.12.1 *Geometric tolerance (principles)*

You have just seen that, in some circumstances, dimensions and tolerance of size, however well applied, do not impose the necessary control of form. If control of form is necessary then geometric tolerances must be added to the component drawing. Geometric tolerances should be specified for all requirements critical to the interchangeability and correct functioning of components. An exception can be made when the machinery and techniques used can be relied upon to achieve the required standard of form. Geometrical tolerances may also need to be specified even when no special size tolerance is given. For example, the thickness of a surface plate is of little importance, but its accuracy of flatness is of fundamental importance.

Figure 7.56 shows the standard tolerance symbols as specified in BS 308: Part 3, and you can see that they are arranged into groups according to their function. In order to apply geometric tolerances it is first necessary to consider the following definitions.

Fig. 7.56 *Geometric tolerance symbols*

	Type of tolerance	Characteristics to be toleranced	Symbol	Clause ref.
For single features	From	Straightness	—	3.1.1
		Flatness	▱	3.1.2
		Roundness	○	3.1.3
		Cylindricity	⌭	3.1.4
		*Profile of a line	⌒	3.1.5
		*Profile of a surface	⌓	3.1.6
For related features	Attitude	Parallelism	//	3.2.1
		Squareness	⊥	3.2.2
		Angularity	∠	3.2.3
	Location	Position	⊕	3.3.1
		Concentricity	◎	3.3.2
		Symmetry	⩦	3.3.3
	Composite	Run-out	↗	4
Maximum material condition			Ⓜ	2.6
Boxed dimension (dimension which defines true position)			☐	2.7.1

*May be related to datum when it is necessary to control position in addition to form.

7.12.2 *Geometric tolerance*

This is the maximum permissible overall variation of form or position of a feature; that is, it defines the size and shape of a tolerance zone within which the surface, median plane or axis of the feature is to lie. It represents the full indicator movement it causes where testing with a DTI is applicable – for example, the 'run-out' of a shaft rotated about its own axis.

7.12.3 *Tolerance zone*

This is the zone within which the feature has to be contained. Thus, according to the characteristic that is to be toleranced, and the manner in which it is to be dimensioned, the tolerance zone is one of the following:

- A circle or a cylinder.
- The area between two parallel lines or two parallel straight lines.
- The space between two parallel surfaces or two parallel planes.
- The space within a parallelepiped.

Note: The tolerance zone, once established, permits any feature to vary within that zone. If it is considered necessary to prohibit sudden changes in surface direction, or to control the rate of change of a surface within this zone, then that should be additionally specified.

7.12.4 Geometrical reference frame

This is the diagram composed of the constructional dimensions that serve to establish the true geometric relationships between positional features in any one group, as shown in Fig. 7.57. For example, the symbol indicates that the hole-centre may lie anywhere within a circle of 0.02 mm diameter that is itself centred upon the intersection of the frame. The dimension indicates that the reference frame is a geometrically perfect square with sides of length 40 mm.

Fig. 7.57 *The geometric tolerance diagram: (a) drawing requirement; (b) reference frame*

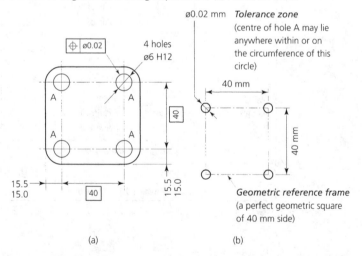

7.12.5 *Applications of geometric tolerances*

Figure 7.57 shows a plate in which four holes need to be drilled. For ease of assembly, the relationship between the hole-centres is more important than the group position on the plate. Therefore, a combination of dimensional and geometric tolerances has to be used.

Using the dimensional tolerances and the boxed dimension that defines true position, it is possible to determine the pattern-locating tolerance zone for each of the holes, as shown in Fig. 7.58. That is, the reference frame (which must always be a perfect square, with sides of 40 mm in this example) may lie in any position providing its corners are within the shaded boxes. The shaded boxes represent the *dimensional tolerance zones*. The hole centres may then (in this example) lie anywhere within the 0.02 mm circles that are themselves centred upon the corners of the reference frame.

Fig. 7.58 *Interpretation of geometric tolerances*

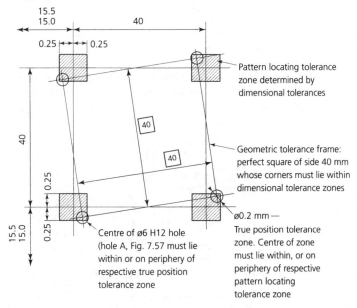

Figure 7.59 shows a further example. It is a stepped shaft with four concentric diameters and a flange that must run true with the datum axis. The shaft is to be located within journals at X and Y and these are identified as datums. Note the use of leader lines terminating in solid triangles resting upon the features to be used as datums. The common axis of these datum diameters is used as a datum axis for relating the tolerances controlling the concentricity of the remaining diameters and the true running of the flange face. Figure 7.59 also shows that the dependent diameters and face carry geometric tolerances as well as dimensional tolerances. The tolerance frames carry the concentricity symbol, the concentricity tolerance, and the datum identification letters. These identification letters show that the concentricity of the remaining diameters is relative to the common axis of the datum diameters X and Y. The tolerance frame whose leader touches the flange face indicates that this face must run true, within the tolerance indicated when the shaft rotates in the vee-blocks supporting it at X and Y, as shown in Fig. 7.60.

These two examples are in no way a comprehensive résumé of the application of geometric tolerancing but stand as an introduction to the subject and as an introduction to BS 308: Part 3. This British Standard details the full range of applications and their interpretation, and includes a number of fully dimensioned examples.

Fig. 7.59 *Datum identification*

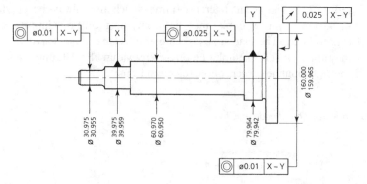

Fig. 7.60 *Testing for concentricity. (Note: Geometric tolerance shows maximum permissible eccentricity, whereas DTI will indicate 'throw' (2 × eccentricity) when the shaft is rotated)*

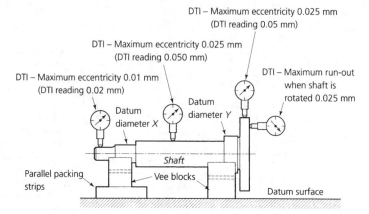

7.13 Virtual size

To understand this section, it is necessary to introduce the terms *maximum metal condition* and *minimum metal condition*.

- The *maximum metal condition* exists for both the hole and the shaft when the least amount of metal has been removed but the component is still within its specified limits of size, i.e. the largest shaft and smallest hole that will lie within the specified tolerance.
- The *minimum metal condition* exists when the greatest amount of metal has been removed but the component is still within its limits of size, i.e. the largest hole and the smallest shaft that will lie within the specified tolerance.

Figure 7.61 shows the effect of a change in geometry on the fit of a pin in a hole. In Fig. 7.61(a), an ideal pin is used with a straight axis. It can be seen that, as drawn, there is clearance between the pin and the hole. However, in Fig. 7.61(b) the pin is distorted and now becomes a tight fit in the hole despite the fact that its dimensional tolerance has not changed.

Figure 7.62(a) shows the same pin and hole, only this time dimensional and geometric tolerances have been added to the pin. For simplicity, the hole is assumed to be geometrically correct. The tolerance frame shows a straightness tolerance of 0.05 mm. This means that the *axis of the pin* can bow within the confines of an imaginary cylinder of 0.05 mm diameter. The dimensional limits indicate a tolerance of 0.08 mm. The worst conditions for assembly (tightest fit) occur when the pin and the hole are in their respective maximum metal conditions and, in addition, the maximum error permitted by the geometric tolerancing is also present. Figure 7.62(b) shows the fit between the pin and the hole under these conditions.

Fig. 7.62 *Virtual size: (a) pin with geometric and dimensional tolerances; (b) virtual size under maximum metal conditions (pin diameter = 25.00 mm); (c) virtual size under minimum metal conditions (pin diameter = 24.92). Dimensions in millimetres*

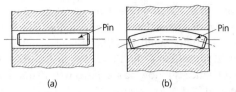

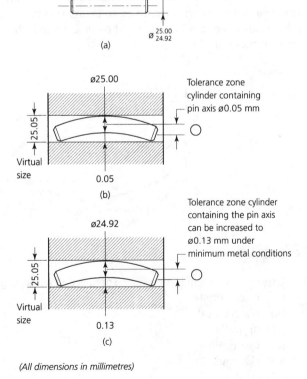

(All dimensions in millimetres)

Under such conditions the pin will just enter a truly straight and cylindrical hole equal to the pin diameter under maximum metal conditions (25.00 mm) plus the geometrical tolerance of 0.05 mm, i.e. a hole diameter of 25.05 mm. This theoretical hole diameter of 25.05 mm is referred to as the *virtual size for the pin*.

Under minimum metal conditions, the geometrical tolerance for the pin can be increased without altering the fit of the pin in the hole. As shown in Fig. 7.62(c), the geometric tolerance can now be increased to 0.13 mm with no increase in the virtual size or the change of fit.

7.14 The economics of geometric tolerancing

Figure 7.63 shows a typical relationship between tolerance and manufacturing costs for any given manufactured component part. As tolerance is reduced the manufacturing costs tend to rise rapidly. The addition of geometric tolerances on top of the dimensional tolerances aggravates this situation still further. Thus geometric tolerances are only applied if the function and the interchangeability of the component are critical and absolutely dependent upon a tightly controlled geometry. To keep costs down, it can often be assumed that the machine or process will provide adequate geometric control and only dimensional tolerances are added to the component drawing. For example, it can generally be assumed that a component finished on a cylindrical grinding machine between centres will be straight and cylindrical to a high degree of geometric accuracy. However, current trends in quality control and just-in-time (JIT) scheduling tend to increase the demand for closer dimensional and geometric tolerancing to ensure that no hold-ups occur during assembly and that quality specifications are met first time without the need for corrective action or rejection by the customer.

Fig. 7.63 *Cost of tolerancing*

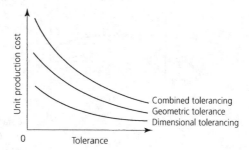

EXERCISES

7.1 Explain how magnification can be obtained in measuring instruments:
(a) mechanically
(b) pneumatically
(c) optically
(d) electrically
(e) electronically

7.2 With the aid of a sketch, describe the measurement of the *major diameter* of a screw thread and calculate that diameter, given:

calibrated diameter of the setting standard = 20.00 mm
reading over setting standard   = 18.003 mm
reading over screw thread   = 17.509 mm

7.3 With the aid of a sketch, describe the measurement of the *effective diameter* of a screw thread and calculate that diameter, given:

pitch   = 3.5 mm
calibrated diameter of the setting standard = 25.000 mm
reading over the setting standard   = 28.003 mm
reading over the cylinders (wires)   = 29.003 mm

7.4 With the aid of a sketch, describe the measurement of the *minor (root) diameter* of a screw thread and calculate that diameter, given:

$D_s = 29.978$ mm
$R_5 = 12.577$ mm
$R_6 = 14.499$ mm

7.5 With the aid of sketches, explain how an optical profile projector can be used to measure thread pitch and flank angles.

7.6 (a) With the aid of sketches, explain how Taylor's Principle of Gauging is applied to screw-thread gauges.
(b) With the aid of sketches, explain how an external screw thread can be checked using gauges.

7.7 With the aid of sketches and typical calculations, explain how **two** of the following dimensions applicable to involute gears are calculated and checked:
(a) tooth thickness at the pitch line
(b) constant chord
(c) pitch measurement
(d) base tangent

7.8 Discuss the need for the geometric tolerancing as well as the dimensional tolerancing of engineering components.

7.9 With the aid of sketches, explain what is meant by the following terms as applied to geometric tolerancing:
(a) tolerance zone
(b) geometric reference frame
(c) virtual size

7.10 (a) Fully dimension a component of your choice, using dimensional and geometric tolerancing.
(b) Discuss the economics of geometric tolerancing.

7.11 (a) Describe the essential differences between touch-trigger probes, analogue (scanning) probes, laser probes and optical probes and describe a typical application for each type.

(b) Describe the principle of operation of setting and operating a 3D coordinate-measuring machine.

7.12 Research manufacturers' literature and describe how the indexing accuracy of a rotary table can be checked using:

(a) a photoelectronic autocollimator and accessories

(b) a laser interferometer and accessories

7.13 Research the Renishaw website given earlier in the chapter and explain, with the aid of sketches, the essential principles of the following devices:

(a) retro-reflector

(b) polarising beam splitter

7.14 Research the Renishaw website given earlier in the chapter and explain what is meant by:

(a) polarisation of light

(b) deadpath error.

8 Advanced manufacturing technology

The main topic areas covered in this chapter are:

- The need for advanced manufacturing technology (AMT).
- CNC programming including: main programmes, nested loops, canned cycles and macros, subroutines, scaling mirror imaging, zero shift.
- CAD/CAM.
- Simulation of automated systems.
- Robot programming.
- Automated assembly.
- Cost of automation and assembly.

8.1 The need for advanced manufacturing technology

The Department of Trade and Industry has defined advanced manufacturing technology (AMT) as *the application of computers to manufacturing operations*. The importance of AMT cannot be overstated. For example, when the Advisory Council for Applied Research and Development (ACAR) published its report on *New Opportunities in Manufacturing* in 1984 it stated that: 'Companies not adopting AMT may not exist in 10 years' time.' The main elements of AMT as defined by the DTI will be covered here and in the following chapters. These elements are:

- Computer numerical control (CNC).
- Computer-aided design and manufacture (CAD/CAM).
- Flexible manufacturing systems (FMS).
- Computerised management systems, i.e. inventory control, production control, and automated stores and parts issue.

Figure 8.1 shows the interrelationship between these elements so that they become a wholly integrated system.

Fig. 8.1 *Areas of advanced manufacturing technology (AMT)*

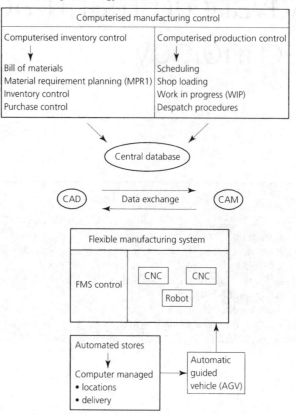

In recent years the manufacturing sector of the UK has been under intense pressure due to a number of factors that threaten the very existence of many companies. These factors are:

- Increased market competition from local and overseas competitors.
- Shorter product life cycles.
- Shorter working week and increasing labour costs.
- High cost of capital investment in new plant and equipment.
- Unpredictable economic situations.
- Greater sophistication of consumers who demand higher quality, reliability and variety of products.
- Pressure to produce goods within tighter environmental constraints that, in turn, add to the cost of the product and perhaps the use of more sophisticated processes.

Let's now look at what a company must do to overcome these factors in order to survive:

- Reduce the lead time from the design inception stage to producing the finished product.
- Increase the flexibility of the manufacturing process both in terms of handling part variety to accepting design alterations and scheduling changes.

- Increase response to changes in the market demand.
- Reduce the product costs.
- Increase its overall efficiency.

Where companies have accepted and exploited the introduction of computers into industry they have been able to meet some of the survival criteria. The two most important computer applications in industry are computer-integrated manufacture (CIM) and advanced manufacturing technology (AMT). We have already defined AMT, and we can now define CIM as *the unimpeded flow of electronic data within industry*. These computer applications are mainly concerned with increasing the flexibility and control of the manufacturing environment, while reducing the duplication of information and effort. Let's now start by considering the computer numerical control (CNC) of manufacturing equipment.

SELF-ASSESSMENT TASK 8.1

Research and describe how the car industry has adopted AMT to remain competitive in a global market.

8.2 CNC programming

CNC programming is the process of converting the information on the component drawing into a format that the machine control software can recognise and act upon. The principles of programming were introduced in *Manufacturing Technology*, Volume 1, Chapter 5. Let's now consider some further part programming techniques.

The part program for any machine tool or system must be designed with *safety as the main priority*. Very large machine tools operating at high speeds and driven by powerful motors are potentially dangerous and must be treated with the utmost respect. For this reason the programs and programming techniques discussed throughout this chapter will follow the essential rules to ensure safe operation. Let's commence by considering a very simple program structure in which only one tool or cutter is used, and all the necessary cutting data can be loaded at the start. The structure of such a program will contain the following information.

8.2.1 *Start-up information*

This is the first part of the program where any defaults such as English or metric units are set. It is also necessary to state whether positional information will be in an absolute or incremental format. A safe tool change position must be set with the associated tool offsets, speeds and feed values. Failure to associate speeds and feeds with the correct tool could result in the tool using the last defined (modal) speed and feed value stored in the machine register. This could be dangerous and result in personal injury and also damage to the machine. Remember that *modal* commands continue to be active until they are cancelled or changed.

8.2.2 *Machining information*

This is the information that controls the relative movements of the cutter and/or the workpiece and thus generates the required shape and features. The workpiece may have a simple two-dimensional profile or it may have more complex three-dimensional contoured surfaces. The machining information must also contain such data as the depth of cut, macros, and canned cycles for roughing and drilling, etc. As shown in *Manufacturing Technology*, Volume 1, the cutter must always be introduced to the work either on a ramp or a curved path so as not to leave a dwell-mark on the workpiece. This technique introduces the cutting forces gradually as the cutter is fed into its correct depth.

8.2.3 *Close-down information*

This is where the cutter returns to the tool-change position away from the workpiece. At the same time the spindle drive is turned off and the program returns to the start position ready for the next component. This will allow the operator to unload or inspect the workpiece under safe conditions, away from any sharp cutters and, if necessary, to change the cutting tool if the machine is configured for manual tool changing.

8.3 Structure

Computer programmers will tell you that a well-structured program is easy to read, is easy to debug, and is thus easier for everyone to understand. This, in turn, results in increased operational safety, particularly if companies adopt a 'house style' enabling operators to know where to look instinctively for any particular program feature. We can summarise the requirements of a correctly structured program as follows:

- Start-up information.
- The main program.
- Any features such as subroutines, nested loops, canned cycles, etc., branching from the main program.
- The close-down information.

The use of standard macros, subroutines, loops and operational skips, reduces the complexity and length of the program, thus aiding readability and reducing the program memory capacity required. A typical program structure is shown in Fig. 8.2.

Fig. 8.2 *Basic program structure*

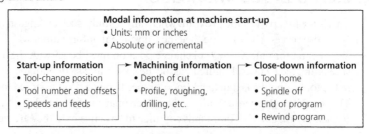

8.3.1 *Canned cycles*

An example of a *canned cycle* is the drilling cycle, as introduced in *Manufacturing Technology*, Volume 1. This previous example used the G81 code which does not include a 'pecking' infeed. *Pecking* is the term used when the drill is fed a short distance into the work and then retracted to clear and break up the swarf before making the next feed increment. The number of 'pecks' will depend upon the depth/diameter ratio of the hole. The deeper the hole the greater the number of 'pecks' required to ensure that the flutes of the drill do not become clogged. This could lead to breakage of the drill. Where parameters such as hole depth, retract height, feed rate and *number of pecks* are entered on one line we use the G83 code, i.e.

G83 R3 D10 N5 F200

This would produce a hole 10 mm deep, drilled in 5 *pecks*, at a feed rate of 200 mm/min with a retract height of 3 mm. Other examples of canned cycles include:

- Tapping screw threads.
- Pocket and slot milling.
- Area clearance within a boundary.
- Turning screw threads.

8.3.2 *Turning example using a subroutine with a loop and a canned cycle*

When a component design results in its program containing fixed sequences of frequently repeated patterns, these sequences can be stored in the control memory as a subprogram, called a *subroutine*, to simplify the task of programming. A subroutine is simply a set of instructions that can be called up and inserted repeatedly into the main body of a program by entering an identifying code.

It is sometimes useful for a subroutine to be repeated a set number of times. The letter P is used to identify the subroutine number. In addition, when using a Fanuc or similar controller, the letter L is used to specify the number of repeats, for example:

N400 M98 P2030 L2

means go to subroutine number 2030 and repeat it twice before returning to the main program. This technique can be illustrated by using a turning program to produce the component shown in Fig. 8.3(a).

The operation is to turn down the 28 mm diameter to 20 mm diameter in two equal passes of the tool for a length of 50 mm. Because of the simplicity of this example it would normally be performed by a canned cycle on most machines. However, for this example, we will stay with a subroutine. The following program segment assumes that earlier program blocks have set appropriate speed and feeds, set the *tool offsets* (tool–nose radius compensation) and have established the appropriate *absolute positioning mode*, and established the *radius programming mode*. This is because it is essential to tell the controller at the start, whether program dimensions refer to the *radius* of the workpiece relative to the

Fig. 8.3 *Turned component requiring a subroutine with loop: (a) component to be turned; (b) axes for turning. Dimensions in millimetres*

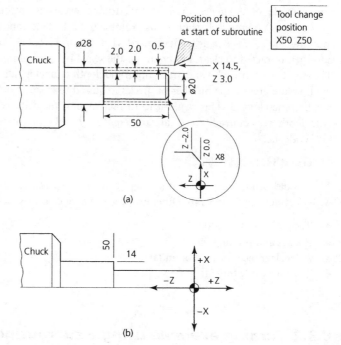

(a)

(b)

spindle axis or whether they refer to its *diameter*. As a reminder, Fig. 8.3(b) shows the position of the X and Z dimensions and movements when turning. Remember that movements of the cutting tool *into* the workpiece are always prefaced by a *minus* sign.

Subroutine	
:3030	Colon followed by identification number specifies the start of subroutine number 3030
N3040 G91	Incremental programming
N3050 G01 X−2.5	Feedrate move to cutting depth whilst the tool point is still 3 mm clear of the end of the job
N3060 Z−53	Feedrate move to turn diameter
N3070 X0.5	Feedrate move of tool 0.5 mm away from work at end of cut
N3080 G0 Z53	Rapid return with tool 0.5 mm clear of previously turned diameter
N3090 M99	End of subroutine

1. Since main program will call for the subroutine to be *repeated twice*, the component will be turned to the finished size with 2 mm being taken off the radius (4 mm off the diameter) at each pass of the tool.
2. The 2.5 mm infeed may cause some confusion. Remember that the tool is retracted 0.5 mm at the end of each cut so that it does not mark the workpiece as it returns to the start position ready for its next cut. Therefore the tool point is 0.5 mm clear of the work at the start of each pass and this must be added to the required depth of cut.

Let us now see how this subroutine is related to the main program.

Main program
N70 _____
N80 _____
N90 X14.5 Z3   Rapid positioning in absolute to start of routine
N100 M98 P3030 L2   Call subroutine 3030 and do twice
N110 G90   On rapid return from subroutine, reset to absolute
N111 G0 X7.0 Z1.0   Rapid to start of chamfer
N112 G1 X11.0 Z−3.0   Machining chamfer at modal feed rate previously set
N120 G0 X50 Z50   Rapid return to tool-change position.

Note that the program blocks (lines) are normally numbered in multiples of 10. This allows for additional programming information to be inserted after the program has been written. In the above example it was decided to include a chamfer on the end of the component to aid a subsequent screw-threading operation. Hence the use of the numbers 111 and 112.

8.3.3 *Turning a screw thread using a canned cycle*

Let's suppose that we need to cut a screw thread on the bar we have just turned down to 20 mm diameter. When cutting threads on the end of a bar it is necessary for the tool to start on a chamfer and finish in a groove (a 'landing' groove). This gives the clearance necessary for the cutting tool to begin and end the thread. Remember that the traverse mechanism cannot start and finish instantaneously, therefore there must be room for the tool to accelerate to the correct traverse speed by the start of the thread and to decelerate to a stop at the finish of the cut. Whilst cutting is taking place the rotational speed of the work and the traverse speed of the tool must be synchronised in order for a thread of the correct lead to be cut (see Section 6.10).

The landing groove at the shoulder end of the thread should be equal to or very slightly less than the minor (core) diameter of the thread being cut. This groove is also useful as it allows any associated component, such as a nut, to be tightened up to the shoulder of the bar. Following on from the previous example we now need to change the tool to one suitable for cutting the groove and also change the speed and feed rates to an appropriate value for this operation. The appropriate program segment to machine the landing groove is shown in Fig. 8.4.

The procedure for cutting the screw thread is shown in Fig. 8.5. Although the relative rotational movement of the workpiece and the linear movement of the cutting tool *generate* a thread of the correct depth and pitch, a *form tool* is used to create the crest radius on the thread and give it the correct form. Note how the cutting tool thread form gets progressively deeper. Figure 8.5 also includes the appropriate program segment for cutting the screw thread based on the Fanuc 0T threading cycle. It follows on from the previous program segment, as can be seen from the block numbering, after allowing for a tool change to the threading tool and also changing the speed and feed rates to suit. Note that the pitch of screw thread to be cut governs the feed rate 'F'. Where F1.5 would produce a thread with a 1.5 mm lead (lead = pitch for a single-start thread, and lead = pitch × number of starts for multi-start threads). Note that the thread gets deeper with each pass

Fig. 8.4 *Program segment for turning the landing groove*

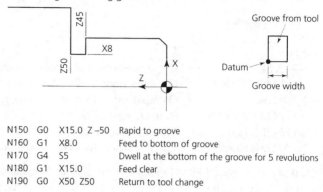

N150	G0	X15.0 Z –50	Rapid to groove
N160	G1	X8.0	Feed to bottom of groove
N170	G4	S5	Dwell at the bottom of the groove for 5 revolutions
N180	G1	X15.0	Feed clear
N190	G0	X50 Z50	Return to tool change

Fig. 8.5 *Program segment for cutting a screw thread*

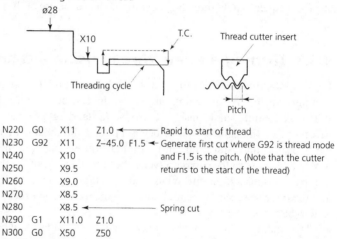

N220	G0	X11	Z1.0 ◄————— Rapid to start of thread
N230	G92	X11	Z–45.0 F1.5 ◄ Generate first cut where G92 is thread mode
N240		X10	and F1.5 is the pitch. (Note that the cutter
N250		X9.5	returns to the start of the thread)
N260		X9.0	
N270		X8.5	
N280		X8.5 ◄————————— Spring cut
N290	G1	X11.0	Z1.0
N300	G0	X50	Z50

of the tool. This is due to the incremental infeed that occurs with each loop of the cycle. Also note how the last cut is termed a *spring cut*. It is at the same depth as the previous cut and it is there to accommodate any deflection of the cutting tool and the workpiece caused by the cutting forces.

The main advantage gained from the use of subroutines and canned cycles is the time saved in programming, particularly if there is a need for the programmer to write out repetitive blocks of programming that involve the tool in a series of identical moves at different stages in the machining.

8.3.4 *Milling example using subroutines*

Figure 8.6(a) shows a fixture plate that consists of 24 holes in a plate 120 mm by 210 mm by 20 mm thick. Each hole is tapped with an M10 screw thread. Figure 8.6(b) shows the program structure for producing the holes, and you can see that it contains *subroutines* that branch from the *main program*. By this method the programmer does not need to repeat a lot of the coded information.

Fig. 8.6 *Drilling and tapping a matrix of holes in a fixture plate: (a) component to be drilled; (b) program structure*

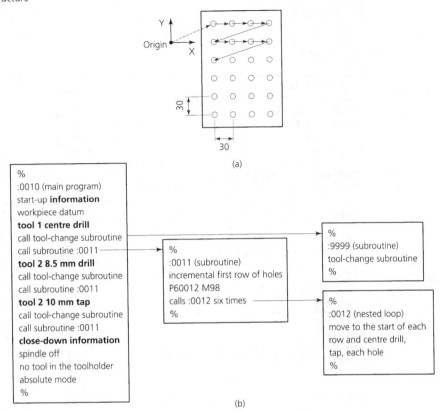

(a)

```
%
:0010 (main program)
start-up information
workpiece datum
tool 1 centre drill
call tool-change subroutine
call subroutine :0011
tool 2 8.5 mm drill
call tool-change subroutine
call subroutine :0011
tool 2 10 mm tap
call tool-change subroutine
call subroutine :0011
close-down information
spindle off
no tool in the toolholder
absolute mode
%
```

```
%
:0011 (subroutine)
incremental first row of holes
P60012 M98
calls :0012 six times
%
```

```
%
:9999 (subroutine)
tool-change subroutine
%
```

```
%
:0012 (nested loop)
move to the start of each
row and centre drill,
tap, each hole
%
```

(b)

8.3.5 *Nested loops*

A nested loop is a repetitive section of program contained (nested) within a subroutine. This is often referred to as *a loop within a loop*. For example, when using linear incremental programming, it is very useful for drilling a matrix of holes where each hole is drilled relative to the previous position. Returning to our fixture plate with its matrix of holes, we can see from the program structure in Fig. 8.6(b) that there is a *nested loop* within the second subroutine.

Finally, to conclude our introduction to subroutines and nested loops, Fig. 8.7 shows the full program for our fixture plate. This has been written for a *Fanuc 0MA* controller. *Note*: programs written for one make and type of controller will not work correctly on any other make and type of controller. *In the interests of safety do not attempt to use this program on a machine with any other controller*. This also applies to all other programs or segments of programs included in this chapter. They will only work correctly and safely in conjunction with the controller for which they were written.

Fig. 8.7 *Full program for fixture plate*

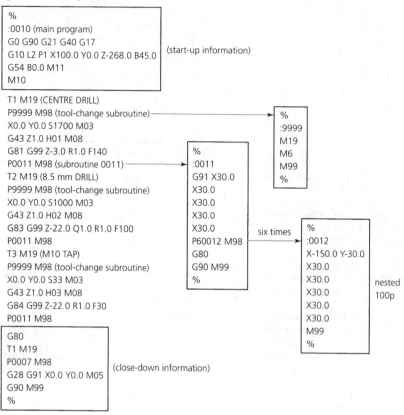

```
%
:0010 (main program)
G0 G90 G21 G40 G17
G10 L2 P1 X100.0 Y0.0 Z-268.0 B45.0        (start-up information)
G54 B0.0 M11
M10
```

```
T1 M19 (CENTRE DRILL)
P9999 M98 (tool-change subroutine)                    %
X0.0 Y0.0 S1700 M03                                   :9999
G43 Z1.0 H01 M08                                      M19
G81 G99 Z-3.0 R1.0 F140                               M6
P0011 M98 (subroutine 0011)            %             M99
T2 M19 (8.5 mm DRILL)                  :0011          %
P9999 M98 (tool-change subroutine)     G91 X30.0
X0.0 Y0.0 S1000 M03                    X30.0
G43 Z1.0 H02 M08                       X30.0
G83 G99 Z-22.0 Q1.0 R1.0 F100          X30.0              %
P0011 M98                              X30.0   six times  :0012
T3 M19 (M10 TAP)                       P60012 M98         X-150.0 Y-30.0
P9999 M98 (tool-change subroutine)     G80               X30.0
X0.0 Y0.0 S33 M03                      G90 M99            X30.0     nested
G43 Z1.0 H03 M08                       %                 X30.0     100p
G84 G99 Z-22.0 R1.0 F30                                  X30.0
P0011 M98                                                X30.0
                                                         M99
G80                                                      %
T1 M19
P0007 M98
G28 G91 X0.0 Y0.0 M05       (close-down information)
G90 M99
%
```

8.3.6 *Macro language*

Most CNC systems have a macro programming language. Fanuc 0M controllers have Macro A or Macro B. The macro language allows the user to store variables within registers, i.e. #100, and use mathematical functions such as sine, cosine, square root, etc. This language is like any other conventional computer language, such as BASIC, and can be used for more sophisticated applications such as:

- Positioning holes on a pitch circle (PCD).
- Elliptical path machining.
- In-cycle gauging.
- Setting up tools and workpieces using probes.

For example, let us suppose that a company has a frequent need to machine bolt holes on a pitch circle with equal hole spacing but that the radius of the pitch circle and the number of holes varies from job to job. If the machine controller does not have a pitch circle canned cycle, then the company may find it worth while to write a suitable *macro* and store it permanently in the controller memory.

Figure 8.8 shows a typical bolt-hole pitch circle with its parameters. This macro would enable calculations to be performed for the coordinates of each hole on the pitch circle for any particular values of A, R and H, needed for a given job, within the machining program. The programmer would merely need to enter a single line statement containing the required values of A, R and H to cause the coordinate values to be calculated. A suitable canned cycle would then be used to perform the actual machine movements. The macro call statement takes the following form:

G65 P9400 R30 A10 H12

where: G65   code calls for the macro
          P9400   is the macro number (same format as the subroutine)
          R30   is the radius of the pitch circle (mm)
          A10   is the angle for the first hole in degrees of arc from the origin
          H12   is the number of holes

Fig. 8.8 *Holes on a pitch circle: (X0, Y0) = coordinate value of bolt-hole circle centre; R = radius of bolt-hole circle; A = start angle*

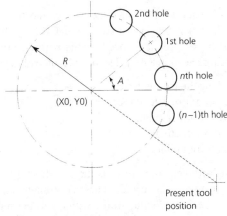

Let us now consider the complete macro P9400 for Fig. 8.8:

```
:9400
#30 = #101;                       Store reference point
#31 = #102;
#32 = 1;
While [#32 LE ABS(#11)] Do 1;     Repeat by number of holes
#33 = #1 +360 * [#32−1] / #11
#101 = #30 + #18 * cos[#33];      Hole position
#102 = #31 + #18 * sin[#33];
X#101 Y#102;
#100 = #100 + 1;                  Increase hole count by 1
#32 =#32 + 1;
End 1
#101 = #30                        Return to reference point
#102 = #31
M99                               End of macro
```

Note that numbers with # in front are variables in the normal computing sense, where variables have the following meanings:

#100	Hole number counter
#101	X coordinate reference point
#102	Y coordinate reference point
#18	Radius R
#1	Start angle A
#11	Number of holes H
#30	Storage of X coordinate reference point
#31	Storage of Y coordinate reference point
#32	Counter for the nth hole
#33	Angle of the nth hole

8.3.7 *Probing (further use of macros)*

Probing was introduced in Chapter 7 in connection with three-dimensional measuring machines. The touch-trigger probes used on CNC machines are somewhat different and are more robust in construction since they are permanently mounted in the machine carousel or magazine (see Section 8.4.1). The benefits of probing can be listed as:

- Reduced machine downtime.
- Reduced operator attendance.
- Approved product quality.
- Increased productivity.
- Increased profitability.

The touch-trigger probe records the coordinates of the position of the probe stylus at the moment it touches a surface. These coordinate values are saved within a register and compared to the desired value stored within another register. The difference between these two values is the *error value* that may be used to update and correct a workpiece datum or a tool radius or length offset value. The stylus of the probe is connected to a tri-filar suspension system that is supported on three rollers. It is the change in electrical resistance as the rollers commence to separate, and not the actual breaking of contact between the rollers, which triggers the probe. The signal is transmitted from the probe by either an optical, inductive or radio transmission systems. The signals from the probes are picked up by means of appropriate receivers. Like any other tool, the probe is called up by the program as and when required. It is then activated by being spun in one direction to switch on (P9001) and, after use, spun in the opposite direction to switch off (P9002) before returning the probe to its storage position in the carousel or tool magazine.

Let us now consider an example of a macro segment to probe a Z datum, as given in Fig. 8.9.

:0001	Program number one
Call up the probe	
P9001	Spin the probe to turn it on
P9005	Clear the macro registers

Fig. 8.9 *Use of probe to establish height of a surface above the Z datum*

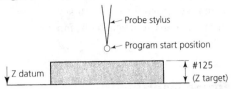

Move the probe to the start position

G65 H1 P#100 Q21   Start of macro: assign metric units

G65 H1 P#125 Q10000   Target at Z10.00 (*note*: Q dimensions in microns)

G65 H1 P#119 Q12   Update the workpiece datum P2 from the reference datum P1

P9018 M98   Call up the Z measure macro

Move the probe away

P9002   Spin the probe to turn it off

The reason for using this macro is to determine automatically the *actual* datum surface (Z = 0) of the workpiece. This is necessary when components of different thicknesses have to be machined using a common program. No matter what the thickness of the workpiece, once the probe has touched its upper surface, that surface then becomes the new datum.

8.3.8 *Scaling*

The scaling feature allows the X and Y coordinates in milling and drilling and the X and Z coordinates in turning to be increased or decreased by a scaling factor from their stated values in the program. For example, Fig. 8.10 shows a component requiring two slots to be machined with their widths equal to the cutter diameter. To simplify the programming, a subroutine could be written to machine the inner slot, and this could be called twice. Once *without* the scaling facility active to machine the inner slot, and once *with* the scaling facility

Fig. 8.10 *Scaling*

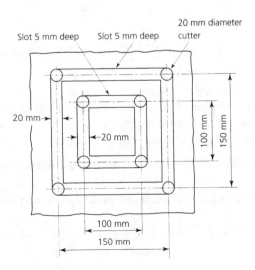

active, with a scaling factor of 1.5 (150/100), to machine the outer slot. Alternatively, the subroutine could be written for the outer slot and this could be called a second time with a scaling factor of 2/3 active to machine the inner slot.

8.3.9 *Mirror imaging*

Mirror imaging, available with some control systems, is a useful feature that allows either a whole program or a subroutine to have the signs of its coordinate data selectively reversed. The facility is applicable to milling and drilling operations, and either X coordinates or Y coordinates or both may be reversed.

Let us consider Fig. 8.11. This shows a cavity to be machined into a die block at four symmetrical positions. If the program datum (point of symmetry) is defined in the position shown, the subroutine to produce the cavity would normally result in the top right-hand quadrant being machined first. If the subroutine is then called with the X reversal facility active, the top left-hand quadrant would be machined. If the subroutine is called with the X and Y reversal facilities active, then the bottom left-hand quadrant would be machined. Finally, if only the Y reversal facility is active, then the bottom right-hand quadrant would be machined. Clearly, a significant amount of programming time can be saved using the mirror imaging facility and there is also less opportunity for the introduction of programming errors.

Fig. 8.11 *Mirror imaging*

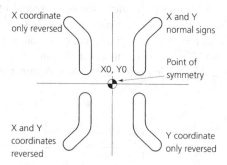

The ISO code for mirror imaging is G28, but control systems vary greatly in their adherence to standards and other methods of activating the feature may be encountered. Assuming that the control system does use G28, the following program lines illustrate how mirror imaging may be called up. For example:

N110 G28 X   This reverses the sign of the X coordinates only for the subsequent subroutine.

N190 G28 Y   This reverses the sign of the Y coordinates only for the subsequent subroutine.

N290 G28 XY   This reverses the signs of both the X and the Y coordinates for the subsequent subroutine.

8.3.10 *Rotation*

Rotation, which is available with some control systems, is a useful feature that allows the whole coordinate system to be rotated by a stated angle. This facility is applied to milling and drilling operations and is particularly useful when a machine feature is repeated at various angular positions around a common centre, as shown in Fig. 8.12.

Fig. 8.12 *Rotation*

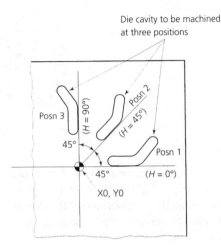

The most efficient way to prepare a program to machine this component would be to write a subroutine to carry out the machining of the cavity and then to use the *rotation feature* to rotate the coordinate axis system each time the subroutine is called. The ISO word address code for rotation is G73 with the H address used to store the rotation angle. However, these standards are not universally applied and other methods of activation may be met. Assuming that the ISO system is to be used, the general form of program to machine the component shown in Fig. 8.12 would be:

N200	call subroutine to machine cavity	Causes cavity to be machined at position 1.
N210	G73 H45	Rotates coordinate system by 45° from angle zero.
N220	call subroutine to machine cavity	Causes cavity to be machined at position 2.
N230	G73 H90	Rotates coordinate system by 90° from angle zero.
N240	call subroutine to machine cavity	Causes cavity to be machined at position 3.
N250	G73 H0	Sets coordinate system back to normal (0°).

8.3.11 *Zero shift*

The zero shift facility is commonly available on most control systems. It allows the program datum to be changed within the program. The component shown in Fig. 8.13 will be used to explain this feature. A program has to be written to machine the large hole A and then the smaller holes B arranged in a pattern about a centre point.

Fig. 8.13 *Zero shift (dimensions in millimetres)*

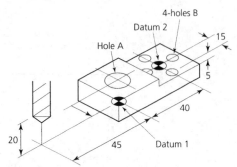

The most convenient datum for hole A is datum (1) and a convenient tool-change position is X−45, Y0, and Z20 with reference to datum (1). The first line of the program would therefore define the start position of the tool (the tool-change position) using a G92 code and the machine operator would position the tool at this point at the start of the program. For example:

N100 G92 X−45 Y0 Z20	Define tool-change position relative to datum 1.
⋮	Program lines for machining hole A.
N1200 G00 X−45 Y0 Z20 M06	Return to tool-change position and change tool.

In order to minimise the calculations for the pattern of holes B, the adoption of datum (2) would be most convenient. Therefore, the program would continue with a G92 code to tell the control system where to locate the existing tool-change position relative to datum (2). This continuation of the program would read:

N1250 G92 X−85 Y-15 Z20	Define tool-change position relative to datum (2).
⋮	Program lines to machine the four smaller holes.
N3000 G00 X-85 Y-15 Z20 M02	Return to tool-change position at end of program.

We have already drawn attention to the fact that a program for one controller will not work on another. For example, when discussing screw-thread cutting we used a G92 code for the appropriate canned cycle. This was because the program segment was written to work with a Fanuc 0T controller. For this milling example we have assumed the use of a controller operating with ISO codes; therefore, we have used a G92 code for position preset.

8.3.12 *Block delete*

The block-delete feature is used to omit parts of programs that may not be required. For example, two components may be identical except for a single hole. Sometimes the hole will be required and sometimes it will not. Rather than write two programs, a single program incorporating the hole can be used. When the hole is not required, the operator can press the 'block skip' button, the block of code preceded by '/' will be skipped, and the hole will not be inserted.

SELF-ASSESSMENT TASK 8.2

1. List the important points for creating a safe working part program.

2. State how block delete would be useful for in-cycle inspection.

8.4 Tooling systems for CNC machines

One of the most useful features of CNC machine tools is their flexibility. However, a wide variety of cutting applications can only be carried out if a correspondingly wide range of cutting tools is available. Manufacturing companies will often have a range of machines of different ages and varying degrees of technical sophistication. The machines will have different tool-mounting methods, spindle nose tapers, draw bar arrangements, etc., and this can lead to each machine needing a unique set of tools and holders. This, in turn, can lead to an unacceptably high investment in tooling and also introduce storage and retrieval problems.

Fortunately, tooling manufacturers offer a variety of solutions to the problem of locating and clamping a range of cutting tools in a variety of machine tools. These 'tooling systems' consist of a range of standard cutting tools and tool holders together with appropriate adaptors for each type of machine used. This allows the inventory of standard tools and holders to be minimised since all of them can be used on any machine with some general characteristics. Therefore using a well-designed tooling system offers the following advantages:

- Maximum tool interchangeability between different machines.
- Easier process planning and tool layout planning.
- Reduced overall tooling costs, since the tooling inventory is minimised.
- Downtime due to tooling shortages minimised.
- Good repeatability of location accuracy between tool and adaptor also between adaptor and machine.
- Standard holders and adaptors make presetting of the machine easier.
- Tooling systems are compatible with automatic tool-changing devices.
- Tooling systems are compatible with touch-trigger probe systems.

Tooling systems may be either manual or automatic. Manual systems are quite satisfactory for jobbing and low-volume production. However, the higher the volume of production the more economical it becomes to consider investment in automatic tool-changing systems. Thus it is advisable to purchase a tooling system that can adapt to automatic tool changing even if it is slightly more costly in the first instance. Figure 8.14 shows a typical tooling system that is equally applicable to manual or automatic tool changing.

Fig. 8.14 *Tooling system for machining centres*

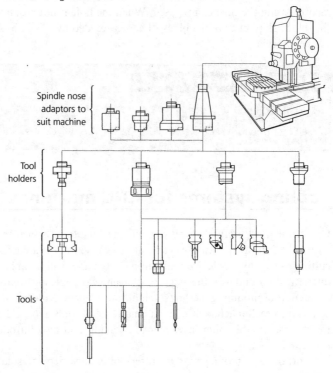

Spindle nose adaptors to suit machine

Tool holders

Tools

8.4.1 *Tool changing*

Tool changing can be performed manually or automatically. Since automatic tool-changing facilities add considerably to the cost of a machine, manual tool changing is still widely used for small quantity batch work. When manual tool changing is used it is essential to minimise the changeover time. This can be achieved by using quick-change toolholders and preset tooling. The tools are kept in a 'crib' placed conveniently beside the machine. The 'crib' is a stand in which the tools are not only stored, but are also located in the order in which they are to be used and in such a position that they can be easily grasped and taken from their location.

Automatic tool-changing systems are classified according to the way in which the tools are stored.

Indexable turrets

These can be programmed to rotate (index) in order to present the tools mounted in the turret in the sequence in which they are required. Indexable turrets are widely used on turning centres and also on some milling and drilling machines. In the latter case, the drive to the cutting tool is also transmitted through the turret. Such a system lacks the rigidity of a conventional machine head and spindle assembly and is usually only used on comparatively light-duty machines.

Tool magazines

These are indexable storage facilities and are used only on machining centres. Two systems are shown in Fig. 8.15. The tool magazine is indexed to the tool-changing position. An arm removes the current tool from the machine spindle and inserts it into the empty socket in the magazine. The magazine then indexes so that the next tool to be used is presented to the tool-change position. The arm then extracts the toolholder and tool from the magazine and inserts it into the machine spindle ready for use.

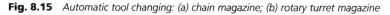

Fig. 8.15 *Automatic tool changing: (a) chain magazine; (b) rotary turret magazine*

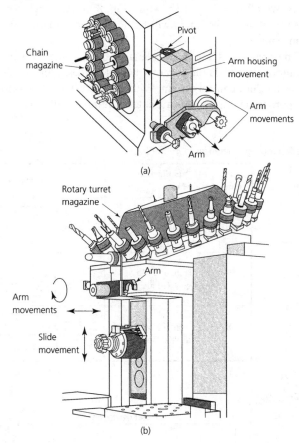

8.4.2 *Tool replacement*

When tooling has to be replaced due to wear or breakage, one of the following situations will arise.

- A suitable 'non-qualified' tool will be fitted; that is, a tool will be used which is suitable for the job but which is not a direct replacement for the previous tool. Therefore the operator will have to reset the tool-nose radius and the tool-length offset file in the tool offset register before machining can recommence. This is obviously time consuming and expensive and is unacceptable if 'just-in-time' production is being employed.

- A 'qualified' replacement tool will be substituted. This is a tool that has identical dimensions to the one it is replacing, within known tolerances. This will allow machining to continue without having to re-datum the tool and with a minimum loss of time. Figure 8.16 (ISO 5608: 1989) shows how different dimensions of a tool can be qualified.
- A 'preset' replacement tool will be used. This is a tool which is not necessarily a direct replacement for the worn tool, but one that has been preset in a setting fixture to a known offset value, as shown in Fig. 8.17. Although the operator/setter will need to edit the offset file, there will be no need to re-datum the tool before recommencing machining.
- The final possibility, only available on the more advanced machines, is the use of automatic tool datum and offset edit facilities, using touch-trigger probes interfaced to the machine control system. With this system, the tool is touched onto a probe and the appropriate offset values are entered automatically into the offset file.

Fig. 8.16 *Qualified tooling (tolerances in millimetres)*

Letter symbol	Qualification of tool	Sketch
Q	Back and end qualified tool	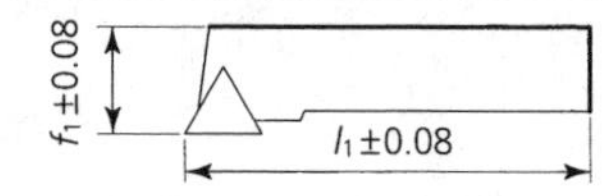 $f_1 \pm 0.08$ $l_1 \pm 0.08$
F	Front and end qualified tool	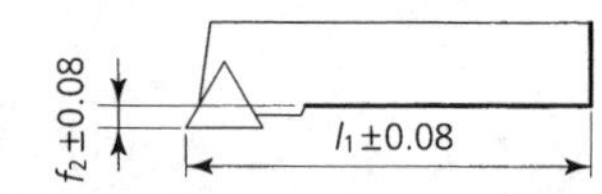 $f_2 \pm 0.08$ $l_1 \pm 0.08$
B	Front, back and end qualified tool	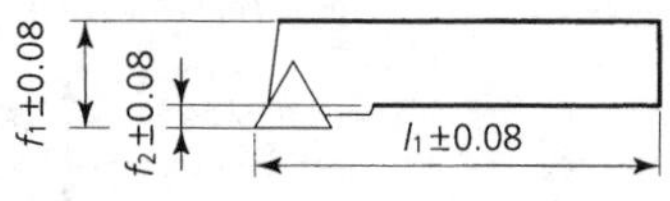 $f_1 \pm 0.08$ $f_2 \pm 0.08$ $l_1 \pm 0.08$

Fig. 8.17 *Preset tooling*

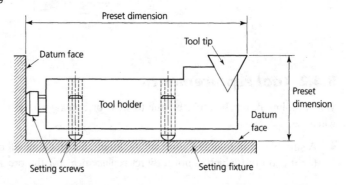

It is no longer considered economical to regrind worn cutting tools made from cemented carbides; nor is it desirable in the case of coated carbides since the coating would be destroyed. Further, regrinding alters the size of the tool and this is not acceptable when 'qualified' tooling is being used – the exception being twist drills of various types. Normal practice is to use disposable tip tools as described in Section 6.6. Figure 8.18 shows a typical disposable tip system. The tips are designed to enable them to be indexed round to a new cutting edge several times before finally being discarded.

Fig. 8.18 *Disposable-tip system*

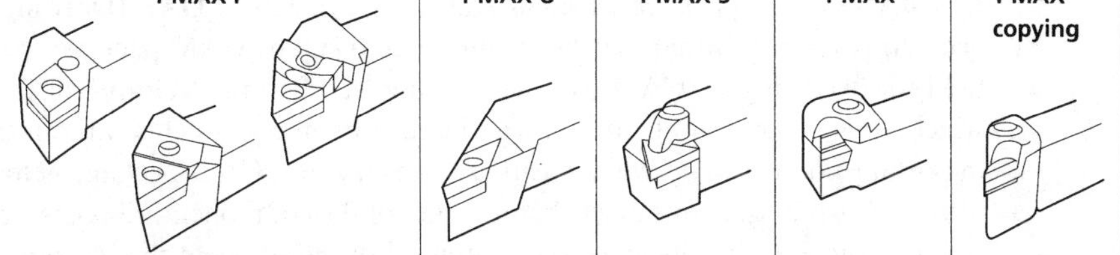

T-MAX P	T-MAX U	T-MAX S	T-MAX	T-MAX copying
Lever, wedge, wedge clamp	**Screw clamp**	**S-clamp**	**Top clamp**	**Top clamp**
• First choice for external turning	• First choice for internal machining and line copying	• Small shank dimensions	• Long-standing system for indexable inserts with adjustable chipbreakers	• Stable clamping
• For one holder a large number of geometries, single and double sided, are available	• Small shank dimensions	• Suitable for internal machining	• Good alternative for stainless and heat-resistant materials	• Excellent for medium to rough copying
• Short indexing time, esp. for lever design	• Quick indexing with the U-lock screw			
• For better accessibility, use 'wedge clamp' for external and 'wedge' for internal machining				

8.5 CAD/CAM

The previous sections of this chapter have shown how CNC machine tools can be programmed to machine a variety of shapes. However, as mentioned in Chapter 1, design decisions must be made with manufacturing implications in mind, i.e. *concurrent engineering philosophy*. At a basic level using the same geometry as created by the designer in a computer-aided design (CAD) package, rather than recreating it in a computer-aided manufacturing (CAM) package, is important in terms of both accuracy and time saving. The different levels of complexity can range from simple two-dimensional (2D) profiles for flame cutters to the complex multi-axis machining of dies from three-dimensional (3D) surface models. The advent of rapid prototyping technologies is a very important development where concept models can be grown directly from the design geometry or machined directly on a desktop milling machine. Figure 8.19 shows the integration of each area.

Fig. 8.19 *Relationship of CAD/CAM functions*

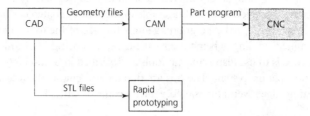

8.5.1 *CAD systems*

Industry standard packages, such as AutoCad, can export DXF (Drawing Exchange File) geometry files in a format that can be imported by many CAM packages. Other file formats used for exporting by CAD packages include IGES (Initial Graphical Exchange System) which is also a neutral file format (can be read by many systems without translation). The important point to note when creating geometry in a CAD package is to ensure that the *world origin* is at the bottom left-hand corner of the component. This ensures that the world origin is aligned with the workpiece datum of the component that is going to be simulated in the CAM package.

8.5.2 *CAM systems*

Computer-aided manufacturing (CAM) systems include PEPS (Production Engineering Productivity System) that can import DXF files. These, in turn, can be defined as part boundaries (outer profiles and inner pockets or slots) called K curves. These boundaries are usually defined in a clockwise direction for milling as these are used to indicate the direction of the milling cutter. The language used in PEPS is very powerful, with one-line commands defining pockets or profiles. Once the program has been proved, i.e. the simulation of the cutter paths is correct, then the simulation can be *post-processed* into any CNC language, i.e. Heidenhain, Fanuc, Allen Bradley, Phillips, etc.

8.5.3 *PEPS programming language*

Let us consider the example shown in Fig. 8.20. This has four curves (K1, K2, K3 and K4) and one group of holes (G1). The geometry was created by importing a DXF file for the component drawn in AutoCAD. In the PEPS geometry menu the DXF lines can be converted to a K curve simply by selecting the line to 'K curve command' and selecting each line in turn, going in a clockwise direction around each boundary. Finally, select E to end and close the boundary. The group of holes is defined in a similar way. Select 'Group' from the geometry menu and then select each individual point defined within the group. The final geometry is defined thus:

    K1 is the outer profile
    K2 is the long inner pocket
    K3 and K4 are square pockets with a circular island
    G1 is a group of three holes to be drilled

Fig. 8.20 *PEPS example drawn in AutoCAD*

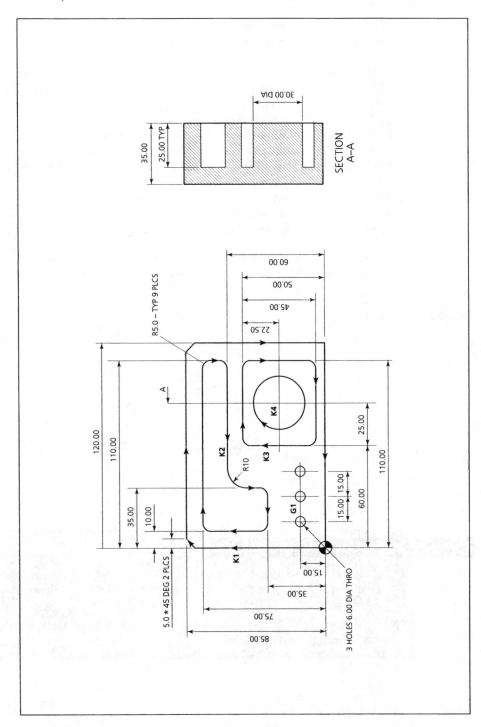

The machining commands to produce K1 are defined as follows:

TOOL 1 D20	is a 20 mm diameter cutter
FRO X37.5 Y28.5	is the tool-change position
SPI 1250	is the spindle speed in rpm
FED V150 H200	is the vertical and horizontal feed rates in mm per min
CLE 3	is a clearance plane to which the cutter moves rapidly up and down
RAP	is rapid feed rate
RET	retracts the tool to the clearance plane
DES−15	descends at feed rate 15 mm below the part surface
OFF L0	offsets the cutter to the left of the profile by the cutter radius
PRO TK1	profiles K1 in the direction it was defined (tangential), in this case a clockwise direction
RET	retracts to the clearance plane
SPI 0	spindle off
GOH	means go home to the tool-change position

Note: The start-up and close-down information are defined in the same safe manner as conventional CNC programming. This simulation is shown in Fig. 8.21.

Fig. 8.21 *Profiling cutter path*

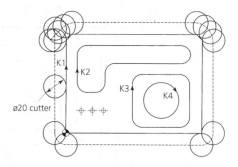

SELF-ASSESSMENT TASK 8.3

1. If the command to pocket K2 is PKT X15 Y40 Z−5 R3 N1 K2 F0 A, write the full program to machine pockets K2 and K3 with K4 as an island, as shown in Fig. 8.22.

2. If the command to drill the group of holes G1 is DRI G1 Z−6 R3 N1 D0, write the full program to produce the holes.

Fig. 8.22 *Self-assessment task*

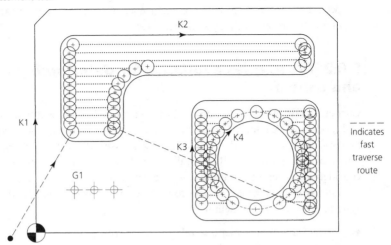

8.6 Simulation of automated systems

Simulation of tool paths is useful before machining a component on a very expensive CNC machine tool, for example:

- The operator can be certain that the tool is not going to make any unpredicted moves.
- Programs can be prepared off-line to avoid wasting valuable machine time.
- Safe clamping positions can be determined.
- Programmers only need to learn one language irrespective of the number of different machine controllers that may be on site.
- Programming using simulation is simpler and quicker for complex components and does not require calculations that might lead to human error.

The same philosophy is true before any expensive plant or automated equipment is built. The following section shows how complex automation, such as a flexible manufacturing system (FMS) as discussed in Section 9.8, can be computer simulated.

8.6.1 *FMS control characteristics*

Any flexible manufacturing system (FMS) will display the following characteristics:

- *Concurrency or parallelism*   In an FMS many operations take place simultaneously.
- *Asynchronous operations*   Machines complete their operations in variable amounts of time.
- *Deadlock*   In this case, a state can be reached where two parts require each other's current process as the next event.
- *Conflict*   This occurs when two or more processes require a common resource at the same time.

- *Event driven* An FMS can be viewed as a sequence of discrete events. Since operations occur concurrently, the order of occurrence of events is not necessarily unique; it is one of many allowed by the system structure.

8.6.2 *Techniques available for determining FMS layout and control*

Current trends in the layout design and control of FMS have turned towards simulation and modelling. An indication of the importance of simulation and modelling is illustrated by companies such as Citroën and Yamazaki, with the latter company spending 100 000 hours planning prior to the installation. Both companies reported that they derived considerable benefits from using computer simulation.

Let us examine the advantages to be derived from any technique that can model such a system. For example, modelling could achieve:

- Correct layout of elements in relation to one other.
- Evaluation of different designs.
- An opportunity to compare different operating policies.
- The prediction of performance.
- An opportunity for the education of management and operatives.
- Determination of control strategy.
- The ability to explore the above features by simulation and modelling without the need for building, disrupting the operation of, or destroying the real system.

Most systems are subject to random influences, such as the vagaries of human beings; however, since an FMS is computer controlled, many of the random disturbances can be eliminated. This means that there can be greater confidence in the results of an FMS over conventional manufacturing methods.

8.6.3 *Graphical simulation*

This is where the physical design and interaction between elements within an FMS can be visualised, and has the advantage of giving the system designer information to identify excesses and deficiencies within the system and, to some extent, predict how the system will perform. Examples of 3D design packages are GRASP, Workspace and RobCAD. They have all been developed to model robots, conveyors, automatic-guided vehicle systems (AGVS), etc. These features aid the development of workplace layout by manipulation of these entities. Other features include *event processors* that can link events together such as robots picking and placing objects on a conveyor. Also, CNC machines can be automatically started and stopped. These events can then be saved in the computer memory along with their associated parameters, e.g. starting time, duration of actions, delays, accelerations, etc. In this manner complex interactions between components within an FMS may be studied and evaluated. An example of a typical cell containing a robot, conveyors, a CNC lathe and a CNC milling machine are shown in Fig. 8.23.

The following example is based upon the flexible manufacturing cell shown in Fig. 8.23. The GRASP sequence of a part passing through the system is as follows:

Fig. 8.23 *GRASP FMS cell*

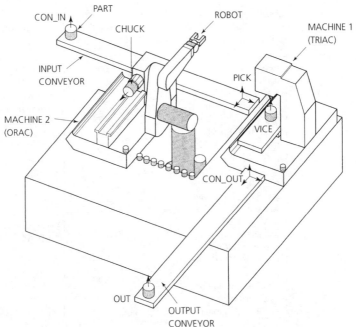

- PART enters on the INPUT conveyor to a sensor at PICK.
- From PICK to VICE.
- PART process on the TRIAC milling machine.
- From VICE to CHUCK.
- PART processed on the ORAC lathe.
- PART exits on the OUTPUT conveyor at CON_OUT.

Using a graphical robot and simulation package (GRASP), we can now write a program for this sequence. The GRASP commands are similar to the VAL 2 programming language, i.e.

POSITION vice shift Z100	means position the TCP 100 mm above vice
GRIP part	means grip the workpiece part
RELEASE part to chuck	means release the part to the location CHUCK
SET ROB_GO	means set the signal ROBOT_GO to *true*
WAIT until TRIAC_STOP	means wait until the signal TRIAC_STOP is *true*

The full GRASP program for the previous sequence is shown in Fig. 8.24.

8.6.4 *Discrete event simulation*

In a discrete model the states of each *entity* within an FMS is modelled individually; for example, a machine working or waiting, or a workpiece waiting or being worked on. There are combined activities such as machine-process-workpiece activity. In other states entities can be in a queue waiting for conditions to change – for example, a workpiece waiting for a

Fig. 8.24 *Full GRASP program*

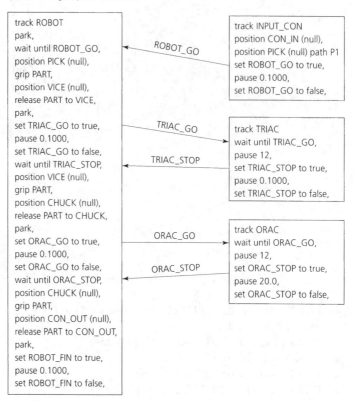

```
global variables
signal ROBOT_GO, ROBOT_FIN, TRIAC_GO, TRIAC_STOP, ORAC_GO, ORAC_STOP;
path P1 straight speed 500.0000;
```

```
track ROBOT
park,
wait until ROBOT_GO,
position PICK (null),
grip PART,
position VICE (null),
release PART to VICE,
park,
set TRIAC_GO to true,
pause 0.1000,
set TRIAC_GO to false,
wait until TRIAC_STOP,
position VICE (null),
grip PART,
position CHUCK (null),
release PART to CHUCK,
park,
set ORAC_GO to true,
pause 0.1000,
set ORAC_GO to false,
wait until ORAC_STOP,
position CHUCK (null),
grip PART,
position CON_OUT (null),
release PART to CON_OUT,
park,
set ROBOT_FIN to true,
pause 0.1000,
set ROBOT_FIN to false,
```

ROBOT_GO

```
track INPUT_CON
position CON_IN (null),
position PICK (null) path P1
set ROBOT_GO to true,
pause 0.1000,
set ROBOT_GO to false,
```

TRIAC_GO
TRIAC_STOP

```
track TRIAC
wait until TRIAC_GO,
pause 12,
set TRIAC_STOP to true,
pause 0.1000,
set TRIAC_STOP to false,
```

ORAC_GO
ORAC_STOP

```
track ORAC
wait until ORAC_GO,
pause 12,
set ORAC_STOP to true,
pause 20.0,
set ORAC_STOP to false,
```

machine to become available. The selection of an entity from a queue depends on its characteristics, such as the type of machine required. Once the entity has been selected its state is changed, e.g. the machine is working. The activities each entity undergoes are considered to begin and end instantaneously and are known as 'events'. Generally these types of simulation are controlled by a timing mechanism known as the 'three phase' method, for example:

- Advance the clock to the soonest event.
- Terminate any activities that are due to finish at that moment.
- Initiate any activities permitted by the conditions built into the model.
- Repeat the process.

From this type of model a variety of performance measures can be considered, such as machine utilisation, work in progress (WIP), throughput times, buffer queue performance data, etc. An example of a discrete event package is PROMODEL and modelling languages are SIMAN and HOCUS.

1. Briefly discuss the benefits of using CNC simulation; however, state why modern CNC machines which use graphics may obviate the use of such systems.

2. Describe the benefits of 3D simulation and why automated guided vehicle systems are better than conveyors in some circumstances.

3. Describe how simulation can accurately define the cycle time of a given flexible or CNC system and state how that can help to determine the cost of using automation.

8.7 Industrial robots

Industrial robots may be defined as computer-controlled, reprogrammable mechanical manipulators with several degrees of freedom, capable of being programmed to carry out a variety of industrial operations. If a robot is to reach for, move, and position a workpiece or tool, it requires an arm, a wrist subassembly and a 'hand' or end effector. The sphere of influence of a robot depends upon the volume (envelope) into which the robot can deliver its wrist subassembly and end effector. A variety of geometric configurations have been developed and the most widely used will now be appraised.

8.7.1 *Cartesian coordinate robots*

These have three orthogonal linear sliding axes, as shown in Fig. 8.25(a). The manipulator hardware and control systems are the same as CNC machine tools. Therefore, the arm positional resolution, accuracy and repeatability will be the same as for a CNC machine tool. An important feature of a cartesian robot lies in its spatial resolution which is equal and constant in all the axes of motion and throughout the work volume. This is not the case for the other configurations.

8.7.2 *Cylindrical coordinate robots*

These have two orthogonal linear sliding axes and one rotary axis, as shown in Fig. 8.25(b). The horizontal arm telescopes in and out and moves vertically up and down the column which, in turn, rotates on its base. The working volume is, therefore, cylindrical. The resolution of a cylindrical robot is not constant and depends upon the radius of the wrist from the rotational axis of the column. When using a standard resolution digital rotary encoder and an arm length of 1 metre, the resolution of the wrist assembly will be of the order of 3 mm. This is poor compared with a cartesian robot where the resolution is constant at about 0.01 mm. Cylindrical geometry robots offer (in theory) the advantage of higher linear velocity at the wrist end of the arm, as a result of having a rotary axis. In practice, this is limited by the moment of inertia of the arm and the workload. In fact it is quite difficult to obtain good dynamic performance from rotary-based robots. The moment of inertia reflected at the base drive depends not only upon the mass of the arm and the workload but also its distance from the pivot point. This is one of the main drawbacks of robots using revolute joints.

8.7.3 *Spherical (polar) coordinate robots*

These have one orthogonal linear sliding axis and two rotary axes, as shown in Fig. 8.25(c). The arm can move in and out, and it can also be tilted up and down on a horizontal pivot. The whole assembly can pivot about a vertical axis since it is mounted on a rotary base. Encoders, built into the pivots, are used to measure the magnitudes of the rotational movements. The working envelope is a spherical shell, and again the resolution is limited not only by the accuracy of the vertical axis encoder but also by the length of the robot arm, the resolution becoming poorer as the distance of the wrist from the vertical axis increases. On the other hand, the movements of a spherical coordinate robot tend to be quicker and more flexible compared with a cartesian robot.

8.7.4 *Revolute (angular) coordinate robots*

These have three rotary axes and no linear axes, as shown in Fig. 8.25(d). The revolute, angular or articulated robot (different names for the same thing) has two rotary joints and a rotary base. Its range of movements closely resembles those of the human arm and it is fast and extremely flexible. Unfortunately, having three revolute movements its resolution is very poor compared with a cartesian coordinate robot. However, this configuration is the most popular for small- and medium-sized robots.

Fig. 8.25 *Robot arm geometry (courtesy of Butterworth–Heinemann Ltd): (a) cartesian coordinate robot; (b) cylindrical coordinate robot; (c) spherical (polar) coordinate robot; (d) revolute (angular) coordinate robot*

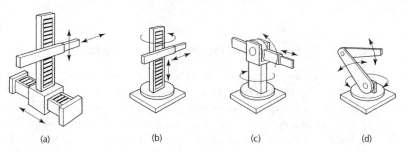

(a)   (b)   (c)   (d)

Figure 8.26 shows a complete revolute-type robot installation and indicates the additional movements of the wrist subassembly. Various types of end effectors can be mounted on the wrist to hold a variety of workpieces and tools. Remember that the greater the number of joints and movements, the lower will be the overall resolution. Further, as wear occurs in service, the greater will be the reduction in accuracy.

Fig. 8.26 *PUMA robot movements*

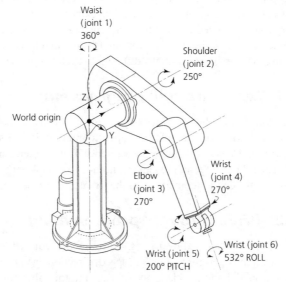

8.7.5 *Robot end effectors*

As has previously been stated, robots can be programmed for a wide variety of tasks and are available in a variety of sizes, shapes and physical capabilities that are reflected in a correspondingly wide variety of end effectors. However, end effectors can be categorised into grippers and special tools:

Grippers

- Mechanical clamping.
- Magnetic or electromagnetic.
- Suction.

Special tools

- Screwdrivers.
- Nut runners.
- Assembly tools.
- Welding tools.
- Inspection probes.
- Various types of sensor.
- Pouring ladles.
- Portable power tools.
- Spray-painting guns.

8.8 Robot-programming methods

As has already been stated, robots share a number of hardware and software similarities with computer-controlled machine tools. They use the same linear and rotary encoders, the same stepper and servo drives, they can have open- or closed-loop control systems, and they are controlled by a dedicated computer which can, in turn, be linked with other computer-controlled devises to build up an automated manufacturing cell. However, the method of programming can be substantially different to that used with CNC machine tools.

8.8.1 'Lead-through' programming

This is done manually by a skilled operator who leads the robot end effector through the required pattern. For example, the operator may hold the spray gun on the end of the robot arm and guide it through the sequence of movements necessary to paint a car body panel. The robot arm joint movements needed to complete this operation are automatically recorded in the computer memory of the robot and can be repeated when required.

8.8.2 'Drive-through' programming

The robot is programmed by driving it through the required sequence of movements, under power, with the operator controlling speed, direction, etc., by means of a teaching pendant, which is a small hand-held keypad connected to the robot controller by a trailing lead. The motion pattern is recorded in the computer memory and can be repeated when required.

8.8.3 'Off-line' programming

'Off-line' programming has the same benefits that CNC simulation enjoys, i.e:

- Checking that the robot simulation does not collide with any object within its cell.
- Expensive robots are not tied up while the operator develops the robot program.
- Only one generic language is required for a number of different robots; the program is post-processed into the required specific robot language.

The GRASP and 'Workspace' simulation languages are examples of off-line programming methods and this software comes with a graphical library of robots, as shown in Fig. 8.27, and a variety of post-processors for a number of different languages. GRASP was discussed previously in Section 8.6.3.

8.8.4 Conventional languages

There are many different robot-programming languages, such as:

- ARLA (ASEA/ ABB Robots)
- FANUC
- FORTH
- KAREL
- VAL 2 (Staubli Unimation Robots, i.e. 'Pumas')
- V+

Fig. 8.27 *Robot simulation using GRASP: (a) robot models from the GRASP library; (b) a welding application being programmed off-line; (c) simulation of the load/unload configuration for assessing the position of the robot in relation to the press brake (from Handbook of Industrial Robotics edited by Shimon Y. Nof; reproduced courtesy of John Wiley and Sons, Inc.)*

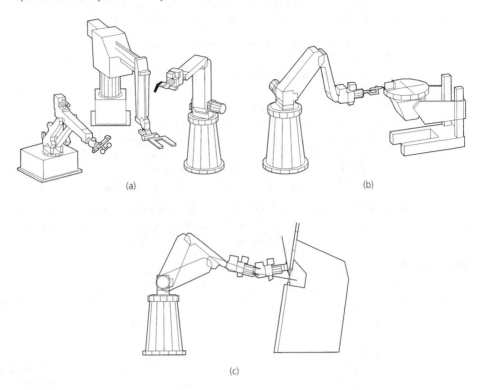

(a)　　　　　　　　　　(b)

(c)

These are but a few of the conventional languages available. However, those mentioned above are very popular and VAL 2 is one of the easiest and most frequently used. The language uses taught locations that are either points stored relative to the robot origin (*transformations*) or recorded positions of each individual joint (*precision*). Each location is given a name, usually in lowercase, which signifies the nature of the location, i.e. '#home' for the precision park position or 'chuck' or 'vice' for locations on machine tools. Simple program commands such as:

APPROACH vice, 100　　means position the gripper tool centre-point (TCP) so that it is 100 mm above the vice

MOVES vice　　means move the TCP in a straight line to the vice

DEPARTS, 100　　means position the TCP 100 mm above its current location

OPENI and CLOSEI　　means open and close the gripper instantaneously. By placing DELAY 0.5 before and after OPENI and CLOSEI will ensure that the gripper pneumatic cylinder has fully actuated and that the robot has actually settled in on the correct position

SIGNAL 1   means activate output signal 1
WAIT SIGNAL 1001   means wait for input 1 to be activated

The example of a robot cell discussed previously in Section 8.6 can be used to show how the following sequence can be programmed in VAL 2. The sequence of a part passing through the robot cell previously shown in Fig 8.23 is as follows:

PART enters on the INPUT conveyor to a sensor at PICK
From PICK to VICE
PART process on the TRIAC milling machine
From VICE to CHUCK
PART processed on the ORAC lathe
PART exits on the OUTPUT conveyor at CON_OUT

PUMA inputs	PUMA outputs
1. Sensor at PICK (1001)	1. TRIAC start signal (1)
2. TRIAC finished signal (1002)	2. ORAC start signal (2)
3. ORAC finished signal (1003)	3. OUTPUT Conveyor GO (3)

If the taught locations are PICK, VICE, CHUCK and CON_OUT, as shown in Fig. 8.23, and if the PUMA Robot is master and all the other machines are slaves, then the VAL 2 program for the above sequence is as given in Fig. 8.28.

Fig. 8.28 *The VAL 2 robot cell program*

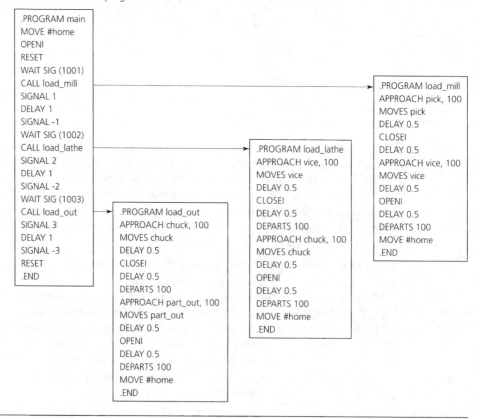

Give reasons why some robots are better at certain tasks than others because of their different degrees of freedom.

8.9 The cost of automation and assembly

In Chapter 1 we discussed the importance of considering assembly and dismantling with care at the design stage. Some basic questions that must be asked when considering assembly are:

- Is the assembly process to be manual or automated?
- Can the components and subassemblies be placed in position easily?
- Do the components need support to keep them in position whilst being fastened together?
- If they do need support, what workholding devices need to be designed and manufactured and how will these affect tooling costs and lead time?
- Do the workholding devices need to be indexable and, if so, do they need to be power operated and linked with the assembly robot control system?
- Does the assembly operator have to make decisions concerning the positioning of components?

In Chapter 1 it was also indicated that any attempt to build an *automated assembly cell* to reproduce *manual assembly processes* is asking for trouble. No automated system can provide the dexterity, sensibility and thought processes of a human being. The cost and complexity of trying to use state-of-the-art technology to approach the skills of human assembly operatives greatly outweighs the results. Therefore, if it is intended to use automated assembly, the product must be designed to suit automated assembly and not the other way round.

The aim should be to keep the assembly cell as simple and as reliable as possible and capable of being set up and maintained at a cost that can be justified. The limitations of robots in terms of speed and accuracy must also be kept in mind. There are no 'brownie points' to be won for using a three-axis robot if a simpler, cheaper and more rigid and dedicated two-axis machine will do the job. Further, it should be remembered that a simple transfer mechanism will operate more quickly and reliably than a robot in many cases. The robot comes into its own mainly when and where flexibility is required.

8.9.1 *Planning assembly*

Modern manufacturing 'systems thinking' suggests that production rates should be matched as closely as possible to customer usage rates (see Section 9.16) to relieve the supplier and the customer of holding stock ahead of need. This means that each product variant should be assembled in levelled quantities over the shortest possible time scale. For example, Table 8.1(a) shows an old-style, large-scale, monthly production schedule, whilst Table 8.1(b) shows a typical small-batch weekly levelled schedule.

Table 8.1 Assembly scheduling

(a) Large-batch monthly production schedule

Product type	Monthly requirement	Weekly production quantity			
		Week 1	Week 2	Week 3	Week 4
A	1100	500	500	100	—
B	300	—	—	300	—
C	200	—	—	—	200
D	100	—	—	—	100

(b) Small-batch weekly levelled schedule

Product type	Monthly requirement	Weekly production quantity			
		Week 1	Week 2	Week 3	Week 4
A	1100	275	275	275	275
B	300	75	75	75	75
C	200	50	50	50	50
D	100	25	25	25	25

The main barriers to the achievement of this type of levelled scheduling are lack of *cellular factory organisation* and long set-up times whenever a new variant is to be made. Cellular organisation is needed because levelled scheduling is only feasible within the small-scale environment of a cell where product variants are few. At the overall factory level, the number of product variants means that the level of complexity is too great for levelled scheduling to be achieved.

Levelled scheduling entails frequent resetting as machines are changed from job to job. Where long set-up times are involved, levelled scheduling can result in too much capacity being lost from production to changeover. Thus long set-up times and levelled scheduling are largely incompatible. However, changeover analysis can invariably be applied to reduce changeover times to insignificant levels. In the longer term, companies aspiring to world-class manufacturing standards seek to move to daily or even hourly levelled schedules where mixed mode production is the norm.

8.9.2 *Required rate of production*

As stated in the previous section, the customer usage rate determines the required rate of production and hence the desired 'cycle time'. For example, a company supplies car rear axle assemblies to a car-manufacturing company who sell 15 000 cars per month. Assuming that there are 20 working days in a month and an 8-hour working day, then:

Number of rear axle assemblies required per day is:
$15\,000 \div 20 = 750$ per day.
Therefore, the cycle time is:
$(8 \times 60)\,\text{min} \div 750\,\text{assemblies} = 0.64\,\text{minutes/assembly}$

8.9.3 *Assembly methods*

The main factors that determine the choice of assembly method are:

- Cycle time required.
- Anticipated product life cycle (and, by extension, the total number of products to be made).
- Design of the product; that is, whether it is simple or complex and its suitability for assembly by automated methods.

For example, consider a high-volume product such as the rear axle assembly mentioned previously. Clearly, this type of product would be more likely to justify investment in dedicated and automated assembly equipment than a low-volume, long-cycle time product. This latter type of product would probably be most economically assembled using general-purpose equipment and manual techniques.

8.9.4 *Estimation of assembly times*

At the planning stage, the assembly system is designed to match the required cycle time as closely as possible. Alternative assembly system designs will be considered at this time and reliable estimating techniques are needed to predict the cycle times that would be achieved. For manual assembly systems, the established work study techniques of predetermined motion time systems, analytical estimating, synthetic time data, and data from existing operations may all have a useful part to play. In addition, desktop manual simulation and computer simulation packages may help to explore the potential interaction and queuing problems between assembly operations.

Automated assembly operations can be estimated on the basis of fixed, machine-paced, transfer and operating cycles. Prototype assemblies, using mock-ups, may also be valuable to confirm assembly cycle times.

8.10 Types of assembly system

A convenient way to classify assembly systems is as follows:

- Manual assembly.
- Dedicated automated assembly.
- Programmable assembly (flexible automation).

8.10.1 *Manual assembly*

This encompasses a wide variety of operations. At its simplest, operators may transport and position the parts and then perform the assembly using hand or power tools. On a more sophisticated level, a basic unit is delivered in a part carrier to the operator on a conveyor or other transfer device and the parts to be added are delivered by automatic feeders or placed in conveniently positioned containers within easy reach of the operator. The final positioning and fixing of at least some of the parts is carried out by the operator as shown in Fig 8.29.

Fig. 8.29 *Manual assembly layout*

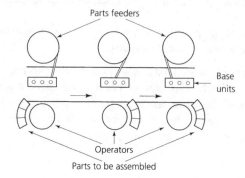

8.10.2 *Dedicated assembly*

Here, the assembly system is dedicated to the assembly of a single product. The assembly system is built around either a linear or a rotary transfer machine. This transfer machine delivers the base units mounted on work carriers to the assembly station. The parts are delivered, positioned on the base unit and fixed in place by automated assembly devices. Figure 8.30 shows, diagramatically, a rotary assembly system.

Fig. 8.30 *Rotary automatic assembly layout*

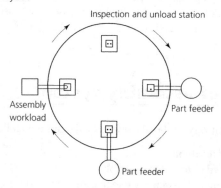

8.10.3 *Programmable assembly*

This type of assembly uses a programmable robot to transfer and position the parts onto the base unit. This gives a degree of flexibility since the robot can be reprogrammed to cope with different parts, locations and assembly sequences. Figure 8.31 shows, diagrammatically, a typical programmable assembly cell.

Fig. 8.31 *Programmable assembly cell*

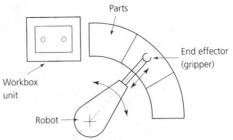

8.10.4 *Advantages of automated assembly*

The actions involved in assembly – including visual selection and sorting of parts, orientation and positioning, compensating for awkward access and less than perfect fit – are all relatively easy for a human operator to accomplish. To achieve the same flexibility in an automated system requires a great deal of attention both to the design of the product for ease of assembly and to the design of the system to compensate for the absence of human intelligence, problem-solving ability and versatility. However, when used with well-designed products, in volumes which justify the investment needed, an automated assembly system gives the following advantages:

- Lower overall cost.
- Increased productivity.
- Improved quality arising from less process variation, and 'in-process' automated inspection.
- Easier integration with other automated equipment.
- Removal of operators from hazardous environments.

8.10.5 *The elements of automated assembly equipment*

The elements of automated assembly equipment include:

- Part-feeding devices.
- Transfer and indexing devices to present the work carrier to the assembly station; part-positioning devices (including robots).
- Workhead mechanism to tighten, rivet, peen, apply adhesive, weld, snap in, or otherwise fix the part to the base unit.

8.10.6 *Part-feeding device*

A typical part feeder uses vibratory action to transfer components from a mass storage hopper onto a feed track that delivers the parts to the workhead with the correct orientation. Figure 8.32 shows a length of feed track that has been fed from a vibratory hopper. It can be seen that simple sorting and orientating devices on the track can ensure that only parts with the correct orientation reach the workhead. Wrongly orientated parts are either corrected or fall back into the hopper to be recirculated.

Fig. 8.32 *Sorting and orientating automatically fed parts*

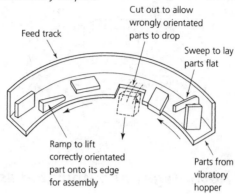

8.10.7 *Transfer and indexing devices*

Most automated assembly is done with the major components of the assembly (called the base unit) being delivered to the assembly station mounted on a work carrier. The other parts are then added to the base unit in sequence and secured. Therefore, the purpose of the transfer or indexing device is to deliver the work carrier to the workhead. A linear transfer device is shown in Fig. 8.33. On each forward stroke of the feed bar, all the work carriers are moved forward by one increment. On the return stroke the spring-loaded pawls are depressed and slide under the work carriers that remain stationary. The cycle is repeated for each forward and return stroke of the feed bar.

Fig. 8.33 *Linear transfer device*

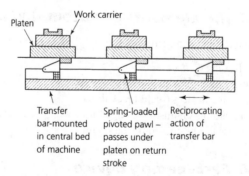

Rotary indexing devices are basically rotary tables that transport the work carriers to the workheads. A typical intermittent drive is the Geneva mechanism shown in Fig. 8.34. This indexing plate rotates with constant velocity and the Geneva plate indexes through a prescribed angle intermittently. Having six slots, the Geneva plate in Fig. 8.34 will index through increments of 60° for each revolution of the indexing plate. The locking plate prevents movement of the Geneva plate when it is not being indexed

Fig. 8.34 *Principle of Geneva indexing mechanism*

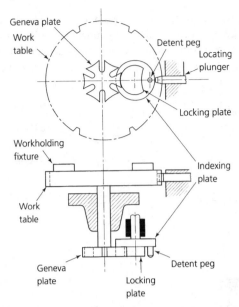

8.10.8 *Part-positioning devices*

These may be of the escapement type, fed by a feed track, or of the robotic pick-and-place type. There are many types of escapement devices successfully in use. In general, they need to be individually designed to suit the component to be fed. An example is shown in Fig. 8.35(a), and Fig. 8.35(b) shows a typical pick-and-place arrangement. The pick-and-place arrangement uses a programmable robot to transfer the components from the feed track and place them on the work carrier or the base unit.

Fig. 8.35 *Part-positioning devices: (a) escapement type positioning device; (b) pick-and-place positioning device*

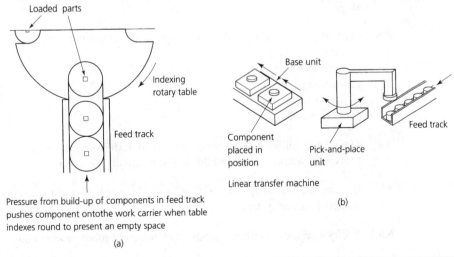

8.10.9 *Workhead mechanisms*

These vary enormously depending upon the application and the fixing action required. They may be operated mechanically, pneumatically or hydraulically. Figure 8.36 shows a typical workhead being used to secure components together by peening over the stem of the component after the previous assembly stations have positioned the components on the base units.

Fig. 8.36 *Assembly workhead*

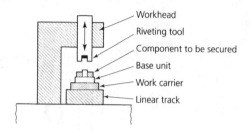

- Workhead
- Riveting tool
- Component to be secured
- Base unit
- Work carrier
- Linear track

EXERCISES

8.1   (a)   Describe the uses and benefits of subroutines in CNC part programming.
       (b)   Differentiate between macros and subroutines in CNC programming.

8.2   Describe a typical machining situation where a macro-programming facility would be useful and explain in general terms how the macro would work.

8.3   Compare the advantages and limitations of computer-aided CNC part programming with manual part programming.

8.4   List the main stages in the preparation of a CNC part program using the PEPS computer-aided part-programming system and briefly explain what happens at each stage.

8.5   With the aid of simple sketches, explain the following CNC part-programming features:
      (a)   zero shift
      (b)   scaling
      (c)   rotation
      (d)   mirror imaging

8.6   Describe the different methods of ensuring that the replacement of a worn or broken tool can be accomplished without loss of machining accuracy.

8.7   Outline the advantages to a CNC machine tool user company of adopting a standard tooling system.

8.8   Briefly describe the different methods of industrial robot programming.

8.9 (a) Compare and contrast the advantages and limitations of industrial robots with human operatives in a manufacturing situation.

(b) Describe three workplace situations where a robot might be used in preference to a human operator.

8.10 With the aid of a sketch, describe the basic layout of a flexible manufacturing 'cell' and discuss the criteria for justifying the high investment involved in setting up such a 'cell'.

8.11 Discuss the criteria influencing assembly and dismantling which must be considered at the design stage in order to achieve efficient manual assembly and routine maintenance.

8.12 Discuss the criteria which must be considered at the design stage when automated assembly is to be used.

8.13 Discuss the advantages of levelled scheduling when planning assembly.

8.14 Briefly compare the advantages and limitations of:
(a) manual assembly
(b) dedicated automated assembly
(c) programmable assembly

8.15 (a) With the aid of sketches, describe the elements of automated assembly equipment.

(b) For a simple assembly of your choice, outline an automated assembly 'cell'.

Part B
Management of manufacture

9 Control of manufacture

The topic areas covered in this chapter are:

- 'World Class' manufacture.
- Management of manufacture.
- Plant layout.
- Group technology.
- Flexible manufacturing.
- Materials requirement planning.
- Manufacturing resource planning.
- Computer-integrated manufacture (CIM).

9.1 Introduction to management of manufacture

The earliest organisations had owner-manager supervision and, even today, many small businesses are operated successfully in a similar manner. One of the first decisions that has to be made in any organisation is which of the many duties must the owner-manager undertake personally. Even in a 'one-man' business materials have to be purchased, products manufactured, invoices prepared, jobs costed and tenders submitted in addition to the raising of capital, and tax considerations. The problems become magnified as the organisation expands and labour has to be hired, paid for, and employment legislation complied with. The high failure rate of small businesses gives testimony to the fact that many people set up in business on the basis of their technical prowess and craft skills but without the necessary organisational skills or without giving sufficient thought to the problems of being one's 'own boss'. Managing an enterprise either alone or, as is more usual, as part of a team requires training and skill in just the same way as operating a machine tool. Small- to medium-sized companies engaged in engineering activities did, and still do, appoint supervisors who are more skilled than the most skilled workers under their control. Such supervisors also have the responsibility for production planning, control of quantity and quality, and the despatch of finished goods. Increasingly, however, even small and medium enterprises (SMEs) need to adopt the principles and practices of scientific management in order to survive.

The scientific approach to management involves the application of the following principles:

- *Forecasting and planning*   This requires an assessment of the future and decision making for future action.
- *Organisation*   The division of labour, the allocation of duties, and the lines of authority and responsibility.
- *Command*   The issuing of instructions to ensure that decisions are activated.
- *Control*   The setting of standards, the comparison of physical events against the set standards, and the taking of any necessary corrective action.
- *Coordination*   The unification of effort to ensure that all activities of the business are pursuing the same objectives.
- *Communication*   The transfer of information between different people and/or sections of the organisation, in particular between management and the workforce.
- *Motivation*   This is the driving force behind all actions. Psychological considerations make it important to recognise the motivation behind the customer buying the goods or services supplied by the organisation and the motivation of the people working in the organisation.

In addition to the principles of management postulated above is the introduction of *behavioural science*. This assists the modern manager in the understanding of human behaviour, for example, job satisfaction, attitude to work, and putting the right person in the right job. Such skills are essential to all managers since 'management is usually about people'.

9.1.1 *'World Class' manufacturing*

Nowadays it is essential for all companies to aim to become 'World Class' if they are to succeed and survive in the global economy. The principles involved apply to all companies large and small. Even small firms can no longer afford to be parochial in their market outlook if they wish to expand and prosper. Let us see how 'World Class' can be defined.

A 'World Class' organisation is one that is continuously more profitable than its competitors. This can only be achieved by being more successful than its competitors in satisfying its customers' needs. *World Class* can therefore be best defined in terms of customer satisfaction – for example, a company that provides high-quality products or services (Q), at less cost (C), and with shorter lead times and meets its delivery promises (D), should be World Class. Since all employees in the organisation may also be viewed as internal customers from the total company's point of view, a high level of worker safety (S) and morale (M) can be added to the definition.

A scoreboard should be developed to measure progress in relation to each of these objectives. The following are examples of QCDSM measurement for a scoreboard:

- *Quality*   Rework; number of customer complaints; defects.
- *Cost*   Productivity; overtime; floor space.
- *Delivery*   Conformance to schedule; lead time; volume of production.
- *Safety*   Number of accidents; number of safety-related suggestions.
- *Morale*   Absentee rate; turnover rate; number of suggestions.

Becoming 'World Class' is a long-term process that requires continuous improvement and innovation to better the current level of achievement. The following action agenda can be followed to implement 'World Class' principles in the manufacturing environment:

- *Understanding the basics* All employees must understand 'World Class' principles. This can be achieved through study sessions and tours of successful 'World Class' plants.
- *Reduce changeover times* Use external instead of internal set up. For example, the tooling is palletised at the side of a press so that the tools currently in use can be rolled out and the replacement tools, externally set, can be rolled in ready for immediate use.
- *Improve plant layout* Minimise transport of materials by changing to U-shaped or parallel lines and cellular manufacturing. Arrange the workplace to eliminate search time.
- *Increase worker training* Multi-skill the personnel to increase their flexibility.
- *Increase worker responsibility* For example, source inspection where each piece received from the previous process is checked by the operator and defects reported immediately.
  - *Problem-solving* Line personnel must be primarily responsible for basic problem solving.
  - *Multiple machines/processes* Make a worker responsible for more than one machine or process.
  - *Data collection* Personnel must record and retain production, quality and problem data at the workplace.
- *Improve process* Make it easy to manufacture without error.
- *Reduce inventory* Cut inventory levels to a minimum with the goal of zero inventory.
- *Reduce lot sizes* Reduce work in progress and flow times by reducing lot sizes.
- *Selective introduction of new equipment* Maintain and improve existing equipment and human work before thinking of new equipment:
  - automate gradually, when process variability cannot otherwise be reduced
  - use machinery with abnormality detection capability
- *Levelled /mixed production*
  - Produce multiple products on the same line and provide upstream processes with balanced loads.
- *Supplier cooperation*
  - Introduce suppliers to 'World Class' principles and help them put it into practice.
  - Cut number of suppliers down to a few good ones.
  - Cut number of part numbers.
- *Introduce kanban system*
  - Use a simple kanban system to move from a push to a pull scheduling system.
- *Improve customer relations*
  - Get to know the customer's needs.
  - Increase make/deliver frequency for each required item.

Many of these requirements will now be considered in greater detail in the following sections of this and the remaining chapters of this book.

9.2 Methods of manufacture

There are four basic methods of manufacture, and these are summarised in Table 9.1.

Table 9.1 *Methods of manufacture*

Method	Description	Examples
Job or unit	Single or very small quantity production of articles to individual customer requirements can involve a single operator or large groups of operators.	Bridges, ships and special components.
Batch production	Involves the multiple production of articles from, say, five units to many hundreds of units; either to customer requirements or in any quantity in anticipation of orders.	Gear boxes, pumps, electronic assemblies.
Mass, flow or line production	The manufacture of very large quantities of products, usually consumer goods made in anticipation of orders.	Cars, household appliances.
Continuous or process production	The plant resembles one huge machine with materials taken in at one end and the finished articles despatched at the other.	Plastic and glass sheet, plaster board.

To differentiate between 'job' and 'batch' production, it is not the number of components which is the deciding factor but the manner in which the production is organised. Consider the manufacture of, say, four components. These could be made by four operators, with each operator making a component outright. This is what normally happens when 'job' or 'unit' production is employed. Alternatively, the components could be passed from operator to operator with each operator specialising in and completing a particular feature of a component. The manufacturing method would then be classified as 'batch' production.

Job and *batch* production have the following characteristics in common: the flow of production will be intermittent; some parts will be for customers' orders, others for stock; schedule control of orders will be necessary to ensure that delivery times are met; there will be a large variety of products.

Flow and *continuous* production techniques have the following characteristics in common: the flow of production is, ideally, continuous; the production is usually in anticipation of sales; profit margins will be smaller than for batch or unit production; close control of costs is required at all stages of manufacture; the rate of flow at each stage of production is strictly controlled to avoid over-production or shortages; there will be a small range of standard products.

Increased production does not necessarily result in increased profits. It could result in a reduction in manufacturing costs but the selling price might also have to be reduced to increase sales volume in line with the increased rate of production.

9.3 Plant layout

Plant layout usually depends upon the method of manufacture. For job and batch production the machines are grouped according to type. For example, all the lathes would be in the *turning section* of the machine shop. For flow and continuous production the machines would be arranged to ensure a smooth flow of work from one operation to the next.

The laying out or rearrangement of machines and/or processes is usually costly in time and labour. Assuming that planning consent can be obtained and environmental constraints overcome, the ideal situation is to layout a *green-field* site. Prior to plant considerations, such factors as the availability of services and human resources, attitudes of the local community, access for transport, proximity of markets, terrain of the new site and, more recently, the availability of local and/or government funding must be taken into consideration.

More often, engineers are called upon to rearrange the layout of an existing plant to accommodate a new product or method of manufacture. Since non-productive *downtime* must be kept to a minimum, the changeover usually has to be achieved during weekends and holiday shutdowns. Specialist firms are available with the necessary equipment and skilled manpower to affect a speedy changeover.

In either of the above cases, the new layout must be carefully planned in advance, in full consultation with the key personal concerned with its operation. The key factors affecting layout can be summarised as:

- The volume of production.
- The type of production and its handling.
- The sequence of the process.
- Monitoring of work-flow.
- Type of building and work area.
- Power sources.
- Temporary stocking area requirements.
- Permanent stocking area requirements.
- Pedestrian areas.
- Service areas.
- Inspection requirements.

9.4 Types of layout

Let's now consider the types of layout. The layout will depend upon the volume of production and the range of product types and will take into account the key factors considered in the previous section.

9.4.1 *Process layout*

In this type of layout the requirements of various manufacturing activities are grouped together – for example, all the lathes in the turning section, all the milling machines in the milling section and all the grinding machines in the grinding section. Whilst such a layout is traditional and still widely used for jobbing and small-batch production, it leads to problems of storage and handling of work in progress. An example of this layout is shown in Fig. 9.1.

Fig. 9.1 *'Process' or 'functional' machine shop layout*

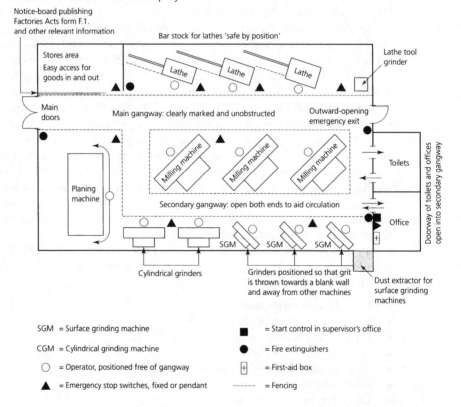

SGM = Surface grinding machine   ■ = Start control in supervisor's office

CGM = Cylindrical grinding machine   ● = Fire extinguishers

○ = Operator, positioned free of gangway   ⊞ = First-aid box

▲ = Emergency stop switches, fixed or pendant   ------ = Fencing

9.4.2 *Product layout*

This is also known as 'line' layout. In this type of layout, the needs of the process take precedence – that is, the machines and processes are laid out in the sequence of operations specified for the production process. An example of this type of layout is shown in Fig. 9.2. The advantages of this arrangement are as follows:

- Greater specialisation and speed of production is possible.
- High capacity dedicated machines can be used.
- The layout lends itself to automation including mechanical handling and robotics.
- Owing to the greater specialisation, it is less labour intensive.

Fig. 9.2 *'Product' machine shop layout*

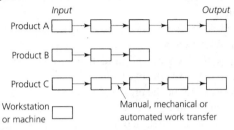

9.4.3 *Group layout*

This is also known as 'cell' layout. In this type of layout, similar machines and processes are grouped together to manufacture small batches of products having family resemblances. An example of such a layout is shown in Fig. 9.3. The advantages of this arrangement is as follows:

- Machine utilisation is high as work can be switched to the first available machine.
- The layout does not have to be changed to suit changes in the product.
- Material handling is reduced.
- In-progress stocking areas are reduced due to the balanced flow of production.
- Production control is simple; delays are quickly apparent.
- There is greater job satisfaction for personnel.
- The manufacturing 'cell' can be automated to accept FMS techniques, as previously discussed in Section 8.6.

Fig. 9.3 *Typical FMS cell*

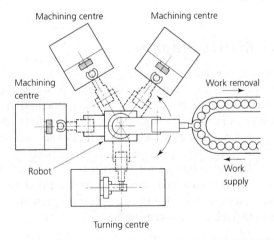

9.5 Methods of plant layout

The basic procedures for laying out equipment in a plant are similar to those used for *method study*. As the process of layout is very much dictated by the unique circumstances in each case, there is no hard and fast set of rules that can be applied. However, the application of *method study principles* will provide a starting point.

1. *Select* the type of layout required – that is, product layout, process layout or group layout.
2. *Record* all the facts concerning the manufacturing method, either the existing method or a new approach.
3. *Examine* the facts critically and in an ordered sequence.
4. *Develop* the most effective layout.
5. *Critically* examine the proposed layout.
6. *Install* the layout.
7. *Supervise* the installation of the layout.

Since it is much easier to move models around than large pieces of equipment, simulation of the process will enable a critique and adjustments to be made easily. The final layout must clearly indicate that the flow of work and information can be monitored. There should be no 'black-holes' down which materials, workpieces and information documents can disappear, only to reappear at suitably embarrassing moments. People, materials and information documents should be on view at all times. It is surprising how often cupboards and small offices can suddenly appear in the layout. Most of the time they are just status symbols and make no useful contribution to the processes and management of manufacture.

Simulation makes use of *flow diagrams* that are scale drawings in detail of the progress of either material or components in relation to the physical environment, i.e. the path such materials or components follow through machines, benches, stores and departments. Using this method the most efficient flow path can be determined. Wasteful journeys and back-tracking can be eliminated and the sequence to provide the minimum delay can be established.

9.5.1 *Matrix diagrams*

Matrix diagrams are scaled work areas marked with a grid to a convenient scale. Scaled cardboard pieces are cut to represent machinery and equipment. These are moved around on the grid until the most suitable layout is obtained. The pieces can then be glued in position or drawn around to produce the final layout drawing. The grid ensures that the width of gangways and the working areas around items of plant are adequate for safety and comfortable working. The maintenance team should also be involved at the planning stage so that adequate room is left for removing and/or replacing plant components. Because of bad planning and lack of foresight, it is not unknown for bricks having to be knocked out of walls so that drive shafts can be removed. The main disadvantage of this system is that it is only two-dimensional.

9.5.2 *Scale models*

Scale models provide a three-dimensional layout enabling any overhead obstructions and lifts to be taken into account. The departure from paper plans permits greater use of colour and enables layouts to be visualised from the point of view of appearance as well as convenience. Colour coding can be used to emphasise any particular storage areas and mechanical-handling equipment.

The scale models are available for a large range of standard machine tools, equipment, benches, storage cupboards, etc., and their bases are prepared to plug into a 'Lego' type base plate which forms a scale grid of the working area. Figure 9.4 shows an example of a typical layout using scale models. Although the models are expensive, a complete scale layout for a large machine shop running into several thousand pounds, it can be claimed that the initial outlay is insignificant compared with the cost savings achieved by getting the layout right first time. The main advantages of using scale models can be summarised as follows:

- They enable dimensions, including overhead clearances, to be accurately measured.
- They enable visual obstructions to be avoided.
- They reduce explanation since the layout is obvious to all concerned, including non-technical staff such as accountants.
- They stimulate the imagination and increase the accuracy and efficiency of the layout.
- They encourage aesthetics and the use of colour.

Fig. 9.4 *Plant layout using models*

In some instances, such as large-scale chemical plants and oil refineries, standard equipment models are not available and highly detailed scale models have to be hand crafted by skilled model makers. Detailed models not only prove the feasibility of the design layout but can also be used for management presentations to such bodies as shareholders and planning authorities. Such models are very costly and invariably end up in glass cases at company headquarters.

9.5.3 *Computer program*

Computer-aided design (CAD) software is available for planning works layouts. The floor area is drawn out to scale and representations of the machines and equipment can be called up from the system database and scaled to match the layout of the working area. One widely used example is 'CRAFT' (Computerised Relative Allocation of Facilities Technique). This uses workpiece handling cost as its computational basis. Other programs are based on the need for similar and associated plant to be grouped together.

Once the layout has been finalised, it should be submitted to the following critique before being put into practice:

- Is the flow of production logical, with minimum back-tracking?
- Is the layout compatible with plant services (e.g. electricity, gas, water, waste disposal) and any mechanical-handling requirements?
- Are there any possible overhead obstructions to clear?
- Are the gangways adequate for the movement of materials and personnel?
- Can maintenance be carried out on every piece of equipment?
- Are storage areas adequate?
- Have all aspects of safety, fire regulations, and personnel comfort been taken into account?
- Has provision been made for monitoring the installation to ensure that it complies with the requirements of the planned layout?

9.6 Process charting in plant layout

In order to understand the requirements when either reorganising the layout of an existing plant or starting from scratch with a 'green-field' layout, it is important to chart the existing or proposed manufacturing process. For this purpose a form of shorthand is available from method study techniques which enables a pathway to be established to scale, indicating the route of a product through the various operations.

This process chart can then be used as the basis of a critique to establish both the optimum method and the optimum layout to meet the production criteria. Figure 9.5 shows the code of five symbols that is used to signify the operations undergone by a product in the course of either manufacture or handling.

Fig. 9.5 *Process-charting symbols*

Activity	Result	Symbol
Operation	Produces, accomplishes, furthers the process	○
Inspection	Verification – quantity and/or quality	□
Transportation	Movement	⇨
Delay	Interferes or delays	D
Storage	Holds, retains or returns	▽

Example 9.1 is concerned with the inspection, stencilling and filling of 250 kg drums, and shows a typical process chart using these symbols. Figure 9.6 shows the plant layout derived from this process chart.

Fig. 9.6 *Plant layout developed from the process chart in Example 9.1*

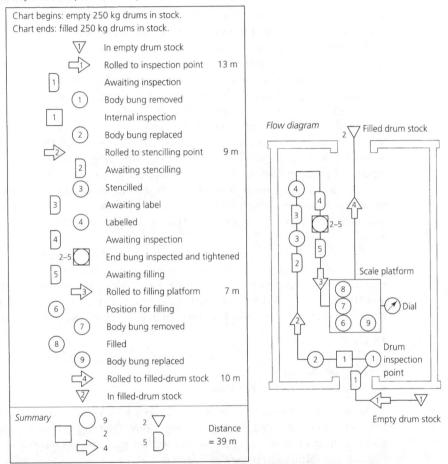

9.7 Group technology

In recent years, manufacturing has had to come to terms with an ever-changing scenario in which the requirement for purpose-specific, or customised, products has replaced the production of large numbers of identical products. In fact, about 90 per cent of the manufacturing base of the UK is involved with batch production rather than continuous production.

Economically successful batch manufacturing, incorporating modern technology, presents problems to companies whose plant layouts were probably designed many years ago. Although these layouts made sense at the time of their inception, they may no longer satisfy the requirements of modern computerised and robotic manufacturing techniques. Such techniques have been forced upon companies by fierce competition from abroad to reduce unit production costs and the need to reduce the time taken to get a new product or component into production (lead time) to satisfy customer requirements. Further, there is a constant demand for better quality with no corresponding increase in cost.

In response to the need for greater flexibility and a more rapid response to the demand for frequent changes in design and manufacturing methods, the *group technology* approach was first adopted in America and continental Europe during the early 1970s. It was adopted shortly afterwards in the UK Group technology (GT) sets out to achieve the benefits of mass production in a batch production environment. It identifies and groups together similar components in order that the manufacturing processes involved can take advantage of these similarities by arranging (or grouping) the processes of production according to 'families' of components. These grouped production facilities are referred to as *cells*.

This allows different 'families' of components and/or assemblies to be manufactured using the cells almost like a number of mini-factories under one roof. These work (or manufacturing) cells have separate planning, supervision, control and even production and profitability targets. With small-batch production in traditional workshops based on process layouts, the non-productive costs can be high for the following reasons:

- Difficulties in scheduling.
- The transfer of work between stations which are widely separated.
- Excessive handling.
- Setting up production processes.

These non-productive costs increase as a proportion of the total cost as the batch size decreases. However, small batches are more easily handled by a cell because of its more closely knit and flexible organisation and the fact that, specialising in a limited range of similar components, set-up times are reduced.

9.7.1 *Design*

New designs should make use of standard components wherever possible as this reduces costs, lead time, and ensures uniformity of quality. Where a new, non-standard component has to be used the designer will often work from scratch rather than spend time searching a retrieval system for a similar, previous design. However, in many cases, the new drawing may turn out to be merely a variant on some previous design. There may only be a change in a single variable such as a different diameter or a change of material specification. To change the design of an existing part would be more cost-effective than starting from scratch every time.

Such an approach would require a retrieval system to be established in which all the components produced in a company would be classified and coded according to various features. For example, geometric form or any manufacturing similarities. Obviously such a system would lend itself to computerisation. Thus the concept of a parts 'family' can be applied to the design process in addition to the manufacturing process. In fact it should, ideally, start in the design process.

9.7.2 *Parts families*

As previously stated, group technology seeks similarities not differences. Therefore, all similar parts can be collected into families. A 'family' or 'part family' is a collection of components which can have similarities in geometric form or can be manufactured by similar manufacturing methods. Once the family has been identified, a composite component can be drawn up which contains all the features of the part family, as shown in Fig. 9.7(a).

The available process machines can then be sorted out into the best mix to produce the composite component. The group of machines selected is then physically moved together to create a *work cell*. The machines in the cell are then set up to produce the composite component, and the actual production components are manufactured by omitting those operations not needed on any particular component. Figure 9.7(b) shows a possible group technology family, whilst Fig. 9.7(c) shows a comparison between a *functional layout* and a *group technology* layout using the equipment for producing a glass container mould, for example, as quickly and as cheaply as possible.

9.7.3 *Management*

With conventional process layouts for batch production, production control demands more and more up-to-date information. This has led to some very complex control systems that attempt to provide management data on a day-to-day basis and even, in some extreme cases, on an hourly basis. Progress chasing then becomes a continuous function. Many of these paper systems have collapsed under the sheer volume of information and have failed to provide the desired information from the shop floor.

Managing a group technology system can benefit the information-gathering process since each group runs as a separate entity and it is only necessary to plan the work into and out of the cell. There is no necessity to plan work through each machine since each cell is self-monitoring. Group self-control is the biggest and most difficult change for conventional management to accept, since responsibility and, therefore authority, is placed firmly at the point of production. External interference by the detailed planning of each operation would effectively destroy the concept of group technology.

Fig. 9.7 *Group technology (parts families): (a) component family; (b) sectionalised (functional); (c) group technology*

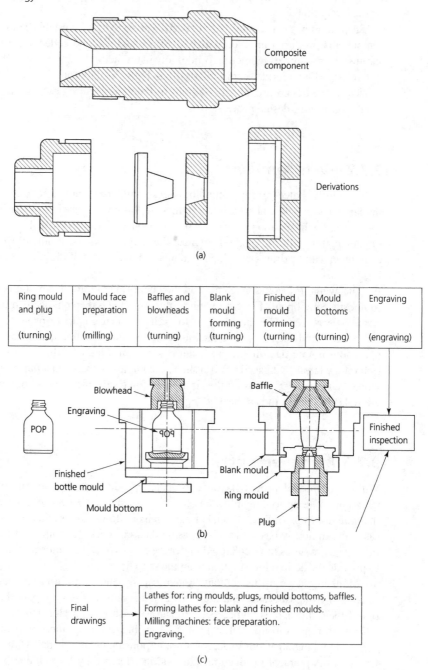

Ring mould and plug	Mould face preparation	Baffles and blowheads	Blank mould forming	Finished mould forming	Mould bottoms	Engraving
(turning)	(milling)	(turning)	(turning)	(turning	(turning)	(engraving)

(a)

(b)

Final drawings → Lathes for: ring moulds, plugs, mould bottoms, baffles.
Forming lathes for: blank and finished moulds.
Milling machines: face preparation.
Engraving.

(c)

9.7.4 *Personnel*

The job-satisfaction level, for many people, has considerably diminished with the advent of flow-line production and the introduction of high-technology machinery. The work becomes de-humanising as operators find themselves repeating the same task every day. The Volvo company in Sweden attempted to overcome the de-humanising effects of the mass-production system of manufacturing cars by creating a group effect. Workers were banded together in teams, with each team producing a complete vehicle rather than simply completing one task repetitively on many cars. Each car carries the group insignia and, in the case of problems, can be traced back to a particular group. All the component parts are delivered to the group rather than to individuals. This job-enrichment provides a positive involvement in the company's products and hence an influence for good on the quality of the finished product. The concept of small-group working coincides with the workers' natural desire for maximum job satisfaction.

9.7.5 *Group characteristics*

John Burbidge, in his book *The Introduction of Group Technology*, Mechanical Engineering Publication 1979, suggests that seven characteristics are shown by an effective group:

1. *The team*   Groups contain a specified team of workers who work solely or generally for the group.
2. *Products*   Groups produce a specified 'family' or set of products. In an assembly department these products will be assemblies. In a machine shop the products will be machined parts. In a foundry the products will be castings.
3. *Facilities*   Groups are equipped with a specified set of machines and/or other production equipment, which is used solely or generally in the group.
4. *Group layout*   The facilities are laid out together in one area reserved for the group.
5. *Target*   The workers in the group share a common product output target. This target output or 'list order' is given to the group at the beginning of each production period for completion by the end of the period. How this is achieved is the group's decision.
6. *Independence*   The groups should, as far as possible, be independent of each other. They should be able to vary their work pace if they so wish during a period. Once they have received materials, their achievement should not depend upon the services of other production groups.
7. *Size*   The groups should be limited to restrict the number of workers per group. Groups of 6 to 15 workers have been widely recommended. Larger groups up to 35 workers may be necessary for technological reasons in some cases. Such large groups have been found to work efficiently in practice.

9.7.6 *Coding and classification*

A system of classification and coding is the foundation to the application of group technology. The grouping of components into part families is the largest problem when considering the change to group technology. One method is to conduct a visual inspection to code components. It is the least expensive and least sophisticated method available. It is also the least accurate. The part families are established by a visual examination of actual components, component drawings or from photographs. Difficulties occur when there are many variations of small details that are not easy to identify.

Another approach to the problem of classification is to use the *production flow analysis* (PFA) system that is based upon the method of manufacture. This system uses route cards as the foundation for sorting components into classes. The main disadvantage of this method is the acceptance of route cards as being the definitive method of manufacture without a critique being carried out. The first sorting is based upon the first operation (set 1, 2, 3, 4, etc.) Then set 1 is sorted into subsets by the second operation (set 12, 13, 14, 15, etc.). Set 12 are those components having their first operation performed on machine type 1 and their second operation on machine type 2. The components are then sorted again by the third operation (set 123, 124, 125, etc.). The procedure is continued until all the route cards have been dealt with. From this sorting, some common characteristics will emerge which will enable the machine loading to be equalised. Where a large number of route cards have to be examined, it is permissible to take random samples of 2000 to 10 000 cards. The PFA system has never found favour in the USA.

The most commonly used method is by parts classification and coding. Though it is the most complex and time-consuming of the available methods, it has the advantage that it lends itself to a computer database. A coding system can be alphabetic, numeric or alpha-numeric. It can be *dependent* (monocode), in which each succeeding symbol is dependent upon the one preceding it. For example the code 12 is a two-digit code where the first digit 1 might indicate a circular workpiece and the second digit 2 would indicate a dimension within a certain range. Because digit 2 is preceded by and dependent upon digit 1, the dimension refers to a circular component and will be a diameter.

Alternatively it can be *independent* (polycode), in which each succeeding symbol in the sequence stands on its own and does not depend upon any preceding symbol. The position of the symbol in the sequence determines its function. For example, the first digit may indicate geometric form and the second digit may indicate overall length within a given range. Thus the second digit of the code will always indicate overall length in this system. Let us now consider two systems that are readily available and commonly used.

9.7.7 The Opitz system

This system was developed in Germany by H. Opitz. It uses an alpha-numeric code 12345 6789 ABC, in which the first sequence of five digits (12345) refers to the design attributes of the parts. The sequence of four digits (6789) refers to supplementary information such as material, dimensions, form of supply of raw material and accuracy, and the alphabet code is available for each company for its own unique purpose.

9.7.8 Miclass

This is the Metal Institute Classification System developed in the Netherlands. The main advantages of this system are that it can be readily computer based and can classify a very large number of parts into their respective family groups. It is an interactive system in which the operator responds to a series of questions (prompts) posed by the computer. The complexity of the part will determine the number of questions. The code can have from 12 to 30 numbers, with the first 12 digits being a universal code applied to any component. The remaining 18 digits are for use by individual companies for their own unique purposes. The universal code is set out as follows:

1	Main shape	2 and 3	Shape elements
4	Position of shape elements	5 and 6	Main dimensions
7	Dimension ratio	8	Auxiliary dimensions
9 and 10	Tolerances	11 and 12	Material

Finally, the case for *group technology* can be summarised as follows:

- Production can be 'customer led', i.e. production is responsive to customer demand in terms of quantity, quality, availability and design variants.
- The use of a coding classification system leads to improved product design and reduced design lead time.
- Standardised machine set ups and tooling can be used.
- More effective use can be made of machines and equipment.
- Handling times and costs are lower owing to less transportation between operations.
- Planning procedures are simplified.
- Buffer stocks are lower.
- Management procedures are simpler.
- Employee job satisfaction is greater.

9.8 Flexible manufacturing systems

The natural extension of *group technology* (GT) is to move to manufacturing cells in which all the functions are externally controlled by computers – that is, a 'people-less' manufacturing cell. In group technology the human beings are part of the process of manufacture but in *flexible manufacturing systems* (FMS) the human effort is confined to component and tooling design, setting up the machines and equipment in the cell, and providing very high level maintenance expertise. It is even possible to install machines that can change their own tools and set themselves.

Again there is a need for part families but, in the case of flexible manufacturing systems, the cell can handle much larger families of parts and deal with them in random order. Other differences are:

- The need for transport (mechanical handling) equipment which can move components from machine to machine in the correct sequence and maintain maximum loading for each machine.
- A computer which can control all the operations and materials handling at the same time.

The outstanding difference between a GT cell and an FMS cell is that the latter involves a very much higher level of capital investment. Before investing in an FMS cell, comprehensive feasibility studies have to be carried out and these must include a five-year market forecast, an estimate of the return on the capital invested, and the level of flexibility required in the system.

Basically, FMS is a group of computer numerically controlled (CNC) workstations connected by a materials-handling system. The individual computers of the workstations and the handling system are themselves under the control of a *master computer*. A hierarchy of control exists in which the master computer deals with the management of the system such as the scheduling of the parts. A *supervisor computer* deals with monitoring the system and acts as the interface between the system and the operator. For example, error conditions are reported and a graphical interface or supervisory control and data acquisition (SCADA) system shows the current state of each part, i.e. where it is within the system, if a machine is busy or if it is waiting for work. In terms of production volume, FMS lies between the high-volume production of dedicated machines used for mass production and individual CNC machines which are best suited for batch production in a group technology cell where there is a large variety of products. This is shown in Fig. 9.8.

Fig. 9.8 *Position of FMS in the production hierarchy*

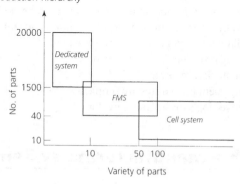

9.8.1 *Concepts*

Where the parts manufactured on flexible manufacturing systems are prismatic in shape, they can be loaded into pallets carrying the necessary fixtures and transported on conveyors or by automated vehicles whose path is determined by guideways in the workshop floor. Such systems are referred to as 'primary conveying systems'. The pallets are coded with a bar code carrying the machining instructions. The bar code is 'read' by sensing devices which route the part to the particular machining destination or to a buffer store conveyor until the required machine is available. Thus, whilst the part is within the system, it is either in transit or being machined. The system has secondary conveyors whose function it is to receive the parts from the primary system and transport them to the individual machines or buffer stores.

The need for buffer stores is due to the different machining times for each operation. Thus buffer stores allow short operations to be accommodated with those requiring longer machining times. The part remains within the system until all the operations on it are complete. Remembering that this is a flexible system, then different parts requiring different operations can be accommodated within the system at the same time providing that the tooling is available. Since the development of *industrial robots*, the range of components that can be handled has grown beyond those with simple prismatic geometry and the overall flexibility has been enhanced. An example of a typical FMS cell is shown in Fig. 8.23.

9.8.2 *Relationships between parts and machine tools*

The need for a parts family is at the root of FMS. Thus there has to be a close liaison between the manufacturing engineers and the designers in order that the machining datum points can be set and agreed. The location points for the fixtures are of equal importance, together with the state of the raw material entering the system. All these facts have to be agreed and established at the planning stage.

The part size will influence the choice of machine within the system (large part: large machine) and the part geometry determines the type of machine tool, as it does in general manufacturing. Where the variety of parts to be manufactured is large, then standard CNC machines should be considered. However, the larger the volume and the smaller the range of components to be manufactured, the greater the tendency to move towards dedicated machines and equipment. This will reduce the flexibility but increase the rate of production. Care has to be taken when considering the part/machine-tool relationship to keep future production requirements in mind. If the market is fluid, then flexibility is the key factor, but if the future demand is more stable, then less flexibility and greater production rates will be required.

The variety and complexity of the workholding fixtures in FMS makes great demand on the ingenuity of production engineers. As in all machining operations, the accuracy of the final product is only as good as the accuracy of presentation of the work to the cutting tools. The concept of the parts family leads to the identification of common datum and common location features. The object, as in all jig-and-fixture design, is to reduce loading time. This is particularly important in FMS, where gains in production times can be easily offset by loading costs. Inspection time and costs must also be reduced if the full benefits of FMS are to be achieved. Automatic inspection stations or probes set within the machine tools themselves must be employed. Actual dimensions are compared with standard dimensions by the computer and adjustments are automatically made to the machine tools. Swarf will be produced in large quantities, so automatic swarf-handling equipment has to be considered. All too often, high-investment systems can be brought to a halt as tools become broken and movements clogged, simply because adequate swarf-removal facilities have been overlooked.

9.8.3 *Materials handling*

Materials handling is the element in any FMS cell that builds flexibility into the system. Its functions are to move the work between machines, buffer stores, inspection stations, etc., and also to present the work to the machines correctly orientated for cutting. The handling system must have the facility for the independent movement of workpieces between machines – that is, such workpieces must be able to flow from one station to another as the loading and routeing demands. There must also be buffer storage facilities for work waiting to be machined.

At the machine tool, the secondary handling systems must permit workpiece orientation and location for clamping within the workholding fixture ready for machining. For geometrically simple, prismatic parts families, palletisation on conveyor systems or automatic guided vehicles (carts) is the most common primary transport system. Pusher bars, guideways or robots transfer the parts between primary and secondary systems and load the workstations.

The development of industrial robotics has increased the flexibility and versatility of flexible manufacturing systems by enabling rotary parts to be included. The robots can transfer parts between systems and load them into machine fixtures under the control of the master computer. A single robot can be installed in the centre of the system and can load each machine with parts at random as the routeing requires (see Figs 8.23 and 9.3). It will, of course, also unload the parts when the process is complete. The machine tools must be laid out so that they are within the operating radius (reach) of the robot's arm. The *end effector* that is mounted on the 'wrist' of the robot must be able to hold all the components within the parts family of the cell.

9.8.4 *People/system relationships*

Whilst the system itself is automatic, people still have a role to play. This role is one of system management rather than machine operating. The supply of raw material to the FMS cell and the removal of the finished components are still manual operations. In the long term, completely automated factories could become feasible but not necessarily desirable.

The major change is in the content of the work available. The need for trade skills are reduced, but more emphasis is placed upon office-based skills and upon design and production engineering skills. Sociologically the implications are profound with a lessening in demand for skilled and semi-skilled operators, a shortening in the working week and the virtual elimination of the need to work 'unsociable' hours. Robots and computers do not have to see, so round-the-clock 'lights out' operation of the factory can achieve substantial savings in operating costs. The manning needs for such a system is for technicians and engineers of the highest skills in the fields of product and tool design using CAD/CAM techniques, tool setting and changing, and for multi-skilled maintenance engineers with expertise in electrical, electronic, mechanical, hydraulic and pneumatic systems. The computer system will make its own demand on programmers and system analysts.

9.8.5 *Benefits*

Having examined the technology and management of FMS cells, let us now consider the benefits of such a system. The manufacturing sector is essential to the UK economy. It absorbs 20.2 per cent of the employed population, provides 80 per cent of all UK exports, and accounts for 20 per cent of the gross domestic product (GDP). Despite its importance to the economy of the UK, the majority of jobs in this sector of industry consist of manufacturing a great variety of parts in low volumes; a figure as high as 80 per cent with batch sizes of 10–50 units is typical. Also, the market life of certain products, for example telephones, has been reduced from 10 years to 18 months. This makes the need to respond more quickly to market needs more important than ever before, and has confirmed the place for FMS to allow changes from old to new designs to be implemented with low investment costs. This view is supported by *Masuyama of Toyota* who states: 'It seems essential to perceive accurately the condition of the market and to supply what is demanded by the market, with a short lead time and at a low cost.'

Companies that have already installed FMS talk of lead times being reduced by 50–60 per cent and scrap levels being reduced from 25 to 5 per cent of output. The companies who invest in FMS are looking for a payback period of two to three years. Also, FMS promises productivity improvements through increased machine utilisation and reduced levels of work in progress, along with a reduction in production cycle time.

An FMS system provides the flexibility needed to make a large variety of components on a continuous basis. The savings accrued by using FMS can be 50 per cent less than the cost of conventional methods, and production can be increased by a multiple of 2 to 3.5. This shows clearly that FMS must continue to be developed and replace conventional methods of manufacture if the manufacturing industry of the UK is to survive in current competitive markets. It should now be apparent that the adoption of FMS strategies has many advantages and we can summarise the commercial benefits as follows:

- A faster response to market changes in design or the creation of new designs.
- Improved product quality.
- Direct and indirect labour costs are reduced.
- Production planning is improved.
- Machine utilisation levels are improved.
- The 'work in progress' inventory is improved since the time spent in the systems is reduced.

It is not possible to buy a ready-made flexible manufacturing system, each installation has to be tailored to suit specific company needs. Therefore, once a company has decided to follow the FMS route, it has to look at every aspect of its production practices, both good and bad. It is in this analysis and during preplanning that the greatest benefit to the company occurs. Even if the decision is made not to invest in FMS, the feasibility study will have revealed flaws within the present methods of production that can be corrected and the company will benefit. Also, the level of management commitment, so essential in the implementation of FMS, will have been explored.

9.8.6 *Problems associated with FMS*

The many benefits accrued by installing FMS are not achieved easily, and there are many problems to overcome before the full potential of the system is realised. The main problems associated when installing FMS are detailed below.

Cost
The capital investment of a FMS is very high, ranging from £100 000 to many millions of pounds. Therefore, poor selection criteria and implementation methods can produce costly mistakes.

Layout
The physical relationship between entities within a FMS is critical. For example, matching the working envelope of a robot so that all the target positions can be reached within the

FMS is very difficult. Also, determining the configuration of the end-effector (gripper) to a workholding device can be very complex. Other physical problems encountered include matching velocities of components on conveyors, automatic guided vehicles (AGVs) and robot arms.

Efficiency

The development of efficient control strategies for the maximum utilisation of a FMS system is most important but, at the same time, extremely difficult. The large capital investment in an FMS cell necessitates the maximum utilisation of the system if an acceptable return on investment is to be made. A survey of 13 Swedish companies indicated that FMS has increased machine utilisation rates by an average of 60 per cent and reduced work in progress (WIP) and 'lead time' by a similar amount. Typical system capital costs ranged from £0.8m to £3.0m in 1986. The payback periods ranged from 2 to 5 years.

Flexibility

The dichotomy of achieving part variety and flexibility whilst, at the same time, maintaining efficiency should be apparent. It is clear that flow-line production must be the most efficient if only one component is being manufactured, since it enables all processes to be fine tuned to achieve maximum efficiency. At the same time, by its very nature, flow-line production has little flexibility. On the other hand, FMS has much greater flexibility and it is because of the variety of parts manufactured in different batch sizes and at different intervals, that the control strategy of FMS is both complex and difficult to plan. For example, how is a critical component to be made ahead of schedule? This will affect the control strategy and efficiency of the system and will require the master computer schedule to be reprogrammed.

SELF-ASSESSMENT TASK 9.3

1. Explain why group technology is important to so many functions within the manufacturing industry.

2. Describe the benefits of FMS and list the important points to note in achieving them.

9.9 Materials requirement planning

Most organisations 'buy in' materials in one of the following ways:

- In the raw state for processing into a finished product; in a part-finished state (e.g. castings or forgings) for processing into a finished product.
- As standard components (e.g. nuts and bolts) ready for assembly.
- As a mixture of one or more of these foundation materials.

The cost of foundation materials can form a large part of the final cost of the finished product. If the elapsed time between the delivery of the foundation materials and the sale of the finished product is small, then the cost of stockholding is low. However, if the elapsed time between the delivery of the foundation materials and the sale of the finished product is high, then the cost of stockholding is significant. Since there is no return on the working capital invested in foundation material, this can adversely affect the profitability of the company. At best it can affect the 'cash flow' of the company since funds tied up in the material represents 'dead money' that cannot be used for more profitable purposes. At worst, if bank borrowing was necessary for the purchase of the material, the interest charges can turn a paper profit into an actual loss.

9.9.1 *Stock control*

The object of the department charged with the function of materials control is to maintain a supply of foundation materials at the *lowest possible cost*. A term that is frequently used in *stock control* is 'lead time'. This is the time taken from the initiation of an order to the delivery of the material into the company stores. If the foundation material is a stock item at the supplier, the lead time will be short. However, if the material has to be manufactured by the supplier, the lead time can become very significant. Thus the problem is that of raising orders at the correct time, both from outside suppliers and from internal production departments. *Materials requirement planning* (MRP) has been in industry for a very long time but with the advances made in recent years in computer technology and software, a computer-based system for materials requirement planning – MRP 1 – has been developed. Before examining MRP 1 in detail, we shall review the more common methods of organising stock control in a company.

The term 'stock control' describes a system of purchasing in which the level of stocks being held in store is used to regulate the raising of buying or production orders. Its function is to ensure that an adequate supply of foundation materials is available at the correct time and in the correct quantities to provide continuous production. Items can be bought in via the purchasing office or manufactured 'in-house'. For the reasons previously mentioned, stocks should be kept to the minimum necessary to maintain production. The amount of foundation material held in stock is governed by the following factors:

- Operational needs.
- Lead time.
- Cost of storage.
- Deterioration of materials whilst in stock.
- The cash-flow position of the company – stock ties up working capital.

Stocks should only be held if:

- Delivery cannot be exactly matched to supply.
- Delivery is uncertain.
- Substantial price reductions provide a cost advantage.
- Bulk buying attracts discounts which offset the cost of storage.
- They provide a buffer of finished products, giving a service to customers.
- In-house production is prone to operational risks (process breakdowns).

9.10 Stock-control systems

The basic feature of these systems is that the stock level initiates any new orders. When the amount of stock falls to some predetermined level – 're-order level' or 'order-point' – then a new order is issued for a further supply. Figure 9.9 shows the basic features of a stock-control ordering system based upon variation of stock over a time period.

- The *order point* (OP) is the level of stock at which a new order is issued.
- The *order quantity* (OQ) is the quantity ordered to restore stocks to the agreed level – also known as the 'batch quantity' or 'batch size'.
- The lead time is the time required either to buy in or to manufacture, in house, a replacement batch.
- If the batch quantity is known, then the lead time can be calculated.
- The order point can be determined by setting off the lead time and batch quantity.

Fig. 9.9 *Simple stock-control ordering system*

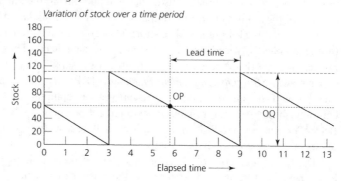

The steps required to set up an elementary stock control system are as follows:

1. Choose the order quantity (OQ).
2. Estimate the lead time to obtain a replacement batch of the required size.
3. Calculate the number of parts that will last, at the normal consumption rate, during the lead time, and thus find the order point (OP).
4. Maintain a permanent record of the stocks of each item.
5. Arrange for a new order to be issued each time the stock level falls to the order point.

9.10.1 *Buffer stock*

The simple system shown in Fig. 9.9 is impracticable because of the imprecision of actual production completion times. If a delay in production occurs and the lead time becomes increased or there is an increase in demand, then stocks would run out causing an interruption in production at some stage. An improved model is shown in Fig. 9.10. The buffer stock represents an insurance against the risk of the exhaustion of stock. Its value is obtained by balancing the cost of a shortage, causing interrupted production, against the working capital locked up in holding additional stock.

Fig. 9.10 *Minimum and maximum stock levels*

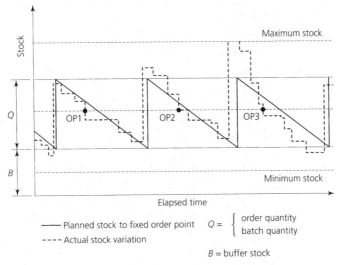

9.10.2 *Stock level*

The *minimum stock* is a control level to reduce the risk of exhausting stocks, as shown in Fig. 9.10. When stocks reach the *minimum stock level*, the fact is reported so that an investigation can take place and corrective action taken. Buffer stock (B) and minimum stock need not be the same amount because the former is a fixed amount of stock maintained as an insurance against high demand or delays in delivery, whilst the latter is a control level.

The *maximum stock* is also a control level and indicates when stock levels are unacceptably high in economic terms (see Fig. 9.10). When this point is reached, production must be reduced or even ceased until stocks have reached a more manageable level. The maximum stock level is set so that it is in excess of buffer stock by an amount sufficient to allow for small variations in the production plan. Only when something is wrong in the system are the maximum and minimum stock levels normally reached. A typical stock control record card is shown in Fig. 9.11.

9.10.3 *Batch quantity*

Although it was originally thought possible to calculate the *economic batch quantity* (EBQ) using mathematical formulae, EBQ is now commonly selected for each case on merit and previous experience. That is:

batch quantity = (total requirement per year)/(batch frequency)

This simple calculation avoids batch quantities being set to suit production convenience without regard to stock costing.

Fig. 9.11 *Stock-control record card*

Batch quantity:	100					
Order point:	56			Description:		
Buffer stock:	40			Bearing Block		
Maximum stock:	160			Part No.: 3890 A		
Minimum stock:	30					

Date	In	Out	Balance	Date	In	Out	Balance
3.6.91	98	–	98	2.9.91	–	12	36
10.6.91	–	12	86	9.9.91	–	2	34
17.6.91	–	16	70	16.9.91	104	–	138
24.6.91	–	14	56	23.9.91	–	12	126
1.7.91	–	14	42	30.9.91			
8.7.91	–	4	38				
15.7.91	–	14	24				
22.7.91	102	–	126				
29.7.91	–	16	110				
5.8.91	–	16	94				
12.8.91	–	14	80				
19.8.91	–	16	64				
26.8.91	–	16	48				

Significant Dates
24.6.91 – Re-order
15.7.91 – Minimum breached; investigate
26.8.91 – Re-order

9.11 Materials requirement planning (demand patterns)

Let us consider an everyday item such as a bicycle. Figure 9.12 shows the relationship between the component parts of the bicycle. The completed assembly is at the highest level and is classified as level 0. The items making up the constituent parts are classified as the lower levels 1, 2, 3 and 4. Therefore, if the items classified at these lower levels are not being sold as spare parts, they can be classified as *dependent* items – that is, the demand is solely dependent upon the number of complete bicycles assembled.

The completed bicycle is an end in itself and it is not used in, or as part of, a higher level item. It is, in fact, the finished product that the customer purchases from the local cycle shop. The volume of production is based upon sales forecasts for that particular model and

Fig. 9.12 *Materials requirement planning*

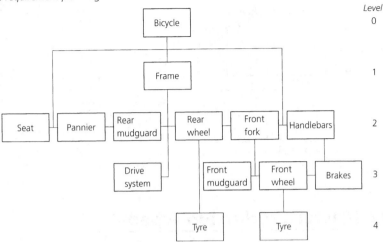

is *independent* of the demand for any constituent parts. Thus the demand for a company's products can be classified into two distinct categories:

- *Dependent demand*, generated from some higher level item. It is the demand for a component that is dependent upon the number of assemblies being manufactured. As such, it can be accurately calculated and inserted into the *master schedule*.
- *Independent demand*, generated by the market for the finished product. The magnitude of the independent demand is determined from firm customer orders or by market forecasting. For MRP purposes it is independent of other items. It is not used in scheduling any further assemblies.

A stock-control system could keep all the parts shown in Fig. 9.12 in stock using either order-point methods or sales forecasts for each item. To do this, each separate item has to be treated as independent. MRP 1 software, however, can calculate the amount of stock of dependent items required to meet the projected demand for the independent higher level item, which, in this example, is a bicycle.

It is claimed that MRP 1 is superior to the standard methods of stock control previously discussed in this chapter. This is because, whilst the demand for the independent items (level 0) cannot be predicted accurately – and are therefore unsuitable for stock-control methods – lower level items (1, 2, 3, 4) are dependent on a higher level item and can thus be calculated with accuracy. That is, although it is only possible to forecast next year's sales of complete bicycles approximately, it is possible to predict that exactly as many tyres will be required as there are wheels.

Further, stock-control methods assume that usage is at a gradual and continuous rate. In practice, however, this is not the case. The call on parts will occur intermittently, with the quantity depending upon the batch size of the final product. This often results in the holding of excess stocks. Material requirement planning assists in the planning of orders so that subsequent deliveries arrive just prior to manufacturing requirements, as shown in Fig. 9.13.

Fig. 9.13 *Comparison: stock control versus MRP 1*

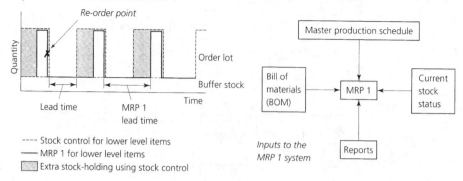

---- Stock control for lower level items
—— MRP 1 for lower level items
▨ Extra stock-holding using stock control

9.12 Master production schedule

This is the schedule or list of the final products (*independent* or *higher level* items) that have to be produced. It shows these products against a time or period base so that the delivery requirements can be assessed. It is, in fact, the definitive production plan that has been derived from firm orders and market forecasts. In order to minimise the number of components specified, care has to be exercised in deciding how many of the basic level 0 assemblies are to be completed. The total lengths of the time periods (the *planning horizon*) should be sufficient to meet the longest lead times.

9.13 Bill of materials

The bill of materials (BOM) contains all the information regarding the build up or structure of the final product. This will originate from the design office and will contain such things as the assembly of components and their quantities, together with the order of assembly. The structure and presentation of a typical BOM is shown in Fig. 9.14. The number of parts should also be shown at each level.

Fig. 9.14 *Structure for bill of materials: S = subassemblies; C = components*

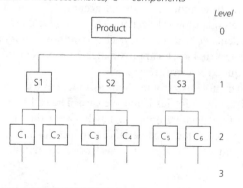

9.13.1 *Current stock status*

This is the inventory or latest position of the stock in hand, pending deliveries, planned orders, lead time, and order batch quantity policy. All components, whether in stores, off-site, or in progress, need to be accurately recorded in the inventory record file.

9.14 MRP 1 in action

Upon receiving the input data, the system has to compute the size and timing of the orders for the components required to complete the designated number of the final product (the independent or highest level item). It performs this operation by completing a level-by-level calculation of the requirements for each component. Since the latest stock information is available to the system, these calculations will be the net value of the orders to be raised:

requirements = net requirements + stock currently available

where

stock currently available = current stockholding + expected deliveries

Inherent in the calculations is the factor for lead time, i.e. ordering and manufacturing times. The system must determine the start date for the various subassemblies required by taking into account the lead times, i.e. it offsets these lead times to arrive at the start dates. A complication that the system has to overcome is the use of some items which may be common to several subassemblies. The requirements of these common items must then be collected by the system to provide one single total for each of the items. Figure 9.15 shows a standard planning sheet for one of the outputs of MRP 1. Other reports required in addition to the order release notices are:

- Rescheduling notices indicating changes.
- Cancellation notices where changes to the master production schedule have been made.
- Future planned order release dates.
- Performance indicators of such things as costs, stock usage, and comparison of lead times.

Fig. 9.15 *Standard planning sheet for one of the outputs of MRP 1*

	1	2	3	4	5	6	7	8
Gross requirements				140				
Scheduled receipts			50					
On-hand	100		150					
Net requirements				100				
Planned order releases				100				

A number of advantages are claimed for MRP 1 software. These include:

- A reduction in stocks held in such things as raw materials, work in progress, and 'bought-in' components.

- Better service to the customer (delivery dates are met).
- Reductions in costs and improved cash flow.
- Better productivity and responses to changes in demand.

9.15 Manufacturing resource planning (MRP 2)

MRP 1 is a large step forward, compared with manual systems, in the procurement of materials and parts in as much as it provides a statement of what is exactly required and when. Unfortunately, it takes no account of the production capacity of a particular manufacturing facility. Most factories have a production output that can only be changed over a period of time either by improved productivity or the injection of capital to purchase more productive equipment.

Clearly, by themselves, the outputs from MRP 1 have limited value to the *master production schedule* except to detail the material requirements. Without the information regarding production capacity, it is difficult to match MRP 1 to actual production schedules. Remember that *capacity planning* is concerned with matching the production requirements to available resources (human resources and plant). It would be impractical to have a master production schedule that exceeded the capacity of the plant to meet its requirements. This could lead to decisions to extend the plant capacity by increasing the labour force, shift working or subcontracting, when what is required is more sensible scheduling.

The shortcomings of MRP 1 led to the development of MRP 2 that brought into account the whole of the company's resources, including finance. Thus the initials MRP also stand for *manufacturing resource planning*. This is quite a different, and more difficult, philosophy than the original *materials requirement planning* when one considers the many variables that can affect shop-floor production. For example, machines may breakdown, staff may become ill or leave, and rejects may have to be reworked. In addition, MRP 2 takes into account the financial side of the company's business by incorporating the *business plan*. MRP 2 is used to produce reports detailing materials planning together with the detailed capacity plans. This enables control to be effected at both the shop floor and the procurement levels. Note that MRP 2 does not replace MRP 1 but is complementary to it and *the two systems are used together*.

As in all computer systems, the information produced is only as good as the information fed into it. Therefore, the system is dependent upon the feedback of information relating to those things under its control, i.e. such data as the state of the manufacturing process and the position of orders. MRP 2 contains enough information, and the power to organise it, to be able to run the whole factory. Figure 9.16 shows a schematic layout of production incorporating MRP 1 and MRP 2 organised around the master production schedule.

MRP 2 is a management technique for highlighting a company's objectives and breaking them down into detailed areas of responsibility for their implementation. It involves the whole plant and not just the materials provision. The key to MRP 2 lies in the master production schedule (also called the *mission statement* – which is a statement of what the company is planning to manufacture) and, because it is the master, all other

Fig. 9.16 *Schematic layout of MRP 1 and MRP 2 organised around the master production schedule*

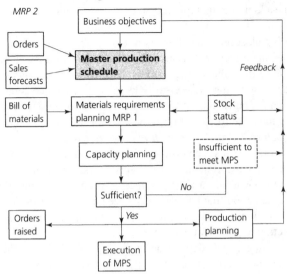

schedules should be derived from it. The master schedule should *not* be a statement of what the company would like to produce, as its own derivation lies in the company objectives. It cannot in itself reduce the company's lead times, but it will provide the incentive for the plant personnel to take action in an area if required. It is important, as in all planning, to monitor adherence to the master production schedule. MRP 2 will not stop over-ambitious planning but it will show up the consequences.

9.16 'Just-in-time'

The phrase 'just-in-time' (JIT) originated in Japan with the production plant of Toyota as the most frequently quoted application. This led to the incorrect conclusion that since Toyota is involved with mass (repetitive) production of cars, then JIT only applies to this mode of manufacture. In fact JIT applies to any mode of manufacture. In the USA the term 'zero inventories' is used but the philosophy is exactly the same as JIT. Perhaps the most apt definition of JIT belongs to R. Schonberger (*Japanese Manufacturing Techniques*, Free Press, New York, 1982):

> Produce and deliver finished goods *just-in-time* to be sold, subassemblies *just-in-time* to be assembled into finished goods, fabricated parts *just-in-time* to go into subassemblies, and purchased materials *just-in-time* to be transformed into fabricated parts.

A second approach is propounded by D. Potts (*Engineering Computers*, September 1986):

> A philosophy directed towards the elimination of waste, where waste is anything that adds to the cost but not the value of the product.

A third approach that appears to combine the other two is put forward by C. Voss (*Just-in-Time Manufacture*, IFS, London, 1987):

An approach that ensures that the right quantities are purchased and made at the right time and in the right quantity and there is no waste.

JIT philosophy, therefore, is not only about minimum stocks being available at the point of manufacture; it is also about reducing costs by eliminating those things that add nothing to the value of the product, such as avoiding interest charges on working capital tied up in servicing unnecessarily large stocks. At the same time, it is necessary that the availability and quality of the product is maintained. However, any manufacturing system should have the foregoing objectives, so it is in the application that *JIT gets nearer to the manufacturing ideal than many traditional systems.*

Toyota uses a system known as *kanban* (meaning = 'card'); so, in this form, their JIT system is a card system. In the kanban system assembly schedules are derived from a master plan which is drawn up to meet a specific 'planning horizon'. The 'planning horizon' is the time period that has previously been agreed as realistic and trustworthy, for example, six weeks may be appropriate. Using the master plan and the time scale, the daily production requirements can be set.

Referring to the dependency theory of MRP 1 and the bicycle example considered earlier, a card (*kanban*) would be issued to the store holding the parts for the final assembly and a container of parts would be despatched to the assembly point. A card would then be issued to the manufacturing centres for replacements to be made for the final parts store. The manufacturing centres would, in turn, issue a card for any replacements that they require. Cards would be passed down through the system to raw materials purchase and subcontractors. Thus, ideally, at each level the replacements arrive 'just-in-time' to maintain production and no stocks are held.

The issue of a card from the master plan or schedule 'pulls' the work through the system, since no production can take place without its authorisation. Figure 9.17 illustrates this point. In such a 'pull-through' system in which everything is dependent upon the arrival of an authorisation card, there must be times when no production takes place. This is because of the over-riding principle that nothing is produced until it is required except, of course, for the minimum (small) inventories kept in the appropriate stores.

The impression might be given that the labour force sits around doing nothing until a card arrives. This is not the case. The major advantage gained from any inactivity is that the reason for that stoppage can be immediately investigated. Everyone – management and production workers – can be engaged upon tackling the problem. In the meantime, those members of the production team not engaged directly in solving the problem can be engaged in quality circle work, training or maintenance activities.

The Toyota system is just one example of the philosophy of JIT because, in any analysis, JIT is simply a philosophy. It covers a range of production-control techniques that have the objectives of eliminating every facet of the business that fails to make a contribution. It focuses minds on achieving production which is not only 'just-in-time' but which produces 'just enough' with no surplus and costly excess stocks. It streamlines the production process by removing the cushion of safety stocks, excess stocks, and/or inflated lead times that can cover up many real problems in a company. It concentrates the minds of the entire workforce on the business of the company. By making every action important, there has to be involvement and commitment from everyone for the successful implementation of JIT. The team approach to problem solving is inherent in JIT.

Fig. 9.17 *Flow diagram for a simplified* kanban *system*

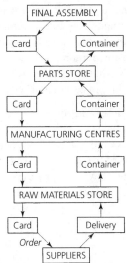

Without a JIT philosophy, 'waste' can be overlooked in a company because it is not recognised and there is often a 'we have always done it this way' mentality. When waste is considered to be any activity that fails to contribute to the value of the product, it can cover such things as the production of scrap and reworking, excessive transporting by poor routeing and poor plant layout, overproduction to compensate for scrap, machine breakdowns and absenteeism. These things can too easily become part of the production scene by the acceptance of their inevitability. The analysis of management and working practices necessary when introducing a JIT philosophy immediately shows up these faults in the system.

It has to be borne in mind that each element in a company is dependent on all other elements and therefore JIT can become a very vulnerable system. The employees and suppliers are all part of the system so that any conflict will halt production very quickly. In a large organisation, for example, an industrial dispute in one link in the supply chain can quickly bring the rest of the chain to a halt.

9.17 Just-in-time and MRP 2

The *kanban* (card) system just described is only one way of applying a JIT philosophy. As production becomes more varied and involved, the amount of data required to run the system increases significantly and computer support is required. To this end MRP 2, which was introduced in Section 9.15, is a computer-based system that has the objective of unifying all the associated functions in a company, from initial planning through to delivery of the finished products. It is applicable to any form of company irrespective of size and diversity of production. It breaks down the company's *business plan* into detailed operational tasks for each part of the organisation. The effective planning element removes any confusion that could exist by concentrating the minds of the workforce – management and productive labour – on what has to be achieved.

Since JIT has the major objective of eliminating waste in all its forms and, as confusion leads to wasteful excesses, then the adoption of MRP 2 promotes the concept of JIT. The key to successful implementation is correct planning and scheduling. Both have the main aim of producing only what is required, when it is required, and are therefore compatible. Thus, the benefits of JIT can be summarised as:

- Better quality by closer control.
- Reduction of lead times.
- Reduction of stocks and levels of work in progress.
- More involvement of the workforce, leading to greater job satisfaction.
- Increased efficiency by the process of continually challenging the existence of everything connected with the production in a never-ending search for greater efficiency.

9.18 Computer-integrated manufacture

The above techniques are all integrated into a seamless system where data is shared between all functions within any department. These systems contribute towards the philosophy of computer-integrated manufacture (CIM), as shown in Fig. 9.18.

Fig. 9.18 *Computer-integrated manufacture (CIM)*

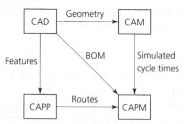

9.18.1 *Computer-aided process planning*

Computer-aided process planning (CAPP) reduces the amount of manual clerical work involved in planning the processes required to manufacture a component, subassembly or assembly. This is achieved by integrating or sharing data with other computer systems such as CAD. Geometrical data or features from such systems can be used to decide which processes will be used. Other computer systems, such as manufacturing databases, can also be used to automate the planning process. For example, databases can contain previous process data on similar components, that share similar family features, i.e. a family of components. Coding systems can be used to identify components, group technology can be used to group components into feature based families. The route each component takes as it passes through each process towards its finished state is then passed to the *computer-aided production management* (CAPM) system.

9.18.2 *Computer-aided production management*

CAPM is another layer of integration with other systems; for example, *bill of materials* data, automatically generated from CAD systems, can be passed to CAPM systems along with routeing data for each component required. The purpose of CAPM is to schedule and prioritise the sequence of events within a manufacturing system. Again, data from tool paths simulated in CAM systems can be used to derive cycle times. This enables the most efficient schedule to be developed so that lines are balanced, capacity is planned, people and machines are fully utilised, and downtime is minimised.

9.18.3 *Product data management systems*

Product data management systems (PDMS) are now used to manage data within a company. Such systems as the *intranet* – that is, an internal web system for a company – enables data to be shared at all levels. Many companies are international corporations and data needs to be shared amongst the many plants of such a company on a global basis. For example, some aircraft companies share design data across the Atlantic using 'intranet' systems. Other benefits, such as developing designs based on previous process information, are useful in a 'design for manufacture' context. Many companies are very close to becoming *closed-loop* systems, i.e. *shop-floor data capture systems* feedback data which can be used to monitor progress.

SELF-ASSESSMENT TASK 9.4

Review the latest trends in manufacturing hence find the meaning of:

- World Class manufacturing.
- Lean and agile manufacturing.
- Reverse engineering.
- Rapid prototyping.
- Logistics.
- Enterprises.
- CIM, CAPP, CAPM.

EXERCISES

9.1 Produce a critique of a factory layout with which you are familiar.
  (a) Discuss the criteria upon which the critique is based.
  (b) List the necessary information that would be required before planning the layout of a new workshop.

9.2 The design function of a product lie between those required for marketing and those required for its operation. Discuss the influence of the marketing and operation functions on the design of the product.

9.3 (a) List and explain **six** control arrangements that make up a manufacturing control system.

(b) Describe how such a control system is implemented in a place of work with which you are familiar.

9.4 With reference to component grouping, define, in your own words:

(a)   (i) classification

      (ii) coding

(b) For any group of components with which you are familiar, design a simple classification and coding system.

9.5 Explain the difference between MRP 1 and MRP 2 and discuss the circumstances in which a company would wish to develop from MRP 1 to MRP 2.

9.6 Prepare a report for senior management outlining the philosophy of JIT and its possible implications for the production process at your own plant.

9.7 For a company with which you are familiar, list your ideas for converting it to a 'World Class' plant and explain how these ideas could become the driving force for improvement.

9.8 In the context of CIM, explain how 'virtual companies' can be formed using the latest computer communication links.

10 Cost control

10.1 Principles of cost control

Some basic principles of estimating and costing were introduced in Section 11.5 of *Manufacturing Technology*, Volume 1, where the processes of estimating and costing were used to arrive at the selling price of a product in advance of that product being manufactured. The purpose of the exercise was to arrive at a price that could form the basis of a 'quotation' or 'tender' when seeking a new order for a product yet to be made. It is relatively easy to tender for a batch of motor cars which are in regular production and for which all the costs are known and the selling price has been fixed. It is quite a different matter to tender for an oil-rig of high complexity to a new design which has yet to be built.

Once the tender has been accepted and the order has been put into work it is necessary to monitor the costs to see if they agree with the estimate. If the manufacturing costs are greater than estimated, then the firm will make a smaller profit or even a loss on the order. If the manufacturing costs are less than estimated then the firm will make an increased profit. Unfortunately, tendering in a highly competitive market leaves little, if any, room for making allowances for cost over-run when estimating and tendering. The object of this chapter is to analyse the costs of manufacture and examine some methods of cost control.

Historical costing

This is the term used when the costs of manufacturing and marketing a product are derived after the product has been completed. It is the true cost of making and selling a product and the various cost elements are determined from records of the cost of materials, bought-in components, the labour used, the machines used, and anything else that directly or indirectly is attributable to that product. Unfortunately, any discrepancy between the true cost and the estimated cost does not become apparent until the product has been completed, by which time it is too late to retrieve the situation. If the product is the first of a batch, then it may be possible to review the manufacturing processes and recoup any loss in the remainder of the batch. However, if a 'one-off' item has been produced, no corrective action can be taken.

Standard costing

This is the term used when estimated cost elements are used to set standards that must not be exceeded at each stage of manufacture. The estimated cost elements are those arrived at during the process of estimating for tendering purposes. Standard costing allows greater management control, and corrective action can be taken as soon as any cost element approaches or exceeds the standard set.

SELF-ASSESSMENT TASK 10.1

1. Discuss the need for cost control in manufacturing companies.

2. Describe the system of cost control used at any organisation with which you are familiar.

10.2 Cost elements

Figure 10.1 shows the basic cost elements that make up the selling price of a product.

Prime costs

These are the actual costs of raw materials, bought-in parts, and labour used in the manufacture of a product. Since they refer directly to the product they are also referred to as *direct costs*.

Production overheads

These are costs which result from the manufacture of the company's products but which cannot be directly charged against any particular product. Hence they are referred to as *indirect costs*. Table 10.1 lists some typical production overhead costs.

Administration, selling and fixed overheads

The sum of the prime costs and the production overheads represent the manufacturing costs of a particular product. In addition, allowance has to be made for the general management of the company, marketing and selling. Also, costs such as rental for the premises, servicing of loan interest, and local taxation (uniform business rate), continue even when no production is taking place; that is, they are *fixed overheads* and are independent of the level of manufacturing activity within the company.

Profit

If the preceding items could be accurately assessed and allocated to a company's products, then those products could be sold without making a profit or a loss. However, a company is primarily in business to make a profit. The profit is needed to build up the company's reserves against any future downturn in activity, to provide a fund for the replacement and updating of plant and equipment, to pay a dividend to the shareholders, and to pay any taxes due on the profits. When setting the profit level, attention has to be paid to the competitiveness of the market and allowance has to be made for inflation. This latter item is particularly important when the period of manufacture is protracted – for example, when building a large ship or an oil-rig. Failure to allow for inflation in such circumstances can eliminate any profit made on an order.

Fig. 10.1 *Cost structure*

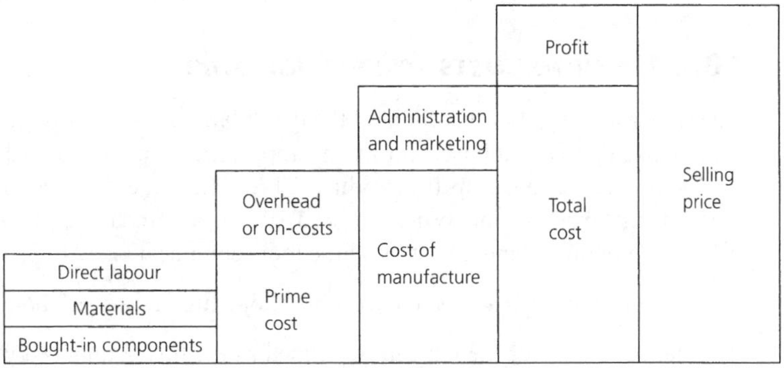

Table 10.1 *Typical production overhead costs*

Workshop premises	Plant
Rent or capital loan charges	Leasing or capital loan charges
Unified business rate	Maintenance and repairs
Depreciation	Energy to operate plant
Insurance	Insurance
Maintenance and repairs	Depreciation
Heating and lighting	
Indirect labour	*Indirect materials*
Supervision	Any materials associated with the activities
Inspection	listed in this table
Stores	
Tool room	
Clerical assistants	
Transport	
Time-office	
Drawing office	

Selling price

A *break-even diagram* for the costs that we have just considered would show how the level of production determines whether a profit or a loss will be made for a given selling price. The use of *break-even analysis* is considered in some detail in Section 10.8. The level of a company's overhead expenses is often taken as an indication of the efficiency of that company, and high overhead costs often indicate low efficiency – but this is not necessarily true, because a highly automated company which is working very efficiently may have minimal direct labour costs so that, in proportion, its overheads appear to be high as a percentage of those labour costs. No matter whether the costing system adopted is simple or complex, the problem still remains as to how to allocate the various costs to any individual product so that it carries its fair share of the costs, yet remains competitive and saleable. Let us now consider some alternative methods of allocating costs.

10.2.1 *Prime costs (direct labour)*

The simplest method of determining the direct labour cost of a product is to multiply the time booked on each job element by the appropriate labour rate and add together all the job elements that make up the product. This is how the direct labour cost is calculated when a historical costing system is used. When a standard costing system is used the calculation of direct labour costs is a little more complicated:

standard labour cost = standard time × operator rating × labour rate for the job

The standard time for the job can be defined as the time a qualified worker would take to complete the job when working to a preset level of performance. Standard time is arrived at by the use of *work-measurement* techniques. Operator rating is a measure of how any individual worker compares with the 'ideal' standard operator. The standard labour cost is not the actual cost, which can only be determined by historical costing, but is a target that needs to be achieved or improved upon if a profit is to be made. Actual work performance must be constantly compared with the standard so that remedial action can be taken early whilst there is still time to prevent a cost over-run.

10.2.2 *Prime costs (direct material)*

The materials used directly in the manufacture of any product must be accurately booked against the job and carefully priced. This applies equally to materials bought in specifically for the job and materials that are drawn from the stores. The cost of materials kept in the stores may vary depending upon market conditions at the time of purchase. Allowance must also be made for any change in cost when the material is replaced. There are a number of ways of allocating cost to stored materials.

- *Market price* The material cost is based upon the market price for the material at the time of issue, irrespective of the price originally paid.
- *Replacement pricing* The material cost is based upon an estimated price that will have to be paid when the material is eventually replaced.

- *Standard cost*   This is a cost set for a material as a matter of company policy and need not be the same as the actual market price for the material.
- *Average cost*   This is where the material has been bought, over a period of time, from a variety of sources at various prices. The material is then charged to the job at the average of all the prices paid.
- *First in : first out (FIFO)*   Where the material has been purchased in batches over a period of time, it is assumed that it will be issued from the stores in the same date order. It will be charged against the job at the original purchase price of each batch.
- *Last in : first out (LIFO)*   Assuming the worst situation of the last batch of material being issued first, all the material is charged against the job at the price of the last batch of material received into stores.
- *Highest in : first out (HIFO)*   This assumes that the most expensively priced batch of material, irrespective of the date of purchase, will be issued first and all the material used on the job will be charged at this price.

10.2.3 *Allocation of overhead costs*

Since overhead costs are indirect costs, their fair allocation against any particular job or product is difficult and a number of systems have been devised to try to arrive at an equitable solution. At present, the most widely used system is *absorption costing*. The procedure is to divide the company into *cost centres*. These cost centres may be *production cost centres*, which are concerned with actual manufacturing such as the foundry and the machine shop, and *service cost centres*, such as the stores, the maintenance department, the sales department, the accounts department, and the design and development department.

Having set up the cost centres, the overhead costs then have to be apportioned in an equitable way between the centres. For example: the rent could be charged against a particular cost centre in proportion to its floor area as a percentage of the total site area; lighting costs could be apportioned according to the number of light fittings and their power consumption; maintenance could be allocated on the basis of the operator-hours spent in a particular cost centre over a set period of time. In the case of service departments, their wages and salaries must also be apportioned in a similar way since they cannot be charged directly against a specific job or product. Thus the total indirect overhead costs of each centre can be built up. Next, the overheads of the service cost centres must be transferred to the producing cost centres according to how much use they make of the service cost centres. Finally, a system has to be devised so that the collective overheads of the production cost centres can be charged fairly against the goods produced by the company.

The methods of allocating overheads and the absorption of collective charges in common use are given below.

Percentage of prime costs
If the total overheads for operating a production cost centre for a year are known, and if the total prime costs generated within that cost centre for a year are known, then the

overheads can be expressed as a percentage of the prime costs. Thus the cost of manufacturing a particular product within that cost centre is:

the prime cost of the product
  + the overhead percentage of the prime cost of production

This method is simple and is used where there is (a) little variation between the different products manufactured and (b) a reasonable balance between material and labour costs.

Percentage of direct productive labour

This method is used where manufacturing of the product is labour intensive and the amount of material is negligible. Here, the total cost centre overheads are taken as a percentage of the total of the cost centres' productive direct labour costs over a set period such as one year. The material is charged at cost. Since the direct productive labour costs usually reflect the skills involved, the allocation of overheads reflects the sophistication of the process and the investment in the plant and equipment used. When this method is used, the cost of manufacturing a particular product is:

material cost for   the direct labour cost   the overhead percentage of
the product   + for the product   + the direct labour cost of the
                                              product for the cost centre

Differential percentage

With this method, the cost centre overheads are loaded onto the direct material costs and the direct labour costs at different rates. Usually the rate loaded onto the material is about a quarter or one-third of the rate loaded onto the direct labour costs. This system is frequently used in small companies as it is not only simple to apply but also allows greater flexibility in pricing strategy.

Operator-hour rate

With this method, the total charges (prime costs and overheads) for the cost centre are divided by the total productive operator-hours of the centre and allocated pro rata to the number of operator-hours spent on each job. This takes into account various rates of pay but ignores the use of different types of equipment. Therefore, it is more suitable for fitting and assembly shops than it is for machine shops where there is a wide diversity of equipment and processes.

Machine-hour rate

The total charges (prime costs and overheads) for the cost centre are allocated to the individual machines in a workshop, taking into account such items as the initial cost of the equipment, the power required, the size and type of product, and routine servicing and maintenance costs. The machine-hour rate is determined by dividing the individual machine costs by the number of productive hours for which that machine works. This is the most suitable system when the production processes used in the cost centre are highly automated.

10.3 Depreciation

Depreciation is the reduction in the intrinsic value of an item of plant or equipment due to normal wear and tear or due to the fact that it has become technologically out-of-date. It must be remembered that each item of plant represents a capital asset in a company's accounts. Since the value of each item of plant is continually diminishing, this represents a reduction of the capital assets of the company and must be shown in the balance sheet if this is truly to represent the financial status of a company. Further, a provision must be made for the eventual replacement of the items of plant under consideration. To avoid repeatedly using the phrase 'item of plant' throughout this section, the term 'asset' will be used instead. The original cost of each asset together *plus* all its additional expenditure (such as servicing and repairs and maintenance) *less* its residual 'trade-in' or scrap value, must be charged against revenue (the money earned by that asset). This charge must be spread over the asset's economic service life as fairly as possible. Remember that an asset's service life is usually very much shorter than its actual useable life. To provide for the replacement of an asset, a sum of money must be set aside annually from revenue earned by the asset in order to accumulate the required amount when the end of its service life is reached.

Allowance must also be made for inflation. It is essential to create this depreciation reserve fund so that new capital does not have to be raised when replacement becomes necessary. From time to time technological innovation may make it necessary to replace plant prior to the planned date and before adequate funds are available. In such circumstances it may be necessary to raise additional capital and the cost of servicing such capital must be included in the estimates of revenue the new plant will earn over its lifetime. There are various methods of writing down the value of an asset over its lifetime and some of these will now be examined. For simplicity, the fact that the depreciation funds can be invested to accrue interest (compound) will be ignored but it can be helpful to some extent in offsetting the effects of inflation.

10.3.1 *Fixed instalment method*

This is also known as the 'straight line' method since the value of the plant is written down in equal amounts over equal increments of time, thus producing a straight-line graph. Let I be the initial cost, R be the remanent (final disposal) value, and n the number of years over which the value of the item of plant is being written down, then the reduction in asset value and the sum set aside each year should be $(I - R)/n$, as shown in Fig. 10.2. This simple method makes no allowance for the fact that the cost of repairs and maintenance is greatest at the end of the life of an asset.

Fig. 10.2 *Fixed instalment (straight-line) depreciation*

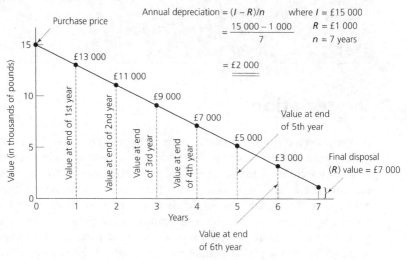

Annual depreciation $= (I - R)/n$ where $I =$ £15 000

$$= \frac{15\ 000 - 1\ 000}{7} \quad \begin{array}{l} R = \text{£1 000} \\ n = 7 \text{ years} \end{array}$$

$= $ £2 000

10.3.2 *Reducing balance method*

This allows for the fact that, financially, most assets depreciate most rapidly in the early years of their life and the depreciation – as a percentage of the original purchase price – becomes less each year. Using the symbols from the previous example, and letting r equal the fixed percentage, then r can be determined from the expression $r = [1 - (R/I)^{1/n}] \times 100$. The application is shown in Fig. 10.3.

10.3.3 *Interest law method*

This provides for depreciation by crediting the asset account with equal yearly amounts. The asset account is then debited with the interest that would otherwise be earned by the capital invested in the asset had that capital been invested in a bank deposit account. The notional interest so deducted is then credited to the *profit and loss account*. The annual increments, less the notional interest on the yearly balance, indicates the amounts to be set aside for replacing the asset at the end of its forecasted service life.

Let S equal the sum set aside each year. Then, using the previous symbols,

$$S = [P(I - R)]/[(1 + P)^n - 1]$$

where P is the notional rate of deposit account interest. This method of calculating the depreciation allowance is frequently used in connection with asset leasing. Whichever of the three previous methods is used, the accumulated reserve at the end of the service life of the asset (nth year) will be $I - R$, as shown in Fig. 10.4.

Fig. 10.3 *Reducing balance (%) depreciation*

Year	Start of year value (£)	Depreciation (£) (n = 32%)	End of year value (£)
1	15 000	32% of 15 000 = 4 800	15 000 – 4 800 = 10 200
2	10 200	32% of 10 200 = 3 264	10 200 – 3 264 = 6 936
3	6 936	32% of 6 936 = 2 221	6 936 – 2 221 = 4 715
4	4 715	32% of 4 715 = 1 509	4 715 – 1 509 = 3 206
5	3 206	32% of 3 206 = 1 026	3 206 – 1 026 = 2 180
6	2 180	32% of 2 180 = 698	2 180 – 698 = 1 482
7	1 482	32% of 1 582 = 474	1 482 – 474 = 1 008

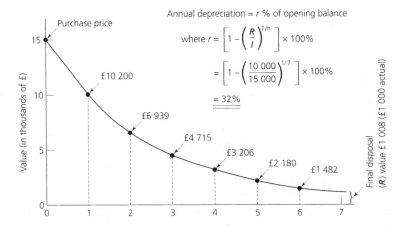

Annual depreciation = r % of opening balance

where $r = \left[1 - \left(\dfrac{R}{I} \right)^{1/n} \right] \times 100\%$

$= \left[1 - \left(\dfrac{10\ 000}{15\ 000} \right)^{1/7} \right] \times 100\%$

$\underline{\underline{= 32\%}}$

There are other systems which allow for depreciation. For instance, in the 'depreciation fund' system the asset is shown at its full value throughout its life but the *profit and loss account* is debited with a fixed annual sum that is paid into a deposit account or into gilt-edged securities. The 'insurance policy' system is similar to the previous example except that the annual sum debited from the profit-and-loss account is credited to an insurance policy that secures a sum sufficient to replace the asset at the end of its service life. The 'revaluation' system requires each asset to be revalued each year. A charge equal to the difference between the book value and the assessed value is then debited against revenue earned by the asset and placed in a sinking fund ready for eventual replacement of the asset, the book value being duly written down. An independent professional valuer is required in the interest of accuracy and impartiality. Finally, in the 'single-charge' system, a single charge is made annually against revenue earned by the asset to cover repairs, renewals and depreciation. The charge may vary and revaluation may also be required from time to time.

SELF-ASSESSMENT TASK 10.3

1. Define what is meant by 'depreciation'.

2. Explain why it is important for a company to make allowance for depreciation.

Fig. 10.4 *Interest law method of calculating depreciation*

Year	Sum set aside each year (£)	Balance carried forward (£)	Accrued interest (£) at 8%	Balance at year end (£)
	(S)			
1	1 569	–	–	1 569
2	1 569	1 569	8% of 1 569 = 126	3 264
3	1 569	3 264	8% of 3 264 = 261	5 094
4	1 569	5 094	8% of 5 094 = 408	7 071
5	1 569	7 071	8% of 7 071 = 655	9 206
6	1 569	9 206	8% of 9 206 = 737	11 512
7	1 569	11 512	8% of 11 512 = 921	14 007

$$S = [p(I - p)] / [1 + p)^7 - 1] \qquad \text{where: } p = 8\% \text{ interest}$$
$$= [0.08 (15\,000 - 1000)] / [(1 + 0.08)^7 - 1]$$
$$= £1569$$

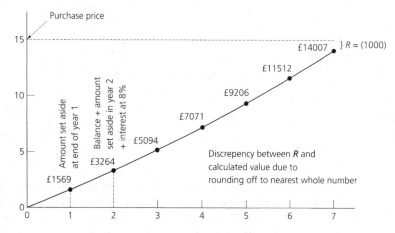

10.4 Costing applications

Having considered the basic elements of all costing systems, let's now consider how these elements can be brought together in a *complete* costing system.

10.4.1 *Job costing*

This is used for costing the production of 'one-off' components or assemblies. It is essentially a historical costing operation with the prime costs of materials and production labour being charged daily against a specific works order and the overheads being allocated on completion. If standard costs have been ascertained then they can be compared with the actual costs on a daily basis and some degree of budgetary control maintained. However, if a cost over-run has occurred, it is very difficult to recoup the situation except by seeking ways to reduce the costs at later stages in the process.

10.4.2 *Batch costing*

This is similar to job costing except that a number of identical components or assemblies are being made against the same works order. If the batch is large enough, then special tooling such as jigs, fixtures and press tools may have to be made. However, the cost of such tools must be recovered, and this cost can either be charged against the order or, sometimes, the customer will pay 'part-cost' towards the tools to reduce the unit component cost. This is often done if a succession of batches is envisaged. Tools wear out and have to be maintained and replaced from time to time and allowance must be made for this in a similar way to asset depreciation, as described earlier in this chapter. Where batches of components or assemblies are being produced, and particularly if repeat batches are being manufactured, closer budgetary control can be exercised than for simple job costing. Standard costing techniques can be applied, and any deviation between actual costs and standard costs can be investigated and corrections can be made to the manufacturing processes employed.

10.4.3 *Process costing*

This is applied to continuous (mass) production and continuous processing (e.g. chemical manufacturing and metal extraction). By its very nature, continuous processing lends itself to standard costing since there is adequate time to compare actual costs with standard costs, to investigate and analyse any discrepancies, and to carry out any remedial action that may be necessary. The aim of process control is to determine the average cost of a product over a period of time. This is achieved by adding together the direct labour costs, the material costs and the overheads, and then dividing the sum of these costs by the number of units produced over a prescribed period of time. Work in progress also needs to be taken into account at the beginning and end of the time period. Data is collected from the various cost centres involved and the more finely these are subdivided the more accurate the analysis becomes. However, the too much subdivision leads to too great a clerical load in collecting and processing the data and, even with computerisation, a balance has to be achieved between the costs saved by accurate cost control and the cost of implementing the system. The service cost centres and the production cost centres require separate costings. A system then has to be found for allocating the service centre *costs* to the production cost centres so that all costs incurred in operating the company may be carried in an equitable manner by the goods manufactured and, ultimately, by the customer. Since all the costs are directly allocated to, and absorbed by, the goods being manufactured, this is known as *absorption costing*.

10.4.4 *Marginal costing*

Overhead costs can be fixed or variable. The fixed overheads are independent of the level of manufacturing activity within the company – that is, all the costs which would remain if the company suddenly stopped manufacturing, dismissed its productive labour force, closed down its workshops, but kept all its offices and non-productive staff intact. The variable overheads are those directly related to manufacturing and which increase or decrease with changes in manufacturing level. There is a school of thought that only the variable overheads should be added to the prime costs to give what are referred to as marginal costs. The difference between the *marginal costs* and the *selling price* contributes to paying all the remaining operating costs, and the fixed overheads and are referred to as *contribution costs*. Marginal costing allows for a differential pricing policy so that items which sell strongly may be loaded with higher contribution costs, whilst items which sell less strongly may be more lightly loaded with contribution costs in order to keep their price down in a competitive market.

SELF-ASSESSMENT TASK 10.4

Compare and contrast the advantages and limitations of the following costing techniques and state the circumstances in which you would use them:
(a) job costing
(b) batch costing

10.5 Break-even analysis

Now that some various cost elements and some various techniques of costing have been considered, the use of break-even graphs, such as the example shown in Fig. 10.5, can be examined. The vertical axis represents *cash* both as revenue from sales and the costs of manufacture. The horizontal axis represents units of production.

Fig. 10.5 *Break-even analysis of profit and loss: (a) analysis of expenditure; (b) analysis of income; (c) break-even analysis of income and expenditure*

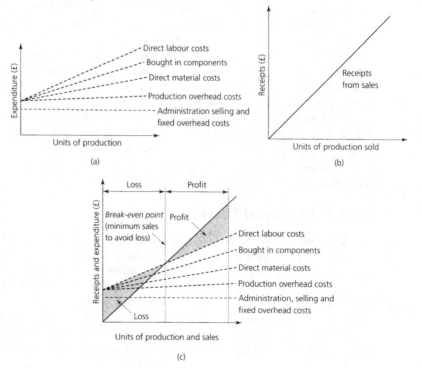

First consider the costs of manufacture. The way these costs build up are shown in Fig. 10.5(a). Since the fixed overheads have already been defined as being independent of the number of units manufactured, it is plotted as a straight line parallel to the horizontal axis. Next, the variable overheads must be added and these increase as the level of manufacturing activity increases. As soon as the decision to manufacture is taken, some expenditure will have to be incurred to prepare for manufacture even before the first item is made. These are production overheads and they will increase as manufacturing becomes more active, more electricity is used and more use is made of the service departments. Finally, the prime costs have to be added. Direct labour and direct material costs are wholly proportional to the number of items manufactured.

Revenue from sales is directly proportional to the number of items manufactured and sold (ignoring discounts and other customer incentives). A plot of the revenue is shown in Fig. 10.5(b). Finally the two graphs are superimposed, as shown in Fig. 10.5(c). The area of the graph where the operating costs are greater than the sales revenue represents a loss. The area of the graph where the sales revenue is greater than the operating costs represents a profit. The break-even point is where the sales revenue and the operating costs are equal. This graph only represents the situation at a particular instant in time. Any active company will be constantly adjusting the slope of the various elements of the graph to maximise its profits. An increase in overheads resulting from the purchase of new and, hopefully, more productive plant should be more than offset by a reduction in direct labour costs. The company should constantly be looking for cheaper and better materials, and the selling price will constantly need to be tuned to the maximum the market will stand.

10.6 Process selection

The first decision that has to be made concerning any new component is to make or buy. Standard components such as nuts and bolts should always be bought in. They are made in large quantities by specialist firms at a lower cost than can be achieved when making a few of such items 'in-house'. To this end, designers should always make as much use as possible of standard components. Where these are made to BSI standards or to DIN standards the quality is also guaranteed. For non-standard items a number of factors have to be taken into account.

- Is suitable plant and labour available 'in-house'?
- Is there sufficient capacity available on the plant in time to complete the components by the required delivery date?
- Can the plant and labour available operate to the required quality standards?
- Can the components be produced at an economic price?
- Will 'buying-in' result in plant and labour standing idle and thus representing a non-productive cost to be recovered?

If the answer to any of the foregoing questions is 'yes', then the decision should be to 'make'. If the answer to any of the above questions is 'no', then the purchase of new plant and/or tooling must be considered. Is there sufficient demand for the component under consideration to warrant the outlay on new plant after allowing for depreciation? If not, can the expenditure on new plant be offset against further, similar work? Again, if the answers are 'no', the decision must be to 'buy-in' the components from specialist subcontractors.

The selection of manufacturing process must also be cost related (as introduced in Section 4.1) as it might appear to be more costly to manufacture a component by powder metallurgy than by drop forging. However, providing the quantity of components required will justify the tooling, the component produced by powder metallurgy will be more accurate, have a better surface finish and will require less secondary machining and finishing compared with forging. Thus, overall the powder metallurgy process may result in a cheaper and better component.

EXERCISES

10.1 Compare and contrast the advantages and limitations of:
   (a) historical costing
   (b) standard costing

10.2 Discuss the cost elements of a typical manufacturing company with which you are familiar.

10.3 Compare the advantages and limitations of any **two** methods from the following list if you were allocating direct material costs:
   (a) market price
   (b) replacement pricing
   (c) standard cost
   (d) average cost

10.4 Compare the advantages and limitations of any **two** methods from the following list if you were allocating direct material costs:
   (a) first in : first out (FIFO)
   (b) last in : first out (LIFO)
   (c) highest in : first out (HIFO)

10.5 Describe **two** methods of allocating direct labour costs.

10.6 Define what is meant by 'overhead' costs and compare and contrast the advantages and limitations of any **two** methods of calculating overhead costs from the following list:
   (a) percentage of prime costs
   (b) percentage of direct productive labour
   (c) differential percentage
   (d) operator-hour rate
   (e) machine-hour rate

10.7 Compare and contrast the advantages and limitations of any **two** methods of calculating depreciation from the following list:
   (a) fixed instalment method
   (b) reducing balance method
   (c) interest law method

10.8 Compare and contrast the advantages and limitations of the following costing techniques and state the circumstances in which you would use them:
(a) process costing
(b) marginal costing

10.9 Describe how 'break-even analysis' can be used to determine the point at which a manufacturing process ceases to make a loss and commences to make a profit.

11 Quality in manufacture

The topic areas covered in this chapter are:

- Quality in manufacture.
- Reliability.
- Quality in design.
- Quality of conformance to design.
- Accreditation of conformance.
- Total quality management (TQM).
- Quality assurance (BS EN ISO 9001/2).
- Quality management.
- Quality control in manufacture, statistical methods.

11.1 Introduction to quality in manufacture

The word 'quality' is bandied about quite loosely, especially in publicity material. As applied to manufacturing it is defined in BS 4778: 1987 as '*the totality of all features and characteristics of a product or service that bear upon its ability to satisfy stated or implied needs*'. Put more simply, it can be defined as 'fitness for purpose'. The survival of manufacturing companies in today's international markets depends upon the achievement of acceptable levels of quality, at minimum cost to the customer. There is no such thing as absolute quality. A customer's idea of quality will change with time and with competition in choice. Therefore a company's attitude to quality must be constantly reviewed in the light of these changes in customer perceptions so that, at all times, that company's products represent 'fitness for purpose', 'value for money' and 'state of the art technology'.

The price of any commodity is negotiable: quality is definitely not negotiable. This is widely appreciated in industry because most manufacturers are, themselves, customers in one way or another. Any apparent savings made by the purchase of inferior materials or components are quickly dissipated in loss of customer confidence. The quality of goods or services goes hand in hand with reputation. This goodwill is a company's greatest asset. It does not just happen, it has to be part of a company's philosophy. That is, it has to be managed so that quality becomes the responsibility of every member of a company. This is the concept of *total quality control* (TQC) and its implementation is *total quality management* (TQM). The concept of TQC has replaced the former system where quality was the sole preserve of the quality control department. Such departments still have an important role to play, but that role has changed with the advent of TQC and TQM.

Figure 11.1 shows the importance of the customer in manufacture. The concept of supplier and customer exists within an organisation to provide a *quality chain* between the external supplier to the company and the delivery of goods to the customer. Each department is the customer of the departments supplying it with goods and services and, in turn, each department becomes the supplier of those departments making use of its goods and services. Similarly, within each department, there is a supplier/customer relationship between the individual members of the personnel of the department. The typist is a supplier of documentation to the manager and the manager, as the typist's customer, has the right to expect a high-quality, error-free typing service.

Fig. 11.1 *The linkage between customer expectations, product design and manufacture*

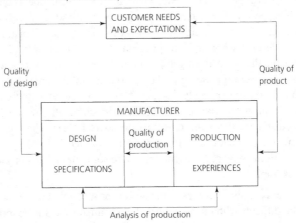

Throughout and beyond all organisations, be they manufacturing concerns, banks, retail stores, universities or hotels, there exists a series of quality chains, as shown in Fig. 11.2. This chain may be broken if one person or one piece of equipment does not meet the requirements of the customer, internal or external.

Fig. 11.2 *Quality chain (from* Total Quality Management – A Practical Approach; *reproduced courtesy of DTI)*

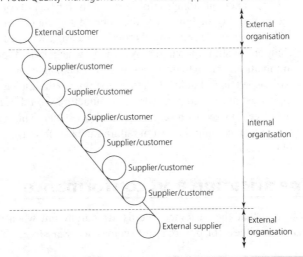

11.2 Reliability

This is the ability to provide 'fitness for purpose' over a period of time. Obviously any component will fail at some stage of its life, although the time span may be such as to give the impression of being virtually everlasting. Failure is considered as being the point at which a product or service no longer meets its fitness for purpose. Increasingly, reliability is a major factor when a customer is making a purchasing decision.

Therefore, there is an increasing need to design reliability into a product from the start. The testing of a new design for its reliability is difficult due to the time factor involved as well as the widely varying environments and conditions under which the product has to operate during its working life. There are many tests in existence specifically designed with reliability in mind. For example, in the aircraft industry, rigs are set up to test the fatigue resistance of wing assemblies. Hydraulically actuated pistons are positioned along the underside of a wing and are programmed to move up and down to simulate the forces acting on the wing during flight conditions. By this method, many thousands of flying hours can be compressed into days as the assemblies are safely tested to destruction. Again, rolling roads and specially designed test tracks are widely used in the automobile industry. In addition, prototype vehicles are test driven under varying climatic conditions over very large distances throughout the world to test their reliability before entering large-scale production and being sold to the motoring public.

From the outset, a designer can take some practical steps to prevent premature failure. For example:

- The use only of components with a proved reliability in assemblies. This can prove to be difficult when a designer is being innovative and is introducing new materials and manufacturing methods.
- The use of uncomplicated designs incorporating as many previously proved features as possible.
- The use, within the cost parameters, of back-up systems and duplicated components where total failure cannot be tolerated, as in the controls of automobiles or aircraft. Duplication decreases the probability of total failure by the square of the chance of a single component failing. Triplication of the number of components in parallel decreases the probability of failure by the cube of the chance of a single component failing. Wherever possible components should be designed to be fail-safe. That is, failure of the component shuts down the system of which it is a part or a back-up system is automatically brought into operation. The pilot light on a modern gas appliance must be so designed that, should it fail, the gas supply to the appliance is automatically cut off.
- Only tried and tested methods of manufacture should be employed until new methods have been proved to be equally reliable or better. This may sound dull to a thrusting young designer, but the cost of failure can be prohibitive both financially and in the loss of human life.

11.3 Specification and conformance

Accepting that the customers' needs are paramount when considering the definition of quality as 'fitness for purpose', it becomes necessary to establish at the outset just exactly what those needs are. Figure 11.1 shows how the customer has to be involved at every stage, from design to delivery. In the production of goods or the delivery of services, there are always two separate but interconnected factors relating to quality:

- Quality in design.
- Quality of the conformance with the design.

Let us now look at these two factors in some detail.

11.3.1 *Quality in design*

The first stage in the development of a product or a service is 'customer needs'. This is the province of the marketing department working in conjunction with the design engineers. This first stage is the most difficult and yet the most crucial as it sets the scene for the whole process of manufacture in the case of goods, and delivery in the case of services. The term *quality of design* is a measure of how the design meets with customers' requirements, i.e. the stated purpose of the *design brief* and the *design specification* derived from it.

The translation of customer needs, as set out in the initial design, into manufactured goods or delivered services is achieved by drawing up specifications. These specifications can best be described as detailed statements of the requirements to which the goods or services must conform. Correct specifications control the purchase of goods or services as well as directing the efforts of the manufacturing processes. Specifications can be expressed in terms of such things as mechanical properties of materials, tolerances on machine parts, and surface finish requirements, all of which are easily recognisable in engineering terms. Specifications may also refer to more abstract concepts such as the *aesthetics* of design. Further, specifications can also apply to the internal supplier/customer relationship. For example, a designer may request the company's legal department to draw up a specification for a product, ensuring that it does not infringe any patent or copyright and does not infringe any current legislation, such as that concerned with product liability.

The translation of customer needs (the *demand specification*) into a *design specification* and then into a feasible *manufacturing specification* is very time consuming, because each demand need must be broken down into sub-needs that are then issued either as purchasing orders or manufacturing requirements. The design specification must contain information concerning the minimum functional, reliability and safety requirements together with the information necessary to achieve minimum cost factors. Thus, the concept of minimum functional requirements is closely related to the achievement of minimum cost factors. Any designer must realise that the demand for higher specifications, such as greater precision, leads to an increase in manufacturing costs. Figure 11.3 shows the influence of the design specifications on the value added to a product. Increased precision often results in increased production costs, leading to an initial rise in the selling price followed by decreasing sales and returns, as indicated by the added value curve. To demonstrate that the 'quality of design' meets customer demand, the design engineers can test materials and components, and build and test prototypes. Customer involvement at this stage is essential in order to secure type acceptance before quantity production commences.

Fig. 11.3 *The influence of added value on design specification (increased precision = an initial rise in the selling price followed by decreasing returns)*

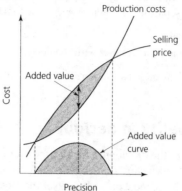

11.3.2 *Quality of conformance to design*

The ability to manufacture goods or deliver a service to agreed specifications at the point of acceptance is referred to as the *quality of conformance*. Many companies now employ conformance engineers whose function is to ensure that goods and services match the design specifications and hence the demand specifications.

Quality cannot be inspected into goods or services. The traditional concept of quality control consisted of inspecting samples of components to ensure that they had been made 'according to the drawings' and rejecting any components that were out of tolerance. Unfortunately the concept that quality is solely the province of the inspection department still persists in many companies.

If the philosophy of customer satisfaction and customer demand-led design has been adopted at the commencement of the process, then this philosophy must be followed throughout the whole system from the basic idea to the finished product or service. This is achieved by regular conformance checks during production to ensure that the specifications are being achieved. *Thus, conformance testing is about checking the quality of the whole process and NOT just about inspection of the finished product.* This avoids the waste of making rejects and the costs of reclamation and rectification. A high level of end-inspection indicates that a company is trying to 'inspect-in' quality. The retrieval and analysis of production data plays a very significant part in achieving specified levels of quality, so that when the product or service reaches the customer and *fully satisfies* all the specified requirements, then *quality of conformance* will have been achieved. Figure 11.4 shows the relationships between inputs and outcomes of quality conformance.

Fig. 11.4 *Quality: inputs and outcomes*

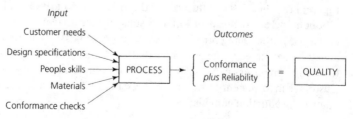

11.4 Accreditation of conformance

In modern demanding and competitive markets, good product design and efficient manufacturing must be underpinned by properly authenticated measurement and testing. For a company to have absolute confidence when issuing a certificate of testing, its own inspection instruments must, themselves, have been certificated by an independent accreditation agency. One such agency is the *National Measurement Accreditation Service* (NAMAS).

NAMAS is a service offered by the National Physical Laboratory (NPL), one of the Research Establishments of the Department of Trade and Industry. NAMAS assesses, accredits and monitors calibration and testing laboratories. Subject to its stringent requirements, these laboratories are then authorised to issue formal certificates and reports for specific types of measurements and tests. NAMAS accreditation is voluntary and open to any UK laboratory performing objective calibrations or tests. This includes independent commercial calibration laboratories and test houses, and also laboratories which form part of larger organisations such as a manufacturing company, educational establishment or a government department. It also includes laboratories who provide a service solely for a parent body.

Every industrialised country requires a sound metrological infrastructure (metrology is the science of fine measurement) so that government, manufacturers, commerce, health and safety and other sectors can have access to a wide range of measurement, calibration and testing services in which they can have complete confidence. Figure 11.5 shows the structure of the National Measurement System that is in place in the UK. It shows that the National Physical Laboratory (NPL) is at the focus of the system and maintains the national primary standards both for the SI base units (mass, length and time) and the SI-derived units, such as force and electrical potential.

Fig. 11.5 *National Measurement System*

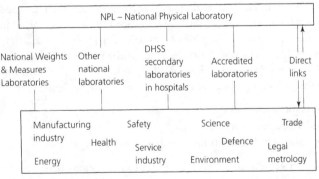

Commerce, industry and other users do not have direct access to the primary standards, but they do have access to secondary standards that have been calibrated against the primary standards. NAMAS provides an essential part of this hierarchical structure through accredited laboratories and test houses which give their customers access to authenticated measurements, calibrations and tests of all kinds. These have formal and certificated traceability to the national primary standards at the NPL. Confidence in the accuracy of measurement and test data is a vital part of ensuring product quality.

Manufacturing industry is constantly being encouraged to improve in every aspect of its quality procedures, and NAMAS plays its part by emphasising the importance of trustworthy measurements and tests.

On an international scale, laboratory, test centres and inspection departments accredited by a body which is trustworthy, and whose competence is beyond question, makes the products and services of a country more readily acceptable overseas. This is particularly important with the advent of the Single European Market. For this reason, NAMAS cooperates fully with other international agencies concerned with laboratory accreditation. Mutual agreements recognising each other's competence are negotiated where appropriate.

SELF-ASSESSMENT TASK 11.1

1. Describe the role of the independent calibration service in the delivery of quality and explain how it could be of benefit to your own company.

2. List the steps back to the NPL reference standard that have been taken for the certification and calibration of the slip gauges kept in the inspection department of your company.

11.5 Total quality management

Quality is not only about 'good design' and 'conformance to design specifications'. Quality should permeate every aspect of a company's business, always focusing upon the needs and hence the care of the customer. The traditional approach to quality has been that 'production makes it' and 'quality control inspects it'. This approach leads to financial loss every time defective products are scrapped or time is involved in reworking and reclaiming defective products. This also leads to late delivery and dissatisfied customers who have to be pacified, and now have the same problems with their customers. It has been estimated that as much as two-thirds of all the effort utilised in business is wasted by this domino effect. Thus, the objectives of any company can be defined as:

- Optimum quality.
- Minimum cost.
- Shortest delivery time accurately adhered to.

To achieve and, more importantly, maintain these objectives will require something better than the traditional approach to quality. It will require more than an attempt to 'inspect-in' quality. It requires the efforts of every resource: the total commitment of every member of the company to satisfying the customers' requirements at all times. The slogan: 'Quality is everyone's business' just remains a slogan if it is not backed by positive action throughout every aspect of a company's activities. It can soon change to 'Quality is nobody's business' without managerial determination from the board of directors downwards. From this approach of total involvement has developed the concept of *total quality management* (TQM). Total quality management can be defined as:

An effective system of coordinating the development of quality, its maintenance and continuous improvement, with the overall objective of the achievement of customer satisfaction, at the most economical cost to the producer and, hence, at the most competitive price.

In theory, since quality affects everyone in the factory, it can be truly stated that 'Quality is everyone's business'.

11.6 Background to total quality management

Shortly after the Second World War, Dr W. E. Deming of the USA introduced the philosophy of total quality management (TQM) to the Japanese. He found a receptive audience for ideas that were based upon the principle of continuous improvement of quality, i.e. quality as a dynamic and not a static process. Western industrialists – particularly in the car industry – began to become aware of the high levels of quality being achieved in Japan and realised that they had to set their own houses in order to survive.

The Deming philosophy of continuous improvement was set out in a 14-point plan. The first point states:

Create constancy of purpose towards the improvement of products and service, with the aims of becoming competitive, staying in business and providing jobs.

This first point is addressed to management, urging them to create an environment in which constant review and improvement can flourish. Quality is not a definitive programme with a finite end, but a continuous process. Philip B. Crosby, also of the USA, finishes his 14-point plan for total quality management with the advice: 'Do it over again.' Thus the clear message is that quality is as much about organising as it is about producing. Total quality management must involve the whole of the organisation, starting at the top with the senior directors and the chief executive.

Total quality management includes not only the entire personnel within an organisation but also the suppliers, because at the heart of TQM lies the internal and external customer/supplier relationships introduced originally in Fig. 11.2. By the introduction of TQM, each person becomes accountable for his or her own contribution to quality. Everyone is motivated to attain the highest quality by being alert to that which is taking place. This cannot be achieved by coercion; TQM has to be 'user-driven' and the challenge to top management is to obtain a change in the culture of the company towards quality without issuing edicts. They must create an environment of job satisfaction where most people want to perform well and be able to take a pride in their own achievements and in those of the company, in addition to feeling a sense of involvement.

Ideas for improvement must come from people with the know-how and experience at all levels of the organisation. Who knows better how to make a product than the person who actually makes it? The objective is to move the pride of the craft possessed by a single operator into the arena where products or services are the result of the contributions of many people of differing skills and abilities. Thus TQM is about changing attitudes and utilising everyone's knowledge and experience in a never-ending quest for the three objectives of:

- Highest quality.
- Minimum cost.
- Shortest delivery time.

The culture of the company becomes that of identifying failures as early as possible and striving to get things *right first time*. Thus TQM is much more sensible than the traditional approach of trying to 'inspect-in' quality by coercion and the efforts of a small band of inspectors; an approach where everyone else's opt-out was 'it's not my job to detect faults'.

We can summarise the advantages of TQM as follows. It enables a company to:

- Identify clearly the needs of their market and hence satisfy the needs of their customers.
- Devise and operate procedures for achieving a quality performance resulting from a clearly defined quality policy.
- Develop good lines of communication throughout the whole of its operation, both internally and externally.
- Critically examine every process which contributes to the final product or service, in order to remove all those activities which are wasteful.
- Measure its own performance and be in a position to continually update and improve its performance.
- Be competitive by understanding all the competition both at home and abroad.
- Provide high-quality goods and services which represent maximum 'value for money' by providing increasing job satisfaction for the workforce through a commitment to the philosophy of teamwork which, in turn, harnesses all the talents of all the people in the organisation.
- Understand *itself* as never before.

11.7 Quality assurance: BS EN ISO 9000

Although, under TQM, everyone from the boardroom down is responsible for the achievement of quality standards, the 'buck stops' at the desk of the company officer who signs the quality assurance certificate. Quality assurance (QA) is defined in BS 4778: 1987 as: '*All activities and functions concerned with the attainment of quality.*' In practice, it is some formal system in which all the operations are set out in written form as procedures and work instructions. The documents are subject to regular review and update and are usually under the authority of one person, the *quality manager*.

Quality assurance is not a new concept. For many years in certain areas of industry customers have demanded certification of quality from suppliers. These certificates of quality were used to verify that the products or services met the specifications demanded by the customer. Some examples are:

- Material specifications in which the results of various prescribed tests are recorded on the acceptance certificate that bears an authorised signature.
- Lifting equipment, including slings and chains.
- Pressure vessels such as boilers, compressed-air receivers and compressed-gas bottles.

The 'new' element of QA is that products and services not usually associated with certification have been brought into line by having written evidence that the company has achieved Nationally Accredited Standards of quality. These standards are such that they are also accepted internationally by keeping the standards of accreditation common amongst all the industrial communities.

In the UK the first standard for quality assurance was published as BS 5750: 1979. However, the interlinking of world trade both within and outside the European Community led to other countries copying the British approach, but with their own particular amendments. This international interest in quality standards led to the introduction of ISO 9000: 1987. This combines international expertise with eight years of British experience in operating quality assurance. It has been adopted without alteration by the British Standards Institution, hence the dual numbering used. The new ISO standard also incorporates the European Standard EN 29000.

11.8 The basis for BS EN ISO 9000

The definition of quality upon which the standard is based is in the sense of *'fitness for purpose'* and *'safe in use'*, and that the product or service has been designed to meet the customers' needs. The constituent parts of this standard set out the requirements of a *quality-based* system. No special requirements are set out that might only be needed or even be achievable by a few companies, for BS EN ISO 9000 is essentially a practical basis for a quality system or systems that can be utilised by any company offering products or services both within the UK or abroad. The principles of the standard are intended to be applicable to any company of any size. The basic disciplines are identified with the procedures and criteria specified in order that the product or service meets with the requirements of the customers. The benefits of obtaining BS EN ISO 9001/2 approval are very real, and include:

- Cost-effectiveness, because a company's procedures become more soundly based and criteria are more clearly specified.
- Reduction of waste and the necessity for reworking to meet the design specifications.
- Customer satisfaction because quality has been built in and monitored at every stage before delivery.
- A complete record of production at every stage is available to assist in product or process improvement.

By the adoption of BS EN ISO 9001/2, a company can demonstrate its level of commitment to quality and also its ability to supply goods or services to the defined quality needs of customers. For many companies it is merely the formalising and setting down of an existing and effective system in documented form so that the validity of the system can be guaranteed by obtaining *external accreditation*.

By having an agreed standard as the basis of supply there can be major benefits to all the parties concerned. A customer can specify detailed and precise requirements, knowing that the supplier's conformance can be accepted. This is because the quality system has been scrutinised by a third party, i.e. it has been accredited. However, successful implementation will be dependent upon the total commitment of the whole management team and, in particular, the person at the top of the organisation.

11.9 Using BS EN ISO 9001/2

Customers may specify that the quality of goods and services they are purchasing shall be under the control of a management system complying with BS EN ISO 9001/2. Together with independent third parties, customers may use the standard as an assessment of a supplier's quality management system and, therefore, the ability of a supplier to produce goods and services of a satisfactory quality. The standard is already used in this way by many major public-sector purchasing organisations and accredited third party certification bodies. Thus it is in the interests of suppliers to adopt BS EN ISO 9001/2 in setting up their own quality systems.

The direct benefits to companies who have been assessed in relation to the standard and who appear in the DTI *Register of Quality Assessed United Kingdom Companies* consists of:

- Reduced inspection costs.
- Improved quality.
- Better use of manpower and equipment.

Where a company is engaged in exporting goods or services, then there is a direct advantage in possessing mutually recognised certificates that could be required by overseas regulatory bodies.

11.10 The quality manager

Responsibility for the implementation of BS EN ISO 9001/2 is best placed in the hands of a professionally qualified quality manager. Such a position must carry all the authority and responsibility for the operation of the agreed quality system. Although such a person may initially have a line management position, it is better that the quality manager is seen to operate across departmental boundaries and report directly to the chief executive, since 'Quality is everyone's business'.

The function of the quality manager is to coordinate and monitor the quality system, ensuring that effective action is taken at the earliest moment so that the requirements of BS EN ISO 9001/2 are met. However, the first task of the quality manager will be to convince the other managers in the organisation that quality must become the vital part of all the company's operations. Any narrow view of quality is totally alien to an environment aspiring to total quality management (TQM). This means that the education and training of all personnel will be at the top of the agenda for any effective quality manager.

1. Prepare a block diagram of the quality–control system for your company or any other major organisation with which you are familiar.

2. Define what is meant by 'reliability'.

3. Define what is meant by 'quality conformance to design'.

4. Explain briefly how accreditation of conformance can be achieved.

5. Explain briefly what is meant by the term 'total quality management' (TQM).

11.11 Elements of a quality system

Having considered the essential principles and background of a quality-assurance system, let us now examine the essential elements to set up such a system.

11.11.1 *Setting up the system*

The quality system must be planned and developed to encompass every function within the organisation and, in particular, the functions of customer liaison, designing, manufacturing, purchasing, contracting and installing. The infrastructure of the installation will have to be fully understood in respect of responsibilities, procedures and processes in order that their functions can be presented in a documented form. Existing quality-control techniques will need to be examined and updated in terms of equipment and trained personnel to ensure that the capability exists for the fulfilment and recording of the quality plans.

11.11.2 *Control of design quality*

The objective of any design is to meet the customer's requirements in every respect. Therefore, it is vital that the input to the design function is established and documented. This input being the result of consultation by the marketing and design staff with the customer. There should be sufficient trained staff and resources available to ensure that the design output meets the customer's requirements, and that any changes or modifications have been controlled and documented.

11.11.3 *Traceability*

All the products needed in the fulfilment of the customer requirements should be clearly identified in order that they can be traced throughout the organisation. This is necessary to ensure that a capability exists for the tracing of any part that could be delivered to a customer. The need for this traceability might arise in the case of a dispute regarding non-conformity or for safety or even statutory reasons. Identification is also important where slight differences exist between the requirements of different customers.

11.11.4 *Control of 'bought-in' parts*

All purchased products or services should be subjected to verification of conformance to previously agreed specifications. Remember that organisations are usually both customers and suppliers. Control is exercised by documentation of the purchasing requirements, together with the inspection (and hence verification) of the purchased product. Many companies insist on details of the quality system used by subcontractors and suppliers. Hence the value of all companies having accreditation under BS EN ISO 9001/2. Even where a customer supplies a product to be processed, the customer must be assured that the receiving company is, itself, 'fit for purpose' to carry out the process. Whilst the product is in the possession of the processing company, they are responsible for its well-being, i.e. that it remains free from defects.

11.11.5 *Control of manufacturing*

Clear work instructions are at the heart of any manufacturing process. They eliminate any confusion by showing, in a simple manner, the work to be done or the service to be provided and also indicate where the responsibility and authority lie. If the customer's requirements, via the design function, are clearly specified on the work documentation, together with quality criteria, then the task of the manufacturing function becomes more easily controlled. BS EN ISO 9001/2 spells out the items that work instructions should include, and emphasises the control of any additional processes, such as heat treatment, thus avoiding expensive errors.

11.11.6 *Control of quality of manufacture*

Prompt and effective action is essential to the maintenance of quality throughout the system. Not only must defective parts be identified but the causes must quickly be rooted out. This could lead to changes in the design specifications or the working methods. The control of the manufacturing process starts with the inspection of the incoming goods for conformance to agreed specifications. Documentary evidence of such conformance should be provided either by the external supplier or by the company's own in-coming inspection. The accreditation of every company's measuring and testing equipment by an external body such as NAMAS provides consistency and avoids disputes between supplier and user.

An essential part of any quality-control system is the means of indicating the status of goods with regard to their inspection; that is, 'Inspected and approved', 'Inspected and rejected' or 'Not inspected'. Any part that does not conform to the specifications must be clearly identified to prevent unauthorised use or despatch. Documentation must be raised to indicate the nature of the non-conformation and of either its disposal or any remedial work performed upon it.

11.11.7 *Summary*

BS EN ISO 9001/2 is about documentation procedures. It is not a panacea of quality but it is a mechanism by which a company can formalise its operation in a standardised way. The written procedure must be unambiguous and clearly state what is to happen, NOT what the company thinks ought to be happening! The system chosen should be compatible with the business and should be capable of being maintained. Communication throughout the organisation is essential. The workforce must know what is happening and how it will affect them. This is particularly important during the assessment period, when the accreditation is being applied for, as the assessors may talk to anyone in the company. It is important, subsequently, to maintain and improve upon the standards achieved during the accreditation assessment. Successful companies who have achieved accreditation under BS EN ISO 9001/2 can display one or more of the marks of quality shown in Fig. 11.6 on their products and advertising media.

Fig. 11.6 *BSI marks of quality (reproduced courtesy of BSI)*

THE REGISTERED FIRM SYMBOL
indicates that the company operates a quality system in line with BD 5750 – the British Standard for quality systems – providing confidence that goods or services consistently meet the agreed specification.

THE KITEMARK
indicates that the product which displays it complies with the appropriate published specification.

THE SAFETY MARK
indicates that the product meets the required safety specification.

THE REGISTERED STOCKIST SYMBOL
indicates that the stockist operates a quality system which ensures the continuity of the quality chain through to the purchaser.
Small wonder, then, that our marks are recognised by purchasers the world over as top marks.

11.12 Prevention-oriented quality assurance

The attainment of quality in general terms means some form of inspection at some stage in the production process. The crucial decisions are:

- WHAT type of inspection procedures to use.
- WHERE in the process the inspection will take place.
- HOW to detect process variations before they become defects.

The need for inspection is rooted in the fact that any process is subject at any given moment to disturbances that lead to variations during a production run. No two parts are produced exactly the same, there are always some differences resulting in deviations occurring either side of the objective. *Variability* can be defined as the 'amount by which the process deviates from a predetermined norm' and, therefore, some variation must be tolerated.

This inherent *production variability* is a characteristic of the process of production and can be the result of any of the following.

Random causes

Certain aspects of production variability occur by chance and it is difficult to ascribe them to any one factor. Examples of random causes are:

- Variations in material properties.
- Variations in process machine conditions such as wear in bearings and slides.
- Temperature fluctuations.
- Differences in the condition and supply of coolant.
- Variations in operator performance.

Assignable causes

These are causes of variation that can be identified. Examples of assignable causes are:

- Tool wear.
- Obvious material differences (e.g. different sources of supply).
- Changes in equipment.

Assignable causes usually result in large process variability and must be kept under control. Quality can be classified into two characteristics that will determine the techniques to be used in their inspection.

Variables

These are characteristics where a specific value can be measured and recorded and which can vary within a prescribed tolerance range between prescribed limits and still remain acceptable. For example:

- Length, height and diameter inspected by measurement.
- Mass.
- Electrical potential (voltage).
- Pressure and temperature.

Attributes

These are characteristics that can only be acceptable or unacceptable. For example:

- Colour (if you want and have ordered a blue car, a red one will not be acceptable).
- Size (a toleranced dimension inspected by gauging will be accepted or rejected on a 'go' or 'not go' basis).

SELF-ASSESSMENT TASK 11.3

1. Briefly discuss the elements of a quality system as related to your own company or to an organisation with which you are familiar.

2. For a machined component, state the causes that adversely affect the quality of that component.

11.13 Types of inspection

There are basically two types of inspection that can be carried out at any stage of a process. These are:

- *Total (100%) inspection*   With 100 per cent inspection every part will be examined. It may be thought that this will guarantee that all defectives will be identified. However, this is not necessarily the case where human beings are the inspectors and arbiters of conformance. This type of monotonous, repetitive inspection procedure leads to loss of concentration and some defectives being missed, which can be as high as 15 per cent at the end of a shift. Some products themselves preclude 100 per cent inspection; for example, the ultimate test of a ballistic missile lies in the actual firing! Therefore, some other means must be found.
- *Sampling*   With sampling, a decision on quality is based upon the close inspection of a randomly selected batch of materials or products. The decision of whether the remaining materials or products are acceptable or not is based upon the outcome of mathematically based statistical procedure. The process involves the collection of data and making decisions about conformance as a result of plotting the information on various control charts.

Whichever system of inspection is adopted, it should be clear that the use of 'end-inspection' is wasteful and uneconomical. In many cases it is far too late, resulting in delayed deliveries and either total rejection of a batch of work or, at best, considerable reworking to reclaim the batch at considerable extra cost. Inspection should be used as a check on a process whilst it is in progress and adding value to a job. It is most effective when it is in the hands of the production operatives with the quality-control section only acting in a validating role. This is the principle that *prevention is better than cure*.

11.14 Position of inspection relative to production

Consideration must be given to the position of any in-process inspection points to ensure that costly time is not wasted on processing parts that are already defective. Such inspection points could be:

- Prior to a costly operation (e.g. there is no point in gold-plating a defective watch case).
- Prior to a component entering into a series of operations where it would be difficult to inspect between stages (e.g. within a FMS cell).
- Prior to a station which could be subjected to costly damage or enforced shut-down by the failure of defective parts being fed into it (e.g. automatic bottling plants, where the breakage of a bottle could result in machine damage or failure in addition to the cost of cleaning the workstation).
- Prior to an operation in which defects would effectively be masked (e.g. painting over a defective weld).
- At a point of no return where, following the operation, rectification becomes impossible (e.g. final assembly of a 'sealed for life' device).

11.15 Areas of quality control

In all manufacturing organisations there are three main areas of quality control. Let us now consider them individually.

11.15.1 *Input control*

It would be foolish and wasteful to utilise resources on the processing of materials or parts that are already defective when received from the supplier. The function of in-coming inspection is to detect any faults that could affect the finished product. An example of this could be the machining of rotor and turbine shafts for the Electricity Generating Industry. The very large forgings for the shafts are inspected internally and externally for cracks before being subjected to extensive and costly machining processes.

One way of effecting input control is to rely upon the supplier's out-going inspection. This could mean the production of documentary evidence in the form of certificates, hence the value of **BS EN ISO 9001/2** in standardising quality documentation. Alternatively, the customer has two other courses of action: (a) to provide supervised inspection at the supplier's plant prior to despatch; (b) to carry out 'in-company inspection' at the customer's plant on receipt of the goods or materials, but before accepting a delivery. The latter option could be 100 per cent inspection or sampling to arrive at a decision regarding acceptance or rejection (see Section 11.19).

11.15.2 *Output control*

This is the last link in the quality-control chain, and is necessary, not because the previous levels of inspection have failed, but because of the fallibility of any system of inspection and the resulting consequences of the customer receiving products or services which do not conform to the agreed specification. Again the choice is between 100 per cent inspection, or sampling.

11.15.3 *Process control*

This is the 'let's see how it is doing' approach and, if it is successful, it reduces the need for output control to a minimum. Considering that any process has an inherent amount of variability, it is important that the amount of this variation is known. Thus a capability study has to be performed and the process has to be kept within acceptable limits. The problem for *process quality control* is, therefore, how to detect unacceptable process variation before it leads to non-conformance with the agreed specification.

11.16 Quality-control techniques

Quality assurance (QA) can be defined as:

> *The provision of documentary evidence that an agreed specification has been met and the contractural obligations have been fulfilled.*

It is a powerful definition indicating that all the legal implications associated with the law of contracts have been covered. Quality control (QC) can be defined as:

The application of a system for programming and coordinating the efforts of the various groups in an organisation to maintain or improve quality at an economical level that still allows for customer satisfaction.

The application of these techniques and procedures is necessary to ensure that any goods or services actually meet the design specifications and hence the customer's contractual requirements.

Thus, QA is the objective to be attained, and QC is the means of achieving that objective. Before any type of control can be exerted in any sphere, ground rules have to be established. In the case of QC, in common with any type of control, those ground rules consist of:

- The setting of standards against which performance can be compared.
- The mechanics for assessing performance against the prescribed standards.
- The means of taking any necessary corrective action where an error exists between actual performance and the prescribed standards, as shown in Fig. 11.7 (the QC loop).

Fig. 11.7 *The quality control (QC) loop*

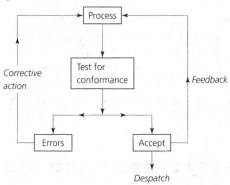

11.17 Acceptance sampling

There is a need for some sort of verification that goods being delivered for use in production, and finished products leaving production, have conformed to the agreed specifications. Where 100 per cent inspection has to be ruled out on the basis of cost and/or impracticality, the decision to accept or reject a batch of work is based upon a sample taken at random from the batch. In this context, a sample can be defined as:

A group of items, taken from a larger collection, that serves to provide the information needed for assessing the characteristics of the population, or as a basis for action upon the population or the process which produced it.

In this context, the term 'population' refers to the batch of components from which the sample is taken. Further, a random sample can be defined as:

One in which each item or collection of items has an equal chance of being included in the sample.

The constant factor in the sampling of goods is that a clear definition of non-conformance must be available. In addition there has to be prior decision on both the sample size and the number of defects found in the sample which would constitute acceptance or rejection of a batch. There are risks involved in basing a decision on a sample. There is the risk, known as *consumers' risk*, of accepting a batch that, overall, contains an unacceptable level of defectives. There is also the risk, known as *producers' risk*, of rejecting a batch that, overall, is acceptable.

The risk involved is a calculated one, based upon the laws of mathematics contained in a statistical analysis. Where risks are concerned, the *probability* of an event occurring has to be taken into account. This probability is measured on a scale between 0 (no possibility) and 1 (a certainty) and the symbols used are:

p = the probability of an event occurring
q = the probability of an event NOT occurring.

therefore $p + q = 1$
where p = (possible outcomes) / (total outcomes)

It is often convenient to convert probabilities into percentages. For example, consider the probability of drawing a king from a full pack of cards. There are 52 cards in the pack and only 4 kings, so the probability is $4/52 = 0.077$ or 7.7%. It must be remembered that if a decision is taken to accept a batch on the statistical evidence deduced from a sample, there is *no absolute* guarantee of the actual quality of the whole batch! Sampling only provides a basis for making a decision that takes into account all the foreseeable probabilities involved.

11.18 Operating characteristics of a sampling plan

To operate a sampling plan, a sample of n items is selected at random from a total batch of N items. The sample is then inspected against the conformance criteria. Usually the inspection is by attributes (i.e. accept or reject), because inspection by variables involves measurement which is usually too slow and costly. Variables such as dimensions can be converted to attributes by assigning them limits of size so that the associated component feature can be accepted or rejected by a simple GO and NOT GO gauge. Dimensions within limits are acceptable, but dimensions that have drifted outside the limits are rejected.

The simplest type of acceptance sampling scheme is one in which a single sample (size n) is taken from a batch (size N) and the decision to accept or reject the whole batch is dependent upon the number of defectives found in the sample. If the number of defectives found in the sample is equal to or less than an agreed acceptance number c, then the whole batch is accepted. But if the number of defectives exceeds c, then the whole batch is rejected.

The effectiveness of any sampling plan in detecting either acceptable or unacceptable batches can be shown graphically by plotting its *operational characteristic* (OC) curve. It can be seen from Fig. 11.8 that the vertical axis represents the percentage of batches accepted and the horizontal axis represents the percentage of defectives in the batch. The ideal sampling plan is one in which any batch having a non-conformance equal to or less than a predetermined percentage is accepted on the basis of a single sample. Similarly, this sample will reject all batches with a non-conformance greater than the predetermined level. This level is known as the *acceptance quality level* (AQL). Figure 11.8 shows the operating characteristic curve for an ideal sampling plan; that is, in this example, 100 per cent of all the batches with 5 per cent or less defectives are accepted and there is total rejection of any batches with a greater percentage of defectives.

Fig. 11.8 *Ideal operating characteristic curve*

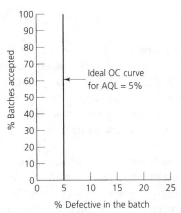

In practice, however, such ideal performance is never encountered since all processes have some inherent variability. A more typical OC curve is shown in Fig. 11.9.

- PAPD is the *process average percentage defective*. This usually coincides with the AQL but is, in fact, the average percentage of defectives being produced when the process is considered to be operating at a level of quality that is acceptable to the consumer.
- LTPD is the *lot tolerance percentage defective*. This is also known as the *consumer* or *customer risk*; that is, the percentage of defectives that the consumer would find unacceptable.

It is usual to locate the two important values of AQL and LTPD on the curve. The calculations for the probability are based on either a binomial distribution or, where the lot total is very large, a Poisson distribution. It is not necessary to perform these calculations because tables of the values are readily available in BS 6000/1/2. The OC curves are also plotted in these standards to assist in the choice of the most appropriate sampling plan for both attributes and variables.

If the sample size is increased, then the sampling plan becomes more sensitive, i.e. it discriminates more reliably between good and bad batches as the sampling approaches 100 per cent inspection. However, as previously stated, increasing the sample size increase the

Fig. 11.9 *Typical operating characteristic curve. (Note: Each individual sampling plan has its own unique operational characteristic curve)*

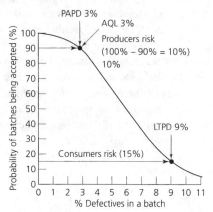

inspection costs proportionately. Similarly, if the acceptable number (c) in a sample is lowered to zero, so that batches are rejected if a single defective is found in a batch, then the sampling plan becomes less discriminatory. It may be thought that a plan with $c = 0$ would be a desirable method of sampling. However, whilst many bad batches would be rejected, many batches with economically viable, and normally acceptable, defective rates would also be rejected. Further, $c = 0$ does not guarantee that all batches that are accepted are free from defectives.

The size of the batch has little effect on the OC curve provided that the batch size is at least five times the sample size. However, the batch size can become important if it is small enough to be affected by withdrawing a sample from it. The smaller the batch, the nearer the sample approaches 100 per cent inspection so that the cost saving of sampling becomes too small to be viable. As previously stated, any sampling plan only gives a basis for making a decision to accept or reject a batch, it does not offer any guarantees: it is a way of weighing up the probabilities. When selecting a sampling plan, consideration must be given to the cost of inspecting an individual part, the cost of rejecting a batch containing a high level of acceptable parts, and the cost of accepting a batch containing too many defectives.

11.19 Classification of defects

In many manufactured goods, more than one type of defect may be present on which the decision for acceptance or rejection can be based. Some defects are more significant than others and, therefore, some classification is required.

- *Critical* This class of defect is based upon experience that a failure in service brought about by such a defect could result in situations where individuals would be at risk.
- *Major* This class of defect could result in failure that could seriously impair the intended function of the product.
- *Minor* This class of defect is not likely to reduce the effectiveness of the product, although non-conformance with the specification is present.

Critical defects may require 100 per cent inspection, whilst the sampling plan for major defects would require a small AQL, with a larger AQL being used for minor defects.

Some products have so many quality characteristics that to attempt to grade them by normal means would be difficult. For example, some complex assemblies such as furniture, vehicles, caravans and houses could contain several defects which could have differing effects upon the acceptance of the product. In these cases decisions are based upon a system of points and grading. The more important the defect, the higher the score and the poorer the product quality. For example:

Critical defects: 20 points
Major defects: 8 points
Minor defects: 2 points

The record of such 'quality scores' indicates trends, so that a constant score shows consistent quality at the level of the score – that is, consistently good or consistently bad. Scores drifting higher shows that the quality is deteriorating. The ceiling for scores is usually set at the score of the critical defects.

11.20 Vendor assessment and rating

One of the spin-offs from acceptance sampling is the fact that when applied to in-coming deliveries, a check can be kept upon a company's suppliers (the *vendors*). Their effectiveness can have a marked influence on the quality of the user's products, and it is in the customer's interests to know the capability of all suppliers. This is achieved by investigating all possible vendors and assessing them in terms of:

- Price.
- Reliability of deliveries.
- Lead time.
- Dependability.
- Strike history.
- Quality-control procedures.

Once a vendor has been selected, a performance assessment is made by rating each subsequent delivery. A typical method would be to have 100 points divided between quality, price and delivery, as shown in Example 11.1.

EXAMPLE 11.1

Maximum points for quality = 40
Maximum points for price   = 35
Maximum points for delivery = 25
Total   = 100 points

Rating of a supplier (vendor):

- 98% of batches are accepted on sampling inspection
- Price £100 per batch; lowest market price £90
- 75% of deliveries are on time.

Quality rating $= (40 \times 98)/100 = 39$
Price rating $= (35 \times 90)/100 \quad = 32$
Delivery $= (25 \times 75)/100 \quad = 19$
Total $\qquad\qquad\qquad$ **= 90 points**

All suppliers are rated in a similar fashion and a rating 'league table' is drawn up for use by the purchasing department.

SELF-ASSESSMENT TASK 11.4

1. Briefly discuss the advantages and limitations of total inspection and inspection by sampling.

2. Discuss the areas of quality control as related to a product of your own choice.

3. Explain what is meant by acceptance sampling and discuss the characteristics of a sampling and its impact on manufacturing costs.

11.21 Statistical process control

Despite the advanced technology used in many current production processes, customer quality requirements have also advanced. Therefore, although process variation is of a lower order, it is still a problem and the output from a manufacturing process still needs to be monitored.

Achieving the necessary quality must lie within the capability of the production department as they have the responsibility for output. Yet there are financial constraints that make measuring and testing for quality an overhead cost since inspection does not add to the value of the product – it does not 'transform' anything. It must be remembered that quality is never free, and any company using the TQM route must provide adequate resources to deliver the quality demanded.

The problem with any production process is the detection of unacceptable process variations before they become defects. Statistical process control (SPC) helps to overcome this problem by measuring samples of items during various stages of the manufacturing process. It is then possible, using statistical means, to make reasoned judgements regarding the level of probability that all the items at that stage of production are being manufactured within the required specification. Further, the drift in variation can be seen at any time and this enables adjustments to be made to the process before the variation results in the production of defective products.

Traditionally SPC was – and still is to some extent – carried out manually with inspectors taking measurements from random samples and plotting the results on statistical charts known as *control charts*. From these charts, trends can be detected and the timing of any remedial adjustments can be determined. However, the introduction of computerised SPC has speeded up data collection and processing, thereby reducing the time lag associated with manual SPC during which the process could go out of statistical control.

SPC uses the results from individual samples to focus attention on the whole process, indicating overall performance. It also provides the information needed to make changes and adjustments to the processing system before defective parts are produced as a result of tool wear or other process variables. Figure 11.10 shows a summary of the flow of information when SPC is being used.

Fig. 11.10 *Statistical process control (SPC): (a) backward control; (b) forward control; (c) accept–reject control*

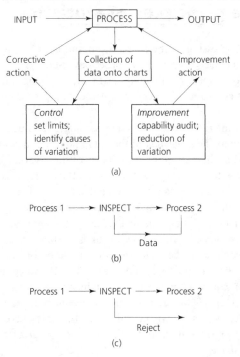

11.21.1 *Backward control*

This is the system shown in Fig. 11.10(a) and its name originates from the fact that the components are inspected following an operation and the results are *fed back* to adjust the process so that it provides maximum quality.

11.21.2 *Forward control*

In this system the components are again inspected following an operation but the results are *fed forward* to the next operation where any defectives can be reworked to produce conformance. This is shown diagrammatically in Fig. 11.10(b).

11.21.3 *Accept–reject control*

In this system the components are subjected to 100 per cent inspection after a given operation. Those components which conform are moved to the next production process, whilst those components which do not conform are rejected. This is shown diagrammatically in Fig. 11.10(c). Because 100 per cent inspection is involved, this system is only economical where fully automated inspection is possible with no bottlenecks to hold up the flow of production.

11.22 Some principles of statistics

Before progressing to the preparation and use of control charts, let's review some of the basic principles of statistical processes. Statistics is the collection, classification and analysis of numerical information. In terms of manufacturing production, the application of statistics provides a very useful tool for predicting the outcomes from a particular process.

The numerical data obtained by sampling spreads over a range of values. This spread is referred to as the *dispersion* of the data and, for control purposes, production engineers are interested in the pattern or *distribution* of these variations about a mean value. The larger the sample the more nearly it represents the characteristics of the total population. In this context, *population* means the total batch from which the sample is taken.

The data obtained from the sample can be shown graphically as a histogram, as shown in Fig. 11.11(a) – the numerical values of the items are plotted on the horizontal axis and the *frequency* with which they occur is plotted on the vertical axis. The width of each vertical column of the histogram (as measured on the horizontal axis) is called the class *interval*. The mid-points of the tops of the vertical columns (arithmetic means of the class intervals) can be connected by straight lines to form a frequency polygon, as shown in Fig. 11.11(a). However, it is more convenient to convert the frequency polygon into a smooth curve, as shown in Fig. 11.11(b). This curve is the basis of all the predictions derived from the sample and is the *curve of normal distribution* (also called a *Gaussian curve* or a *bell-shaped curve*).

Fig. 11.11 *Introduction to distribution curves: (a) histogram and frequency polygon; (b) normal distribution curve*

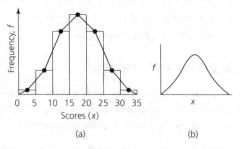

It can be seen from the distribution curve that most of the sample data readings lie on or near to the centre of the variables plotted on the horizontal axis, with progressively fewer points further away from the central point. A single value, the *arithmetic mean* ($\bar{x}$), can be calculated to represent the whole of the distribution:

$$\bar{x} = \frac{\text{sum of all readings } (x)}{\text{number of readings } (n)} = \frac{\Sigma x}{n}$$

For example, if a machine is set up to cut off lengths of tubing, then the mean value would coincide with the nominal length, and the actual lengths of the cut tubes would vary slightly about this size. The distribution of the variations in size about the mean value represents the process variability and, if plotted graphically, would follow a *normal curve of distribution* as discussed earlier. Control chart theory is based upon the assumption that the measurements follow a normal curve of distribution.

There are a number of measures of variability (dispersion) such as 'range', 'mean deviation' and 'standard deviation'. These all have their own particular advantages and limitations but the most satisfactory for control purposes is *standard deviation*, and this will now be considered in some detail.

Standard deviation is, in fact, the distance from the mid-point of a normal distribution curve to the point on the curve where the direction of curvature changes (stops curving downwards and starts curving outwards). There are several variations of the formula for calculating standard deviation depending upon how the data has been collected and collated. The symbol for standard deviation is σ (sigma), and:

$\sigma = \sqrt{[1/n\ \Sigma(x - \bar{x})^2]}$ if individual values of x are given

$\sigma = \sqrt{[1/n\ \Sigma f(x - \bar{x})^2]}$ if frequencies (f) are given for various stated exact values for the variable

$\sigma = \sqrt{[1/n\ \Sigma f(m - \bar{x})^2]}$ if frequencies are given for the values falling in various stated ranges or class intervals.

11.23 Properties of normal distribution

One of the most important properties of normal distribution is that the proportion of a normal population which lies between the mean ($\bar{x}$) and a specified number of standard deviations (σ) away from the mean is always the same irrespective of the shape of the curve. The total area under a curve of normal distribution = 1 (i.e. 0.5 either side of the mean), when the mean = 0 and the standard deviation (σ) = 1. The values of the areas under the curve between the mean and various multiples of the standard deviation are:

Between 0 and 1 standard deviation: area = 0.34135
Between 0 and 2 standard deviations: area = 0.47725
Between 0 and 3 standard deviations: area = 0.49865 (approximately 0.5)

Figure 11.12 shows two examples of normal distribution curves with different shapes. The areas are those bounded by the mean, 2 deviations, the horizontal axis and the curve. The relationship between the mean and the normal distribution curve is shown in Fig. 11.13, and it can be seen that:

Fig. 11.12   *Typical curves of normal distribution: (a) x = 100 and σ = 10; (b) x = 120 and σ = 5*

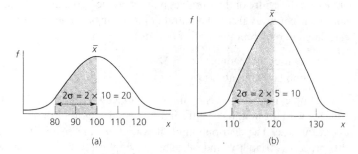

(a)   (b)

- 68 per cent of the total population is contained under the curve between +1 and −1 standard deviations about the mean value (see Fig. 11.13(a)). *Note*: For a symmetrical curve the mean, median and mode are all the same value.
- 95 per cent of the total population is contained under the curve between +2 and −2 standard deviations (see Fig. 11.13(b)).
- 99.73 per cent of the total population is contained between +3 and −3 standard deviations (see Fig. 11.13(c)). That is, for all practical purposes, +3 deviations will determine the upper limit of the distribution and −3 deviations will determine the lower level of the distribution.

Fig. 11.13   *Relationship between mean, standard deviation and normal distribution (see text for details of (a), (b) and (c))*

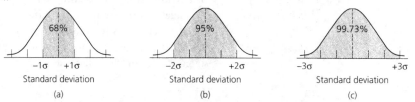

(a)   (b)   (c)

A process with a distribution showing a mean value which is equal to the nominal or target value is classed as being *accurate*, whilst a process with a distribution showing only a small spread, or scatter, is said to be *precise*. High quality is associated with high accuracy and high precision.

Whilst the usual objective of any process is to produce items which are the same, some variation has to be tolerated. However, providing that limits to the variation are set, then items within the limits will conform.

The relationship between the variability of a process and the design tolerances is known as the *process capability* (C_p). For a process to produce items that conform to design tolerances, the spread of the distribution must be equal to or less than the prescribed tolerances.

Where only random variability exists (when the process is under statistical control), the capability index can be calculated from:

$$C_p = \frac{\text{upper limit} - \text{lower limit}}{6\sigma}$$

Note: For values greater than 1, variability is contained within the limits; for values less than 1, variability is outside the limits.

11.24 Control charts for variables

Rather than plot a distribution curve for each sample as it is taken from the process, it is more expedient to plot the distribution of the mean values of the samples to obtain information about the process.

Consider repeated samples of size n drawn from a process, giving a normal distribution about a mean $\bar{x}$ and a standard deviation σ, then the calculated means $\bar{x}_1, \bar{x}_2, \bar{x}_3$, etc., can themselves be arranged into a frequency distribution known as the *sampling distribution of the mean*. This distribution will also be normal and have its own mean value and its own standard deviation. To avoid confusion, the standard deviation of the sample means is referred to as the *standard error of the mean*, where:

process mean $= \bar{x}$
standard error $= \sigma_n = \sigma/\sqrt{n}$

Figure 11.14 shows a comparison between the distribution of a single sample and the distribution of the sample mean. It can be seen that the spread of the latter is much less than for the single sample, i.e. the variation is less for the sample means.

Fig. 11.14 *Comparison between distribution of (a) single sample and (b) sample means. (Note: The spread is narrower for the plot of sample means)*

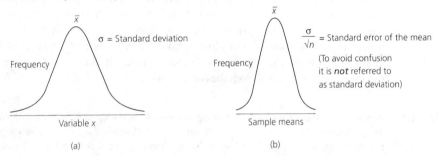

Control charts are graphs onto which some measure of the quality of a product is plotted as manufacture is proceeding. The measure of quality is taken from small samples and recorded on the control chart. Trends can be observed from the plots and corrective action taken before the process comes out of 'statistical control', i.e. before items are produced that do not conform to the agreed specification. Control charts are designed to give two signals:

- A *warning signal*, which alerts the production controllers that a drift is taking place and appropriate corrective action should be planned.
- An *action signal*, which indicates that the process is about to run out of statistical control and the preplanned corrective action must be implemented immediately.

Let's consider the control of production for a batch of 10 mm diameter steel rollers. Sampling shows that the diameters achieved during production are normally distributed about a mean of 10 mm with a standard error from the mean of 0.01 mm. Periodic samples are measured and the sampling means of their diameters are plotted onto a chart, as shown in Fig. 11.15.

Fig. 11.15 *Simple control chart*

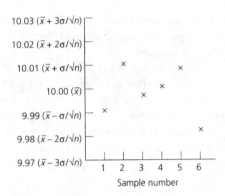

Based on what we have already discussed about normal distribution, we can expect that 68 per cent of all the readings will fall between $\bar{x} \leqslant \sigma/\sqrt{n}$, that 95 per cent can be expected to fall between $\bar{x} \leqslant 2\sigma/\sqrt{n}$, and that almost all the readings will fall between $\bar{x} \leqslant 3\sigma/\sqrt{n}$.

As it is necessary to control the measured diameter of the rollers during a long production run, the control chart provides an easily understood, pictorial representation of the process, and will show clearly any process variation. In Fig. 11.16(a) the last point to be plotted falls outside the $3\sigma/\sqrt{n}$ limit. This indicates a sharp deviation from the standard due to either a random variation or, more likely, a technical fault such as tool failure. Figure 11.16(b) shows a slower 'drift' from the specification. A general drifting of the plot above or below the mean indicates a gradual change in the process settings. The time for resetting can be estimated from the control chart.

Fig. 11.16 *Interpretation of control charts. (a) A sharp deviation from the specification: the plot falling outside the $3\sigma/\sqrt{n}$ limit would indicate a sharp deviation from the standard due to either a random variation or a technical fault. (b) A slower drift from the specification: a general drifting of the plot above or below the mean would indicate a gradual change in the process settings (the time for resetting can be estimated from the control chart)*

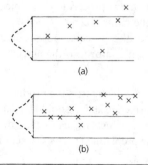

11.25 Warning and action limits

It is important that warnings are available to indicate when the plot, and therefore the process, is drifting out of control to enable corrective action to be planned and implemented. For this purpose, further use can be made of the probabilities offered by the relationship between the curve of normal distribution, the mean and the standard deviation. It can be seen from Fig. 11.17(a) that the probabilities are as follows: 2.5 per cent (1 in 40) of all readings will lie above $\bar{x} + 2\sigma$; 2.5 per cent will lie below $\bar{x} - 2\sigma$; 0.1 per cent (1 in 1000) of all readings will lie above $\bar{x} + 3.09\sigma$; and 0.1 per cent of all readings will lie below $\bar{x} - 3.09\sigma$. It is conventional to use these values as:

> upper warning limit (UWL) at $\bar{x} + 2$ standard errors
> upper action limit   (UAL) at $\bar{x} + 3$ standard errors
> lower warning limit (LWL) at $\bar{x} - 2$ standard errors
> lower action limit   (LAL) at $\bar{x} - 3$ standard errors

where one standard error $= \sigma/\sqrt{n}$. A typical chart layout is shown in Fig. 11.17(b), and the charts can be roughly interpreted as follows:

- If a point falls on or outside the *warning limits*, take a second sample immediately. If the second point also falls on or outside the limits, adjust the process immediately.
- If a point falls outside the *action limits*, assume that the process is out of control; shut it down and take immediate corrective action.
- If points fall within the warning limits, assume that the process is in control and take no action.
- If points fall within limits but are drifting either above or below the mean, and towards one or other of the warning limits, determine the type of action and the time such action should be taken.

Fig. 11.17 *Application of standard deviation to control chart limits: (a) probabilities of process variation; (b) limits applied to a control chart. (Note: Standard error is the standard deviation from the distribution of sample means; $\sigma_n = \sigma/\sqrt{n}$)*

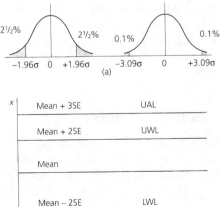

The same principles apply to control charts that indicate the spread of the distribution, i.e. *range charts* (also known as '*w*' charts). The difference between the highest reading and the lowest reading in the sample is taken and plotted as before on a control chart similar to that shown in Fig. 11.18, except that the mean is $\bar{w}$ instead of $\bar{x}$. The lower limits are less important on this chart, since a trend towards the lower limit indicates less spread and, therefore, improved quality.

For both types of control chart the position of the limits is not calculated from first principles. To save much unnecessary calculation, tabulated data is available for establishing the standard deviation and the standard error of the distribution of means (see BS 2564).

Fig. 11.18 *Range chart (w chart)*

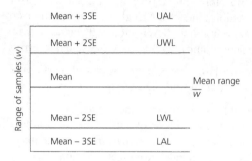

11.26 Control charts for attributes

When inspecting for attributes, the decision making is reduced to 'good' or 'bad', 'right' or 'wrong', and the decision is either to allow the process to continue or to stop it for adjustments. There are two types of control chart based upon defectives – i.e. any item which does not conform to the agreed specification. There are also two types based upon defects – i.e. any feature on an item that is unwanted, for example, blemishes or other surface marks which detract from the appearance of an item but do not render it defective. However, a predetermined number of such minor defects on an item could, in fact, render it defective over all.

11.26.1 *Control of defectives*

Number defective charts (*np* charts)
The sample size is a constant value and there is no requirement for a range chart. As with control charts for variables, the control limits are set at $\pm 2\sigma$ and $\pm 3\sigma$ from the mean. Then:

the standard deviation $(\sigma) = \sqrt{[np(1 - p)]}$

where p = (total defectives/total output) or 'fraction defective'
        n = sample size.

Let us consider the control limits that can be derived from the following process information:

$$\text{total items rejected as non-conformants} = 142$$
$$\text{Number of batches} = 12$$
$$\text{Constant batch size} = 400$$

$$\text{then} \quad p = 142/(12 \times 400) \qquad\qquad = 0.0296$$
$$n = \text{batch size} \qquad\qquad\qquad = 400$$
$$\text{and} \quad np = 400 \times 0.0296 \qquad\qquad = 11.84$$
$$\text{then} \quad \sigma = \sqrt{[400 \times 0.0296(1 - 0.0296)]} = 3.390$$
$$\text{warning control limits} = np \pm 2\sigma = 11.84 \pm 2 \times 3.39$$
$$= \mathbf{18.62} \text{ and } \mathbf{5.06}$$
$$\text{action control limits} \quad = np \pm 3\sigma = 11.84 \pm 3 \times 3.39$$
$$= \mathbf{22.01} \text{ and } \mathbf{1.67}$$

The control limits would be entered on the control chart, as shown in Fig. 11.19, and the number of non-conformants in each of the 12 batches would then be plotted on the chart to give a graphical representation of the production run.

Fig. 11.19 *An np control chart*

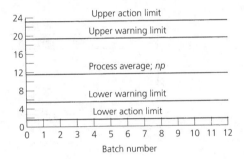

Batch number

Percentage defective charts (p charts)

The sample size for *p* charts can be variable and this allows samples to be taken on a time basis rather than on a batch size basis. However, even *p* charts cannot cater for *wide* variations in sample size. Such charts are only satisfactory where the sample size does not vary by more than 25 per cent either side of the average sample size. The procedure is similar to that for the previous (*np*) charts.

$$\text{process average } \bar{p} = \frac{\text{total number of non-conforming items}}{\text{total number inspected}}$$

$$\text{standard deviation } \sigma = \sqrt{\left[\frac{\bar{p}(100 - \bar{p})}{\bar{n}}\right]}$$

where $\bar{n}$ = average sample size.

Consider the control limits for the following derived process information:

total non-conformants in a day's production = 127
total production for the day = 5050
number of samples taken = 12

Control limits:

$$\bar{p} = \frac{127}{5050} = 0.025 \text{ or } 2.5\%$$

$$\bar{n} = \frac{5050}{12} = 421$$

$$\sigma = \sqrt{\left[\frac{2.5(100 - 2.5)}{421}\right]} = \sqrt{0.579} = 0.761$$

action control limits $= \bar{p} \pm 3\sigma = 2.5 \pm (3 \times 0.579) =$ **4.372%** and **0.217%**
warning control limits $= \bar{p} \pm 2\sigma = 2.5 \pm (2 \times 0.579) =$ **4.022%** and **0.978%**

These control limits can now be entered onto the chart, as shown in Fig. 11.20.

Fig. 11.20 *Control limits p chart*

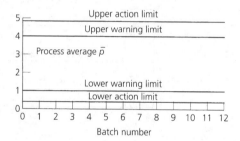

Sometimes this type of chart is referred to as a *proportion defective chart*, the difference being that the numbers are not expressed as percentages but are left as proportions. For example, $\bar{p}$ would be left as 0.025 instead of 2.5 per cent. The standard deviation would then read:

$$\sigma = \sqrt{\left[\frac{\bar{p}(1 - \bar{p})}{\bar{n}}\right]}$$

However, the percentage chart is more easily understood on the shop floor.

11.26.2 *Control of defects*

C-chart

This is used with a fixed sample to count the number of defects contained within that sample. Each defect in itself will not render the item defective but the cumulative effect may result in the item being deemed to be non-conforming. The total number of defects present in each sample is plotted on the control chart.

$$\text{process average} \quad \bar{c} = \frac{\text{total number of defects}}{\text{number of batches inspected}}$$

$$\text{standard deviation} \quad \sigma = \pm(\bar{c})$$

As previously:

action limits occur at $\bar{c} \pm 3\sigma$
warning limits occur at $\bar{c} \pm 2\sigma$

U-chart

This is used with a variable sample size and computes the number of defects per item (it is more practical to compute defects per 100 or even per 1000 items). The operation of these charts is similar to the C-charts – that is, samples are taken at various time intervals instead of batch sizes; the number of non-conformities per item (or per 100 or 1000 items) are computed and the number of non-conformities are plotted onto the chart, where:

$$\bar{u} = \frac{\text{total number of non-conformities}}{\text{total number of items inspected}}$$

$$\bar{n} = \frac{\text{total number of items inspected}}{\text{number of batches inspected}}$$

standard deviation $(\sigma) = \sqrt{(\bar{u}/\bar{n})}$
control limits are at $\pm 2\sigma$ and $\pm 3\sigma$ as previously

As with p-charts, the range of sample sizes has to be within ± 25 per cent.

Note

- It may seem strange with attribute charts that a lower limit is not only set but is actually used by inspectors. Strange because the nearer a reading is to zero defects, then the nearer the process is approaching perfection. However, due to the inherent variability of any process, perfection is greeted with suspicion in quality control and any practitioner will want to check the reading being plotted.
- Some practitioners do not use the warning limits on attribute charts but prefer to use the '7-rule' instead. This states that a change has taken place in the process average if seven consecutive readings are plotted either above or below the established process average or seven consecutive readings are seen to be either rising or falling with regard to the last point plotted.

11.27 Notes for charts for variables

Warning and action limits do not depend upon design limits; they are totally dependent upon the process variability. Therefore, designers must take account of the capability of a process when allocating toleranced work onto specific machines. However, there is a relationship that can be used to compare process variability with design limits. This is the *capability index* (C_p), introduced in Section 11.24. It is also known as the relative precision index (RPI) and has three classifications (see BS EN ISO 9000):

- Medium relative capability, where the variability just meets the design specification with a C_p between 1.0 and 1.3. The control chart average must be set at the centre of the limits and strict control exercised.
- High relative capability, where C_p is greater than 1.3 and the variability does not fill the design limits.
- Low relative capability, where C_p is less than 1.0. Variability is so great that a high level of non-conformity can be expected.

SELF-ASSESSMENT TASK 11.5

1. Explain how the properties of a normal distribution can be applied to statistical process control.

2. Differentiate between attributes and variables.

3. Differentiate between warning and action limits.

4. Discuss the control of defects.

11.28 Cost of quality

Quality is no different from any other function associated with production; there is a cost that must be taken into account and a high level of quality can absorb valuable resources. Like any other function, the costs must be analysed and budgeted. Where TQM is being practised it can be said that every function in the company is associated with quality but the actual costs can be considered in three broad categories.

11.28.1 *Failure costs (internal)*

This includes any cost within the process of production up to the point at which the customer takes delivery. For example:

- *Scrap* Items which must be discarded.
- *Re-work* The cost of any correction processes.
- *Re-inspection* The inspection of any work which has been corrected.
- *Down-grading* Where items can be sold as seconds at a lower price provided they do not detract from the market for first-class items;
- *Others* Must include any loss of production capacity caused by resetting of the process to regain conformance together with the cost of any investigation into non-conformance even to the point of a redesign.

11.28.2 *Failure costs (external)*

This includes any costs occurring after delivery to the customer. For example:

- Loss of reputation and possible future orders.
- Recalling of goods for rectification or replacement.
- Expenses for specialists to travel to the customer to rectify defective goods.
- Product liability and the possible cost of litigation.
- Maintenance of higher stock levels to cover the replacement of faulty goods.

Loss of reputation is the most damaging as this is soon known by all in the industry and it takes considerable time and effort to regain customer confidence.

11.28.3 *Prevention costs*

These include any costs brought about to prevent defective work being produced in the first place. This means the implementation of a 'right first time' philosophy so that no scrap or reworking is involved. This is also known as a policy of 'zero defectives' in the USA.

There must be financial advantages in the pursuance of such an ideal as 'right first time' and now, with the advent of 'just-in-time' production, the drive for minimum defects is imperative. JIT and low-quality production are not compatible, since the former is not possible without great attention being paid to the quality of production. Prevention costs can include:

- The management costs of setting up TQM.
- The provision of inspection equipment.
- Training programmes for all personnel.
- Vetting incoming supplies and vendor appraisals.
- Method study to ensure that the most cost-effective, foolproof methods are in operation.
- High-quality maintenance of process plant measuring and testing equipment.

The cost of quality, like any other organisational function, must be compatible with the selling price of the goods or services. In the quality control of any production system care must be taken to ensure that there is not more 'harness than horse'.

EXERCISES

11.1 Explain how the profitability of a company could benefit by operating a *supplier quality assurance* scheme.

11.2 Describe an example from your own experience of one defect in each of the classes:
   (a) critical
   (b) major
   (c) minor

11.3 A process machine is given a routine maintenance service during which the average setting is reduced and a decrease in variability is noted. Show on a control chart how these changes would be seen.

11.4 Explain the reasons for using standard specifications in the design process, stating any advantages derived from these specifications.

11.5 Define process capability analysis and relate its use to the product designer.

11.6 Explain how the output of a machine or process can be controlled either by 'first-off inspection' or by 'patrol inspection'.

11.7 For a company with which you are familiar, outline the steps being taken to ensure that the customer receives a 'high-quality product or service'.

11.8 In a company of your choice that is seeking accreditation under BS EN ISO 9001/2, describe how that company is implementing the requirements of accreditation.

11.9 Explain how a system of 'quality circles' could be set up in a place of work with which you are familiar. Describe the membership of the 'circle' that you would recommend and give your reasons for that particular combination of people.

11.10 Using control limits $\pm 2\sigma$ and 3σ from the mean, calculate the warning and action limits for the following process sampling criteria:
   (a)   total number of rejects $= 100$
   (b)   Number of batches $= 10$
   (c)   Constant batch size $= 500$

Part C
Management of health and safety at work

12 Health and safety at work

The topic areas covered in this chapter are:

- Health and Safety at Work, etc., Act.
- Workplace Regulations 1992.
- European Safety Directive.
- Safe working practices (PUWER).
- Fire Regulations.
- Electrical safety.
- The management of safety – BS 8800 (OS&H).
- Workplace risk assessment.
- Environmental management in manufacturing (BS EN ISO 14000).

12.1 The Health and Safety at Work, etc., Act 1974

The Health and Safety at Work, etc., Act was considered in some detail in *Manufacturing Technology*, Volume 1. The Act was put on the Statute Book on 31 July 1974, and it was fully implemented by April 1975. It is an *enabling* Act, which means that it is a broad and general piece of legislation that does not go into specific detail in itself. However, it gives powers to the Secretary of State for Employment (acting through the Health and Safety Commission) to draw up detailed *regulations* and *codes of practice* without further reference to Parliament.

The main purpose of the Act is to provide a comprehensive, integrated system of law dealing with the health, safety and welfare of work-people and the public, as affected by work activities. The Act has six main provisions:

- To completely overhaul and modernise the existing law dealing with safety, health and welfare at work.
- To put *general duties* on employers, ranging from providing and maintaining a safe place to work, to consulting on safety matters with their employees.
- To create a Health and Safety Commission (see Section 12.2).
- To reorganise and unify the various inspectorates (see Section 12.3).
- To provide new powers and penalties for the enforcement of safety laws (see Section 12.4).
- To establish new methods of accident prevention, and new ways of operating future safety regulations.

The enabling provisions of the Act are superimposed on the existing legislation contained in the 31 relevant Acts and some 500 subsidiary regulations. It did not remove, cancel or affect existing legislation such as the Factories Acts, the Offices, Shops and Railway Premises Acts, and others. These and other existing pieces of legislation will continue side by side with the Health and Safety at Work, etc., Act until they are replaced by new legislation.

This chapter is mainly concerned with later additions to the Act and subsequent legislation that has been brought in to harmonise the industrial and commercial health and safety requirements of the UK with those of the European Union (EU). It is also concerned with the management of risk and the management of the effects of manufacturing on the environment.

12.1.1 *Administration of the Health and Safety at Work, etc., Act*

- *The Health and Safety at Work Commission*   The Commission has taken over the responsibilities previously held by various government departments for the control of most occupational safety and health matters. The Commission is also responsible for the *Health and Safety Executive.*
- *The Health and Safety Executive* (HSE)   Since 1975, the unified inspectorate amalgamates the formerly independent government inspectorates such as the Factory Inspectorate, the Mines and Quarries Inspectorate, and similar bodies, into one cohesive executive. This is called the Health and Safety Executive (HSE) Inspectorate. The inspectors of the HSE have wider powers than when working under previous legislation and their prime duty is to implement the policies of the Health and Safety Commission.
- *Enforcement*   Should an inspector find a contravention of any of the provisions of previous Acts or regulations still in force, or a contravention of any provision of the 1974 Act, four lines of action are available:
  (a)   issuing an Improvement Notice
  (b)   issuing a Prohibition Notice
  (c)   prosecution: in addition to serving an Improvement Notice or a Prohibition Notice, the inspector can *prosecute* any person contravening a relevant statutory provision
  (d)   the inspector can *seize, render harmless,* or *destroy* any substance or article that he or she considers to be the cause of imminent danger to health or serious personal injury.

12.1.2 *Responsibility*

The Act now places the responsibility for safe working equally upon:

- The employer.
- The employee.
- The manufacturers and suppliers of goods and equipment.

Every employer must ensure that plant and equipment are maintained in a safe condition, and that adequate instruction, training and supervision in safe working practices are provided, but it is an equal responsibility of every employee to cooperate in making proper and full use of the facilities provided.

To ensure this cooperation, each company must have a *health and safety committee* representing both management and labour to examine working practices, actual and potential incidents, co-opt expert advisers when required, organise health and safety training, and enforce safe working practices. It must look at all aspects of a company's activities and ensure that it is a safe and healthy place in which to work.

Similarly, the Act places duties on persons who design, import or supply articles for use at work. Any plant, machinery, equipment or appliance must be so designed and constructed as to be safe and without risk to health. Plant must be regularly tested and examined to ensure its continuing safety. It must also be erected or installed by competent persons to ensure that it is functioning correctly.

Again, anyone who manufactures, imports or supplies goods and materials for use at work, has the responsibility for ensuring that such goods and materials do not represent a hazard. The same applies to contractors hiring out plant and firms leasing plant and equipment. Such plant and equipment must be supplied in a safe condition and not represent a health hazard.

The safety, health and welfare of visitors is also covered by the Act, as is the safety, health and welfare of persons living and working in the vicinity of any firm. This latter point overlaps somewhat with environmental issues that are the subject of separate legislation and inspectorates.

12.2 The Workplace Regulations 1992

The Workplace Regulations 1992 came into effect on 1 January 1996. They deal with basic workplace requirements which apply to all offices and workshops. These do not replace any regulations existing under the Health and Safety at Work, etc., Act but are supplemental to that Act. We shall now consider these basic requirements, but as this is only a summary of the regulations, The Workplace Regulations 1992 should be consulted.

- *Interpretation* The definition of a workplace embraces more than factories, shops and offices and includes all non-domestic premises used by any persons as a place of work.
- *Maintenance of workplace, and of equipment* There should be a planned, preventative maintenance system rather than proactive breakdown maintenance.
- *Ventilation* Ventilation must provide sufficient quantities of fresh pure air with no build-up of dust or fumes. Also, an effective warning device must be used to indicate the failure of any ventilation equipment.
- *Temperature* A temperature of 16 °C is considered normal but can be reduced to a minimum of 13 °C where severe physical effort is required. Also steps must be taken to prevent the temperature becoming too high.
- *Lighting* Emergency lighting must be available in case the normal artificial lighting system fails, particularly in cases where persons are exposed to danger if the lighting fails.

- *Workstations and seating* Workstations and seating must be ergonomically correct to prevent unnecessary physical strain.
- *Miscellaneous regulations* These include:
  - (a) the conditions of floor and traffic routes
  - (b) risks associated with falls and falling objects
  - (c) organisation of traffic routes, doors and gates
  - (d) accommodation for clothing and facilities for changing clothes
  - (e) facilities for rest and eating meals.

12.3 European Machinery Safety Directive

The European Machinery Safety Directive became law in the UK on 31 December 1992. It stipulates that machinery manufactured and/or sold within the EU, including imported machinery, must satisfy wide-ranging health and safety requirements including those relating to noise levels. A new approach was used in preparing the Machinery Safety Directive by coupling it with European Standards, as agreed by CEN and CENELEC. The British Standards Institution participated in this activity, together with the professional engineering institutions.

Barriers to trade within the European Union (EU) have resulted from the difference in safety legislation that existed within the various member states. Therefore, in May 1985, EU Ministers agreed on a new approach to harmonise the technical standards to overcome this problem. The Machinery Safety Directive (Directive 89/329/EEC), as passed by the EU Ministers, sets out essential safety requirements (ESRs) that must be met before machinery is placed on the market anywhere within the union. Thus *compliance* with the Directive ensures the free movement of goods and services relating to machinery within the EU. The Directive represents 'Best Practise Methods' and its implementation and enforcement are aimed at guaranteeing an adequate level of health and safety amongst those persons using machinery. Thus all manufacturers and importers of machinery and/or their agents will be required to certify that their products conform to the Directive. It must be remembered that, ultimately, it is the Courts of Law that will decide upon the meaning and interpretation of the Directive.

The ESRs are expressed in general terms and it is intended that *European Harmonised Standards* should fill in the detail so that designers, manufacturers and importers have clear guidance on how to conform to the Directive. To this end, the harmonised standards are those standards produced by CEN and CENELEC under a mandate from the Commission and which are subsequently approved by the 83.189 Directive Committee and published in the *Journal of European Communities*. The harmonised standards are, in themselves, not mandatory but are a way of showing conformity with the Directive. They form a 'benchmark' against which other ways of achieving conformity will be judged.

The Directive (first published in June 1989) is divided into three parts, all of which have significant implications for persons and institutions responsible for drawing up standards.

- The first part sets out the overall objectives of the Directive and provides guidance on the interpretation of the subsequent Articles.

- The next part sets out the 14 Articles that lay down the scope, exclusions and duties contingent upon designers, manufacturers, importers and member states in implementing the Directive.
- The final part consists of a series of annexes detailing specific parts of the Directive and containing most of the technical details required for its implementation.

A detailed study of this comprehensive document is beyond the scope of this book – in fact, it would more than double its size. For more detailed information, readers are referred to the Directive itself. For assistance in interpreting the Directive, the proceedings of the one-day symposium held in June 1991 and published by the Institution of Mechanical Engineers, the Institution of Electrical Engineers and the Health and Safety Executive are useful.

SELF-ASSESSMENT TASK 12.1

1. Safety regulations are continually being improved; research and list the latest amendments to the Health and Safety at Work, etc., Act.

2. Discuss the main points to note when addressing safety issues in your company or any large institution of your choice.

3. Describe the marking(s) that machines and equipment must now carry to prove that they conform to international safety standards and recommendations.

12.4 Safe working practices (PUWER)

Safe working practices and guarding were introduced in *Manufacturing Technology*, Volume 1. We must now take these topic areas further by considering the *Provision and Use of Work Equipment Regulations 1992* (PUWER) that came into force on 1 January 1993. These Regulations implement the European Community (EC) Directive 89/655/EEC that required similar basic laws throughout the EU on the use of work equipment.

The PUWER lay down important health and safety laws for the provision and use of work equipment. The primary objective of the Regulations is to ensure the provision of safe work equipment and its safe use. Work equipment should not give rise to health and safety, irrespective of its age or place of origin.

12.4.1 *Interpretation of 'use' and 'work equipment'*

- The definition of the word 'use' is wide and includes all activities involving work equipment such as stopping or starting the equipment, repair, modification, maintenance and servicing.
- The scope of 'work equipment' is also extremely wide and includes: single machines such as a guillotine, circular saw bench, or a photocopier, tools such as a portable drill or a hammer; and laboratory apparatus (bunsen burners, etc.).

12.4.2 *Regulations*

The Regulations are too extensive to consider in detail in this chapter but the main headings are listed below. It is in the interests of all persons engaged in any form of industry, either as employee, employer, or self-employed, to be familiar with these Regulations, and should refer to HSE Guidance on Regulations L22 (PUWER 1992, ISBN 0-11-886332-0).

- *Regulations 1 to 3* deal with the preliminary issues, i.e. date on which the Regulations, scope and limitations come into force.
- *Regulation 4* defines who has duties, i.e. employers, persons in control, and the self-employed.
- *Regulation 5 – Suitability of work equipment* deals with the suitability of work equipment in terms of:
  (a) its integrity
  (b) the place where it will be used
  (c) the purpose for which it will be used
- *Regulation 6 – Maintenance* deals with the maintenance that is necessary for the safe and efficient working of all equipment and the maintenance of an up-to-date log where this is required.
- *Regulation 7 – Specific risks* deals with specific risks and the associated restricted use of such equipment to those persons designated competent to use and maintain it.
- *Regulation 8 – Information and instructions* deals with the provision of information and instructions and stipulates that all persons who supervise, manage or use work equipment must have adequate health and safety information available to them. Where appropriate, they must also have written instructions concerning the use of the work equipment.
- *Regulation 9 – Training* deals with workforce training and states that all persons who supervise, manage or use work equipment must have received proper and adequate training for all purposes of health and safety. Such training must include the correct use of the equipment, awareness of any risks entailed and the precautions to be taken.
- *Regulation 10 – Conformity to European Union requirements* states that all work equipment brought into use after 31 December 1992 must conform to the appropriate legislation and comply with any relevant EU Directive. From 1 January 1995 only work equipment products complying with 'essential safety requirements' and bearing the 'CE' mark can be placed on the market and used in the workplace.
- *Regulation 11 – Dangerous parts of machinery* deals with the guarding and safe operation of machinery, and will be considered in Section 12.7.
- *Regulation 12 – Protection against specific hazards* deals with protection against, and the elimination of exposure to, specific hazards such as:
  (a) articles or substances falling or being ejected from work equipment
  (b) rupture or disintegration of work equipment components
  (c) work equipment overheating and/or catching fire
  (d) unintended or premature discharge of any article or substance (gas, vapour, liquid, dust) that is produced, used or stored in work equipment
  (e) unintended or premature explosion of work equipment or any article or substance produced, used or stored in it.

- *Regulation 13 – High or very low temperatures* deals with work equipment and any articles or substances produced, used or stored in it resulting in temperatures that are sufficiently high or sufficiently low to cause injury. Adequate protection must be provided. To comply with this Regulation the risk assessment (Sections 12.5 and 12.10) must consider the hierarchy of controls.

- *Regulation 14 – Controls for starting or making a significant change in operating conditions* deals with the provision of controls that are used in the starting of equipment or in making significant changes in the operating conditions such as speed, pressure, temperature and power. Starting or restarting after a stoppage must involve the use of a control (no-volt release on motor starters). Such controls should be designed and positioned to prevent inadvertent operations (Section 12.5).

- *Regulation 15 – Stop controls* states that stop controls must be readily accessible to bring the equipment to rest in a safe manner and render it safe.

- *Regulation 16 – Emergency stop controls* states that emergency stop controls must be provided on work equipment where other safeguards are deemed to be inadequate when an irregular event occurs. Emergency stop controls are no substitute for adequate guarding. Such emergency stop controls must be conveniently positioned, easily identified (coloured RED) and easy to use (buttons to be mushroom headed).

- *Regulation 17 – Controls* deals with work equipment controls and states that they must be clearly visible and carry identifying marks. No control must be positioned so that its operation exposes the user to risk. Where there would be a risk to a person operating a control, such as equipment starting, then audible and/or visual warnings must be automatically given for a sufficient period of time for the person to avoid risk.

- *Regulation 18 – Control systems* states that control systems must not create any increased risk. Any loss of power source to a control system must not result in any increased risk. Control systems must be 'fail safe'.

- *Regulation 19 – Isolation from sources of energy* states that all work equipment must have an adequate and safe means of isolation from all its sources of energy. Provision must be made so that the reconnection of the sources of energy does not put any person at risk (see Regulation 14). Further, the release of energy after isolation must not put any person at risk.

- *Regulation 20 – Stability* is concerned with the stability of work equipment where there is a risk that the equipment might collapse, slip or overturn. It stipulates that such equipment must be fixed to the ground, fastened, clamped or tied to a secure struture to render it safe.

- *Regulation 21 – Lighting* stipulates that lighting must be adequate and suitable.

- *Regulation 22 – Maintenance operations* stipulates that work equipment must be so constructed that maintenance operations involving a risk to health and safety can be carried out whilst the equipment is shut down. If the equipment needs to be running during maintenance and this presents risk, suitable measures must be taken to enable the risks to be minimised.

- *Regulation 23 – Markings* states that work equipment must be marked in a clearly visible manner with any markings appropriate for reasons of safety and health, e.g. the safe working load (SWL) of lifting equipment. Such markings should conform to the latest release of BS 5378.

- *Regulation 24 – Warnings* is concerned with warnings and warning devices. The warning given by a warning device must be unambiguous and easily perceived and understood, otherwise it must not be used.

12.5 Hazards associated with the working environment

Let's now consider the main points concerning safety in the working environment that should be observed by employers, managers and operational workers when engaged in manufacturing processes. It is intended as a reminder that one can never become complacent where health, safety and welfare are concerned and that the principles of good operating practice can never be reiterated too often.

We shall now consider the hazards associated with the processes discussed in the earlier chapters of this book. The management of risk applies equally to the processes described in Volume 1. Let us commence with the casting processes.

12.5.1 *Casting processes*

As with all processes involving high temperatures, casting processes are potentially hazardous. This is particularly the case when large quantities of molten metal are involved. The structure and lining of all furnaces, crucibles and ladles should be regularly inspected to prevent failure and accidental spillage. It is essential to maintain the casting shop itself in good repair, since water dripping from a leaking roof or water pipes into molten metal can cause an eruption of violent and dangerous proportions. Further, the casting shop floor should be so designed that any spillage should be contained locally and the flooring material should not break up and shatter if hot metal is spilt on it. (For this reason it must never be concrete.) Moulds are often placed on sand beds for pouring.

Casting-shop personnel must wear suitable protective clothing when preparing and pouring molten metal: a face visor, leather apron, leather gauntlet gloves, leather spats and industrial footwear to protect the lower part of legs and feet. Adequate ventilation is also required. This not only removes fumes associated with the fluxing and degassing of the molten metal and the burning out of the resin adhesive used in cores and shell moulds, it also provides a cool environment with an adequate supply of fresh air so that personnel can remain alert and work comfortably.

Where automatic and semi-automatic processes, such as pressure die casting, are used the machine should be fully guarded with the casting area totally enclosed during the injection cycle. This is necessary to contain any molten metal that could escape under very high pressure should the dies burst or should the machine fail to close the dies properly. Such guards should be interlocked with the machine so that it cannot function whilst the guard is open. This prevents the machine operating and trapping the operator whilst the casting is being removed from the dies.

12.5.2 *Work equipment hazards*

The remaining processes involve the use of power-driven machines, so we shall commence by examining the hazards associated with machinery in general, which can be categorised as:

- *Mechanical hazards*, including traps, impact, contact with moving parts, entanglement or through the ejection of materials or machine parts due to structural failure or incorrect operation.
- *Non-mechanical hazards*, including electrical faults, pressurised fluids, exposure to chemicals, extremes of temperature, noise, vibration and radiation.
- *Control hazards*, such as operator (human) error, corruption of software, faulty programming of computer-controlled systems, and failure of computer-controlled systems.

Also, it is important to distinguish between ongoing hazards associated with working with unguarded or inadequately guarded machinery and hazards arising from catastrophic failure of machine components and/or failure of the safety equipment and systems. Hazards associated with machinery can be categorised as:

- *Traps* A person's body is trapped between closing or rotating elements of the machine. This causes crush injuries and can result in the loss of a limb or death. Where the body or limb is drawn into the machine, it is referred to as an 'inrunning nip'.
- *Entanglement* A person most usually becomes trapped in a machine by the entanglement of hair, jewellery items of clothing with the rotating parts of the machine.
- *Contact* A person can suffer serious injuries by coming into contact with the sharp edges of cutting tools and components, ragged swarf, and abrasive surfaces (grinding wheels). In addition, injuries can be caused by coming into contact with excessively hot or excessively cold surfaces, and electrically live components.
- *Impact* Injuries can occur by being struck by moving parts of machinery or by components and broken parts being ejected violently from the machine because of structural failure.
- *Ejection* Injuries can occur from the ejection of sparks, swarf, chips, molten metal as well as the broken components mentioned above. Safety goggles should always be worn in a workshop situation.

12.5.3 *Controlling the risk*

Before moving on to consider the hazards associated with specific manufacturing processes, let us first review some of the methods of controlling the risk.

1. The operation of machinery can be made safer by implementing the following objectives:
   (a) eliminating hazards at the design stage (see Section 12.10)
   (b) reducing or eliminating the need for persons to access dangerous parts of the machine (e.g. the cutting zone) by using automatic feed mechanisms or providing interlocked guards so that access cannot lead to injury
   (c) making dangerous areas inaccessible (out of reach) so that they are 'safe by position'
   (d) providing the operator with personal safety equipment such as safety goggles and ear protectors
2. The safety of persons working with or in the vicinity of the machines also needs to be considered:
   (a) safety awareness training for all persons working with or in the vicinity of machines and other potentially hazardous work equipment
   (b) redesigning equipment to make any potential hazard more obvious and/or warning signs
   (c) providing enhanced skill training in the avoidance of injury
   (d) improving motivation in the use of safety devices and procedures so as to avoid accidents

12.6 Safety by design

As stated above, machine safety should start at the design stage. All work equipment should be designed so as to eliminate hazards. Dangerous areas should be inaccessible, and the need to handle work into and out of the cutting zone should be reduced by incorporating automatic feed mechanisms wherever possible.

12.6.1 *Controls*

The design and positioning of controls can greatly influence the safe operation of machines, for example:

- Correct positioning.
- Easy identification by distinguishing features such as size, colour, feel, direction of movement, etc.).
- Use of correct type of control – no risk of accidental start up.
- Logical directional movement – control and linked machine element move in the same direction.

12.6.2 *Failure to safety*

All machines and work equipment should be designed with failure to safety in mind. This is often referred to as 'fail-safe'. For example, if a pilot light on a gas heater fails, the fuel supply to the appliance should be automatically cut off. Similarly, if the electrical supply to a machine fails, the machine should not restart on its own when the supply is reconnected. The start button should have to be manually activated by the operator.

12.6.3 *Isolation of the supply*

In addition to the 'on' and 'off' controls provided for routine operation of the machine during normal working, provision should also be made for total isolation from all sources of energy supply. The isolating devices should have positive 'lock-off' facilities so that the maintenance fitter can keep the key, thus preventing unintentional restoration of the energy supply whilst maintenance work is being carried out. The machine should be isolated when:

- Changing tools;
- Changing workholding devices.
- Maintenance requires the removal of interlocks, inspection covers and guards.
- Dismantling and reassembly of replacement parts.

Machines should be designed so that, as far as possible, routine maintenance and lubrication can be carried out without the removal of guards, interlocks and inspection covers. A central lubrication system feeding inaccessible parts is also recommended.

12.6.4 *Safety by position*

Those hazardous parts of work equipment and machines that are out of reach and which are kept out of reach are said to be 'safe by position'. However, designers must always be alert to the unexpected, and hazard areas that are normally out of reach may become accessible from a ladder during cleaning and decorating of the workshop premises.

12.6.5 *Machine layout*

The layout of workshops can also influence safe working conditions (see also Section 9.3 *et seq.*). The layout should take account of:

- *Spacing* should be adequate to facilitate access for operation, cleaning, adjustment, maintenance and supervision.
- *Lighting* – both general lighting levels of the workplace (natural and/or artificial) and localised lighting for specific operations and machining – must be better than just adequate, but without glare. Machine lights (often high-intensity halogen quartz spotlights) should operate from low-voltage supplies for safety.
- *Cables, pipe, and conduits* should be placed to allow safe access and avoid tripping. They should be positioned so that they are free from the risk of accidental damage. Wherever possible they should be colour coded to identify their contents.

12.6.6 *Selection*

In view of the litigious age in which we live, new machinery, plant and work equipment should be purchased and installed with the above points in mind in order to avoid the cost of accidents, civil claims and criminal prosecution. The correct selection and layout of machines and equipment in the first place reduces costly alterations and replacements at a later date.

12.6.7 *Machinery guarding*

The guarding requirements of machines and their tooling will vary according to the type of machine and the work being performed on it.

1. Where access to a hazard area (such as the cutting zone) is *not* required during normal operation, then use a:
   (a) fixed guard, interlocked doors or canopy
   (b) distance guard
   (c) trip device

2. Where access to a hazard area (such as the cutting zone) *is* required during normal operation, then use:
   (a) an interlocked guard
   (b) an automatic guard
   (c) a trip device
   (d) self-adjusting guard
   (e) two-handed control

SELF-ASSESSMENT TASK 12.3

1. Within the context of PUWER, research and explain what is meant by the terms:
   (a) assessment of risk
   (b) conformity with EU Regulations
   (c) specific risks
   (d) dangerous parts of machinery
   (e) protection against specified hazards

2. Within the context of PUWER, explain what is meant by the terms:
   (a) safety by design
   (b) machinery hazards
   (c) failure to safety
   (d) machinery layout

12.7 Types and applications of guards

Let's now apply some of these principles to the processes that have already been discussed in this book and in *Manufacturing Technology*, Volume 1.

12.7.1 *Forging and extrusion processes*

Usually forging presses and hammers are not guarded to the same extent as sheet metal presses, because the operator has to load and unload the work zone with long-handled tongues since the workpiece is at or above red-heat during the process. However, the operator must be protected against accidental contact with the hot metal, and protective clothing similar to that recommended for casting should be worn. In addition ear protectors should be worn as the noise level in a forging shop is a potential hazard. As with all high-temperature processes, adequate ventilation is essential to maintain a comfortable working environment and to remove any unpleasant or toxic fumes.

Impact extrusion has similar hazards to forging. However, because of the batch sizes involved, mechanical-handling devices are frequently used to load and unload the presses and total guarding of the working zone is possible.

12.7.2 *Plastic moulding*

Most plastic moulding operations are now heavily automated and this reduces operator risk. However, where manual loading and unloading of the press or the injection-moulding machine takes place, suitable guards, interlocked with the machine-operating mechanism, must be provided to prevent the operator becoming trapped and injured. An interlocked guard suitable for an upstroke moulding press is shown in Fig. 12.1. In Fig. 12.1(a) the guard is open and, in this position, it holds the operating lever down in the safe position. When the guard has been closed as shown in Fig. 12.1(b), the lever can be raised to set the press in motion. In this position the lever prevents the guard from being opened. Interlocked guards are also considered in Section 12.7.3. Overalls are adequate protective clothing. Care has to be taken if the operator has to handle moulding materials or synthetic adhesives as this can cause skin sensitisation. Suitable barrier creams and gloves should be used. Ventilation of the working area is essential, particularly if solvents are being used as these can be both toxic and highly flammable.

Fig. 12.1 *An interlocked guard: (a) guard open; (b) guard closed*

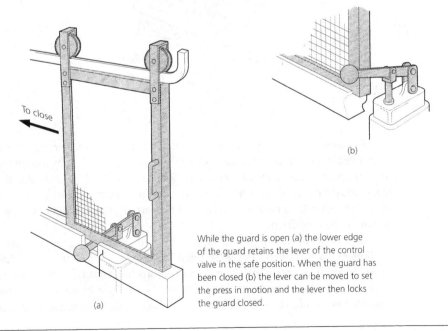

To close

(b)

(a)

While the guard is open (a) the lower edge of the guard retains the lever of the control valve in the safe position. When the guard has been closed (b) the lever can be moved to set the press in motion and the lever then locks the guard closed.

12.7.3 *Presswork (sheet metal)*

Sheet metal pressing operations are amongst the most hazardous processes used because the operator's hands have to enter the working zone of the tools to load and unload the machine. Thus presses must be fitted with a comprehensive guard system that must be *interlocked* with the operating mechanism. This prevents the tools closing and trapping the operator whilst work is being loaded or unloaded. An example of the guard used on a small press is shown in Fig. 12.2. You can see that the flywheel and belt drive are fully enclosed in a fixed guard, and that the interlocked guard (gate) is shown in the open position ready for the operator to load or unload the tools. The machine cannot be operated nor can it operate accidentally, whilst the gate is open.

Fig. 12.2 *Fully guarded power press; the press can only be operated with the 'gate' shut*

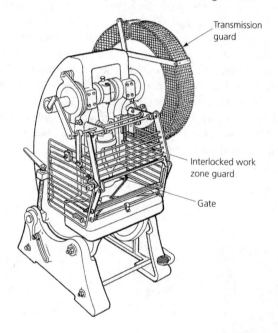

Transmission guard

Interlocked work zone guard

Gate

The guards, and the presses to which they are fitted, must be regularly inspected by an accredited inspector and the guards can only be adjusted and set by a certificated fitter. The use of mechanical-handling devices to load and unload a press greatly reduces the hazardous nature of the operation. No special protective clothing is required by the operator, other than normal overalls and reinforced gloves as the raw edges of thin sheet metal can be very sharp.

Two other devices that can be equally well applied to presses – press-brakes and guillotines – are shown in Fig. 12.3. Figure 12.3(a) shows a photoelectric device fitted to a press brake. If the operator's hands and/or arms interrupt the light beam the machine is immobilised. Upon removing the obstruction to the light beam, the machine can then be operated in the normal way. Figure 12.3(b) shows a *distance guard* fitted to a press-brake.

The guard is set in such a way that the operator cannot physically reach the hazard zone created by the press tools. The side and rear guards are not shown in order to simplify the illustration. They would have to be in position in practice.

Fig. 12.3 *Further guard types for sheet metal working machinery: (a) photoelectric device; (b) distance guard*

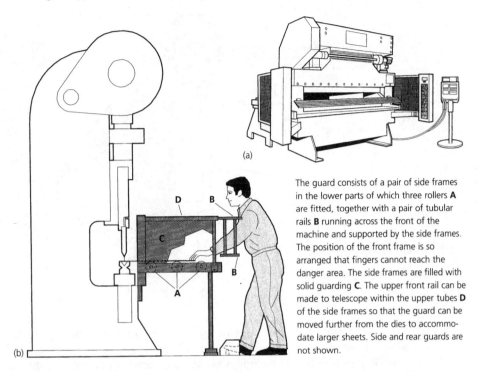

(a)

(b)

The guard consists of a pair of side frames in the lower parts of which three rollers **A** are fitted, together with a pair of tubular rails **B** running across the front of the machine and supported by the side frames. The position of the front frame is so arranged that fingers cannot reach the danger area. The side frames are filled with solid guarding **C**. The upper front rail can be made to telescope within the upper tubes **D** of the side frames so that the guard can be moved further from the dies to accommodate larger sheets. Side and rear guards are not shown.

12.7.4 *Machining*

It is the duty of machine manufacturers to ensure that their products have adequate transmission guards securely fencing the shafts, gears, belts, chains, etc. A typical transmission guard is shown in Fig. 12.4. It is the duty of the employer and the employees to ensure that these guards are kept in place and in good condition. In addition, it is the duty of the employer to ensure that such machines are provided with proper cutter, chuck and splash guards and that these guards are fitted and used correctly. It is the duty of the employee to make proper use at all times of the guards provided, and not to tamper with them or remove them in order to speed up the process and earn increased bonus payments. The type of guard will depend upon the machine and the process. Milling machines are particularly hazardous and a selection of cutter guards for jobbing and small batch work are shown in Fig. 12.5. Where a guard is impractical, 'two-hand control' may be used as shown in Fig. 12.6. The design ensures that both hands are moved to a safe position whilst they operate the shrouded start buttons.

Fig. 12.4 *Fixed transmission guard: a fixed guard constructed of wire mesh and angle section preventing access to the transmission machinery*

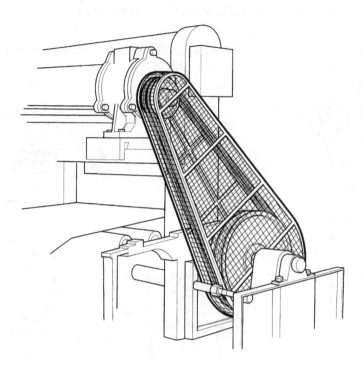

Some guards, such as the chuck and drill guard shown in Fig. 12.7(a), are self-adjusting. As the drill penetrates the workpiece, the guard telescopes so as to allow the down feed of the drill. As the drill is raised from the work at the end of the cut, the guard returns to its original length and safely enclosed the rotating drill. *Trip switches* can also be used and an example is shown in Fig. 12.7(b). The 'mast' of the switch is situated at 75 mm from the spindle. In the event of the 'mast' being deflected by only a matter of a few degrees from the vertical, the machine drive motor is switched off and the brake is applied. The machine will immediately stop.

Automatic and CNC machine tools frequently have the work zone totally enclosed whilst cutting is taking place. This not only protects the operator, but helps to reduce the noise to legally acceptable levels and prevents the coolant being splashed out of the machine. (A considerable volume of coolant is usually flooded over the cutting tools and the workpiece in such machines throughout the cutting cycle.) Loading and unloading can be by industrial robots but, where loading and unloading is performed manually, a separate workstation is frequently provided remote from the cutting zone.

Photoelectric guards are used where mechanical guards cannot be inserted easily. The guard consists of a curtain of parallel light beams that are between the operator and the hazard zone. If a light beam is broken, the machine will be stopped automatically.

Fig. 12.5 *Milling-cutter guards*

12.7.5 *Grinding*

The guards on grinding machines have two purposes:

1. To prevent the operator coming into contact with the rapidly revolving abrasive wheel.
2. To provide burst containment; that is, to contain the wheel should it shatter whilst it is being used. For this reason abrasive wheel guards should be made from steel and should be thicker than is necessary merely to keep the operator out of contact with the wheel. The thickness is specified in the regulations.

Under the grinding wheel regulations of 1970, only certificated persons may change and mount grinding wheels. Because of its relatively flimsy construction and high rotational speeds, a grinding wheel is potentially dangerous. A large wheel, rotating rapidly, stores up kinetic energy that can be released with disastrous results if the wheel bursts. The usual causes of failure are misuse, incorrect mounting or overspeeding. The spindle speed of the machine must be prominently displayed. This speed must be compared with maximum permitted speed marked on the abrasive wheel and, in no circumstance should the wheel be used if the spindle speed is greater than the permitted speed.

Fig. 12.6 *Two-handed control*

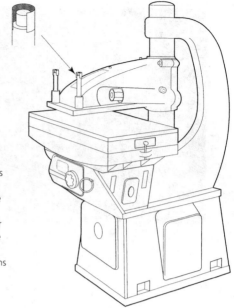

At some machines the use of a guard is impracticable and other measures to protect the operator are necessary. The provision of two-hand controls at this clicking press ensures that the operator has both hands in a safe position while the press head descends. To protect against accidental operation the buttons should be shrouded (see inset)

Fig. 12.7 *Safety devices for heavy-duty drilling machines: (a) self-adjusting drill guard for radial-drilling machine; (b) trip-switch guard for radial-drilling machine*

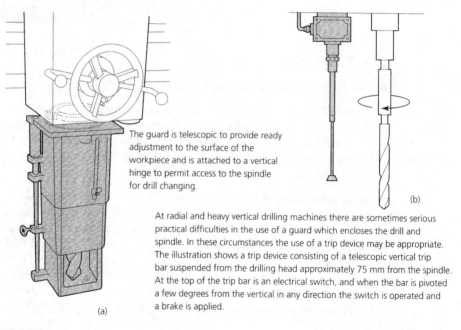

The guard is telescopic to provide ready adjustment to the surface of the workpiece and is attached to a vertical hinge to permit access to the spindle for drill changing.

(b)

At radial and heavy vertical drilling machines there are sometimes serious practical difficulties in the use of a guard which encloses the drill and spindle. In these circumstances the use of a trip device may be appropriate. The illustration shows a trip device consisting of a telescopic vertical trip bar suspended from the drilling head approximately 75 mm from the spindle. At the top of the trip bar is an electrical switch, and when the bar is pivoted a few degrees from the vertical in any direction the switch is operated and a brake is applied.

(a)

12.8 Fire regulations

New fire regulations for commercial and manufacturing premises came into force on 1 April 1989. The employer, manager or owner of a factory, office, shop or railway premises must decide whether or not the premises require a fire certificate. A certificate is needed if:

- More than 20 persons are at work in the premises at any one time.
- More than 10 persons are at work at any one time on a floor other than the ground floor.
- The premises are used as a factory which has highly flammable or explosive materials stored in or under it.

Note: The phrase 'persons at work' means *everyone* including employers, managers, employees, self-employed persons, temporary staff, trainees and apprentices.

The decision of whether a certificate is or is not required rests with the local fire service. From the time of application for such a certificate, all means of escape with which the premises are provided must continue to be kept safe and effective; all means of fighting fires must be maintained in efficient working order; and all employees must receive instruction in what to do in the event of fire. The HMSO publication *Guide to Fire Precautions in Places of Work that Require a Fire Certificate* should be available on the premises and regularly consulted. Regular consultation with the local Fire Prevention Officer is time well spent, particularly if changes are to be made to the premises, plant, processes, working practices, and materials being handled and stored.

Even when the premises are exempted from a certificate by the local fire service or, even if insufficient persons are working in the premises to make a certificate obligatory, it is still the responsibility of the employer, manager or owner of a factory, office, shop, or railway premises to ensure that there is provision for suitable means of escape and also suitable means for fighting fires. Failure to provide proper means of escape or means of fighting a fire is a *legal offence*. In the case of premises not requiring certification, reference should be made to the HMSO publication *Code of Practice for Fire Precautions in Factories, Offices, Shops and Railway Premises Not Required to Have a Fire Certificate*.

12.9 Electrical hazards

The hazards associated with electricity are shock, burns, fire and explosion. Accidents involving *electric shock* are often accompanied by *burn injuries* at the point of contact. An electric shock is an unpleasant sensation producing a convulsive response to an electric current by the body's nervous system. The majority of reported cases are usually from the 240 V domestic supply. An electric current regards the human body as a conductor. Like all conductors, the human body has resistance, capacitance and inductance – that is, it has *impedance*. The value of its impedance can vary with a number of factors such as contact area, contact pressure, skin moisture, etc. The impedance value has been found to vary between 1000 and 2000 ohms. It is the magnitude and duration of the current flow that influences the intensity of the 'shock' that can lead to injury or death. Ventricular fibrillation (extra heartbeats) can occur even at relatively low currents. It has been found from experiments conducted under strict medical supervision, that human beings should be able to tolerate 650 mA for 10–80 milliseconds and should be able to tolerate 500 mA for 80–100 milliseconds. The design of *residual current circuit breakers* has been influenced by this data, and they are set to operate in less than 30 milliseconds.

Table 12.1 shows how even relatively low currents can affect healthy people when sustained for longer than 1 second's duration. From this table it is apparent that sustained shocks, at normal mains voltage, are potentially lethal. However, in alternating current electric-arc welding where the open-circuit voltage is normally 80 V, few accidents related to electric shock have been reported, although it is known that the operators often sustain mild shocks when changing electrodes.

Table 12.1 *Effects of alternating currents on the human body*

If sustained for more than 1 second

Current in mA	Effect
0.5–2	Threshold of perception.
2–10	Painful sensation, increasing with current.
10–25	Cramp and inability to 'let go'. Increase in blood pressure. Danger of asphyxiation from respiratory muscular contraction.
25–80	Severe muscular contraction sometimes involving bone fractures. Increased blood pressure i.e. Loss of consciousness from heart and/or respiratory failure.
Over 80	Burns at points of contact. Death from ventricular fibrillation (uncoordinated contractions of the heart muscles so that it ceases to pump).

In all industry it is recommended practice to use 110 V electric tools supplied from an isolating transformer with its midpoint bonded to earth. This reduces the potential shock voltage to 55 V and there are no reported shock accidents. It would seem that at these fairly low voltages, in dry conditions, the victims are able to 'let go' and the adverse effects of the shock soon wear off, hospital treatment is not required and no accident report is needed.

12.9.1 *Treatment for electric shock*

An electric shock placard has to be exhibited in most industrial premises. This gives simple instructions on remedial treatment when a shock victim is rendered unconscious. Treatment for electric shock is covered in detail in Section 12.18 in *Manufacturing Technology*, Volume 1. Artificial respiration, however, is of no avail against ventricular fibrillation. Usually, this condition can only be treated with a *defibrillator*, which is a device that administers a carefully controlled *direct-current* electric shock to restore the normal heart rhythm. First-aid personnel, trained in the recovery procedures associated with electric shock, should be available.

12.9.2 *Explosions and fires caused by electrical faults and discharges*

A spark from an unenclosed electric motor or even a simple light switch can ignite flammable dust particles and vapours that are present in the atmosphere once the critical concentration has been reached. Electrical equipment used in hazardous environments where there is the possibility of fire or explosion must be flame-proofed to the requirements of BS 5345 and BS 5501 for flammable atmospheres, and BS 6467 for flammable dusts. Other precautions include electrostatic discharge where earth straps must be used to prevent the build up of static electricity.

12.9.3 *Electricity at Work regulations 1989*

These regulations are a revision of the Electricity (Factories Act) Special Regulations 1908 and 1944 and replaced them on 1 April 1990. As these regulations apply to everyone at work, there is thus an overlap with other existing legislation having electrical safety requirements. The new regulations are additional to, but do not supersede, previous regulations. Details of the regulations are too lengthy for inclusion in this chapter, but those involved in the management of manufacturing or the installation and maintenance of manufacturing equipment should make themselves familiar with the requirements of The *Electricity at Work Regulations 1989*.

SELF-ASSESSMENT TASK 12.5

1. Discuss the implementation of the current fire prevention and safety regulations as they apply to your place of work.

2. List the main points in preventing electrocution and state why a properly qualified person of responsibility is present during any electrical work.

3. Briefly describe the life-saving techniques used when saving someone's life if he or she has been electrocuted.

4. Research areas such as PAT testing which are applied to electrical equipment testing.

12.10 Risk assessment

The Health and Safety Executive and Regulation 3(1) of the *Management of Health and Safety at Work Regulations 1992* makes an assessment of risk a legal obligation for companies. Such a *risk assessment* enables the company to identify potential hazards in the workplace and to assess the resultant risk to employees and to anyone present on the premises by invitation. Carrying out such an assessment helps to identify all the protective and preventative measures that have to be taken to comply with the PUWER.

When undertaking risk assessment on work equipment it must be decided whether or not the requirements are already complied with and, if not, what additional action must be taken to ensure compliance. As an aid to completing the form, the HSE has defined five steps:

1. Look for hazards.
2. Decide who might be harmed and how.
3. Evaluate the risks arising from the hazards and decide whether existing precautions are adequate or if more should be done.
4. Record your findings.
5. Review the risk assessment from time to time and revise this if necessary.

A typical risk assessment form is shown in Fig. 12.8.

Fig. 12.8 *Work equipment inventory and risk assessment summary*

Department	Location	Assessor		Date
		Approving manager		
Work equipment	Equipment safety list required	Hazard(s)	Risk LMN	Control(s)	Actioned by (name & date)

SELF-ASSESSMENT TASK 12.6

1. Use a standard risk assessment form to carry out a risk assessment in one area, i.e. a laboratory or workplace at any company or organisation with which you are familiar.

2. Survey your department and report on how effectively it complies with the Management of Health and Safety at Work Regulations 1992.

12.11 Managing safety (BS 8800)

Safety is not only an ethical, moral and regulatory issue, it is clear that there are sound economic reasons for reducing work-related accidents and ill-health and for improving the quality of the work environment. BS 8800, *Guide to occupational health and safety management systems*, came into effect in May 1996, having been prepared by the Technical Committee HS/1. Throughout this section Occupational Health and Safety will be referred to as OH&S. The aim of BS 8800 is to:

> *Improve the occupational health and safety performance of organisations by providing guidance on how management OH&S may be integrated with management of other aspects of business performance.*

There is evidence that employees in small businesses (less than 50 employees) are 40 per cent more likely to have an accident than those in larger organisations (more than 1000 employees).

More than ever, organisations must keep firm control and understanding of what effective OH&S management means to them. At present it is not enough to feel that you have done your bit for OH&S and then forget about it. Pressure for continuous improvement is increasing from a growing number of parties external to an organisation, with a legitimate interest in an organisation's performance.

A successful organisation will integrate OH&S into its day-to-day management arrangements, recognising that there should be one management system providing the procedures and instructions. It will continuously maintain its OH&S management programme, update it and, most importantly, communicate it to those who are most directly affected – employees, contractors, customers, clients, visitors and the public. It is not only essential to those closest to the organisation, but evidence of successful OH&S management is now increasingly sought by other groups that are important to the organisation, such as investors, insurance companies, financial institutions and potential customers. By demonstrating a sound track record, organisations are able to show commitment and a sense of responsibility towards managing OH&S issues on an evolving basis.

Clearly, effective OH&S management will not just happen. From the outset there needs to be commitment at the highest level and a proactive approach from the organisation to address all OH&S issues. This involves taking positive action to introduce and establish an OH&S management system that is supported at every level throughout the organisation.

Initially, it may be perceived that OH&S can be a serious drain on resources, offering little financial return. In practice, it has been shown that the costs can be outweighed by a reduction in accidents, occupational illness, equipment/plant damage, etc. Benefits include:

- Reductions in staff absence.
- Reductions in claims against the organisation.
- Reductions in adverse publicity.
- Improved insurance liability rating.
- Improved production output.
- A positive response from customers wanting to deal with an organisation with a successful OH&S track record and which is unlikely to be disrupted by costly accidents.

Many companies have already achieved BS ISO 9001/2 status and are operating successful quality systems. With the increasing awareness of the importance of managing environmental issues, organisations are also seeking certification to BS EN ISO 14001 (which has superseded BS 7750). BS 8800 completes the trio of business management systems – *quality*, *environmental* and *occupational health and safety*.

For those organisations, OH&S can be seen, at first sight, to be yet another system for the organisation to implement. However, this need not be such an onerous task. BS 8800 embodies the principles upon which BS EN ISO 9001/2 and BS EN ISO 14001 are based, and many commonalities exist between them. An OH&S management system based on BS 8800 allows alignment with these other systems. Organisations can choose for themselves the extent to which they interface or integrate the three business management systems to avoid unnecessary bureaucracy and duplication of paperwork.

BS 8800 provides two approaches to developing an OH&S management system. The first is based on HSE guidance, *Successful Health and Safety Management* HS(G)65 approach, and the second on the environmental standard BS 14001: *Environmental management systems. Specification with guidance for use* Figure 12.9 shows these approaches diagrammatically. The content is essentially the same, the only significant difference being the order of presentation. The material in this section is, necessarily, an all-too-brief overview and the reader is referred to the complete British Standard 8800.

Fig. 12.9 *Elements of a successful health and safety management system (courtesy of British Standards Institution)*

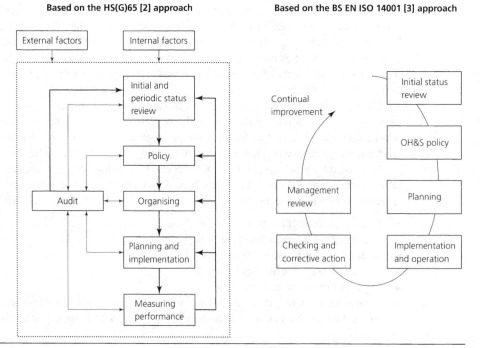

12.12 Environmental management systems (BS EN ISO 14001)

Organisations of all kinds are increasingly concerned to achieve and demonstrate sound environmental performance by controlling the impact of their activities on the environment, taking into account their environmental policy and objectives. They do so in the context of increasingly stringent legislation, the development of economic policies and other measures to foster environmental protection, and general growth of concern from interested parties about environmental matters, including sustainable development. Most companies now recognise the huge public relations (PR) benefits that can result from a transparent sound environmental policy.

Many organisations have undertaken environmental 'reviews' or 'audits' to assess their environmental performance. On their own, however, these 'reviews' and 'audits' may not be sufficient to provide an organisation with the assurance that its performance not only meets, but will continue to meet, its legal and policy requirements. To be effective, they need to be conducted within a structured management system and integrated with overall management activity.

International standards covering environmental management are intended to provide organisations with the elements of an effective environmental management system that can be integrated with other management requirements, to assist organisations to achieve environmental and economic goals. These standards, like other international standards, are not intended to be used to create non-tariff trade barriers or to increase or change an organisation's legal obligations.

BS EN ISO 14001 specifies the requirements of such an environmental management system. It has been written to be applicable to all types and sizes of organisation and to accommodate diverse geographical, cultural and social conditions. The basis of the approach is shown in Fig. 12.10. The success of the system depends upon commitment from all levels and functions, especially from top management. A system of this kind enables an organisation: to establish and assess the effectiveness of procedures; to set an environmental policy and objectives; to achieve conformance with them; and to demonstrate such conformance to others. The overall aim of BS EN ISO 14001 is to

Fig. 12.10 *Environmental management system model for BS EN ISO 14000*

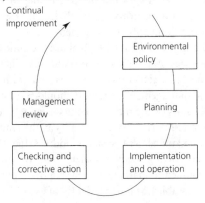

support environmental protection and the prevention of pollution in balance with socio-economic needs. It should be noted that many of the requirements may be addressed concurrently or revisited at any time.

There is an important distinction between (a) the standard specification being discussed that describes the requirements for certification/registration and/or self-declaration of an organisation's environmental management system and (b) a non-certifiable guideline intended to provide generic assistance to an organisation for implementing or improving an environment management system. Environmental management encompasses a full range of issues, including those with strategic and competitive implications. Demonstration of successful implementation of BS EN ISO 14001 can be used by an organisation to assure interested parties that an appropriate management system is in place. Guidance on supporting environmental management techniques will be contained in other international standards.

British Standard EN ISO 14001 contains only those requirements that may be objectively audited for certification/registration and/or self-declaration purposes. Organisations requiring more general guidance on a broad range of environmental management issues should refer to BS EN ISO 14004: 1996, *Environmental management systems – general guidelines on principles, systems and supporting techniques.*

It should be noted that BS EN ISO 14001 does not establish absolute requirements for environmental performance beyond commitment, in the policy, to compliance with applicable legislation and regulations and to continual improvement. Thus, two organisations carrying out similar activities but having different environmental performance may both comply with its requirements.

The adoption and implementation of a range of environmental management techniques in a systematic manner can contribute to optimal outcomes for all interested parties. However, adoption of this international standard will not in itself guarantee optimal environmental outcomes. In order to achieve environmental objectives, the environment management system should encourage organisations to consider implementation of the best available technology, where appropriate and economically viable. In addition, the cost of the effectiveness of such technology should be taken into account.

British Standard EN ISO 14000 is not intended to address, and does not include requirements for, aspects of occupational health and safety (OH&S) management (BS EN ISO 8800 previously discussed); however, it does not seek to discourage an organisation from developing the integration of such management system elements. Nevertheless, the certification/registration process for BS EN ISO 8800 will only be applicable to limited aspects of the environmental management system.

British Standard EN ISO 14000 shares common management system principles with the ISO 9000 series of quality system standards. Therefore some organisations may elect to use an existing management system consistent with the ISO 9000 series as a basis for its environmental management system. It should be understood, however, that the application of various elements of the management system may differ owing to different purposes and different interested parties. Whilst quality management systems deal with customer needs, environmental management systems address the needs of a broad range of interested parties and the evolving needs of society for environmental protection.

The environmental management system requirements specified in BS EN ISO 14000 do not need to be established independently of existing management system elements. In some cases, it will be possible to comply with the requirements by adapting existing management system elements. Remember that it was stated in the previous section that BS EN ISO 9000 (*Quality*), BS EN ISO 14000 (*Environment*) and BS 8800 (*OH&S*) have all been designed with a *common management structure* to avoid duplication of effort and wasteful bureaucracy in an organisation.

12.12.1 *Manufacturing and the environment*

Having looked at the requirements for an environmental management system and the certification/registration of such a system, let us now look at some of the practical issues involved. The potential for damage to the environment resulting from manufacturing activity can be considered under three headings, namely:

- Inputs.
- Processing.
- Outcomes.

Inputs

Protection of the environment must commence at the design stage with the selection of raw materials and natural resources. Wherever possible, either renewable and sustainable materials – or at least materials that are recyclable – should be used. Toxic materials such as the heavy metals (lead, cadmium, mercury, etc.) should be avoided. The raw materials and processes used to manufacture the materials called for in the design must also be considered. Just because a particular material appears to be benign cannot excuse its selection if toxic waste is produced, and high levels of energy are used, in its manufacture.

The product design should also take into account the manufacturing energy requirements. The designer must take into account the energy required to make the raw materials as well as the energy used in processing the materials into the product design. A full energy audit is necessary when choosing the materials to be used. Only hydroelectric and wind-electric energy sources are environmentally safe. All energy production involving the use of fossil fuels causes environmental damage:

- Global warming.
- Air pollution (CO_2, NO_x and sulphur emissions).
- Waste disposal (ash, etc).

The designer must also consider the OH&S implications of the materials chosen for a particular design. However technically and economically desirable a material may appear to be at the design stage, it cannot be considered if the workforce and general public are put at risk by its use and processing. For example, no one would now consider the use of *asbestos* as a raw material or the use of CFCs that destroy the ozone layer.

Processing

The following forms of pollution can stem from manufacturing processes:

- Visual pollution.
- Noise pollution.
- Emissions to air.
- Releases to water.

- Waste management (see 'Outcomes' below).
- Contamination of land.
- Energy useage.

Visual pollution this commences with the building of a new factory on a green-field site. At present the planning authorities expect any new development to blend with its surroundings by careful use of natural contours and landscaping. 'Dark satanic mills' belching black smoke from their smoke stacks are very much a thing of the past. In fact most firms recognise the public relations benefits that can be gained from well-designed and architecturally pleasing modern premises that exhibit an air of competent efficiency. First impressions are important.

Noise pollution Noise is wasted energy and energy has to be paid for. So noise eliminated at source is economically sensible. Further, a company that can live at peace within the community can concentrate on its mainstream activities instead of fighting endless court battles with its neighbours. To this end, factory premises should be sited so that heavy traffic does not have to travel through residential areas. Noise should also be avoided as it has OH&S implications for the workforce and may eventually lead to compensation claims.

Emissions to air Air pollution can come in many forms:

- Soot, dust, fly-ash and other particulates that can dirty the surrounding buildings, open spaces and washing left out to dry.
- Noxious, but not necessarily dangerous, odours that can be unpleasant to live near. For example, the fumes associated with stove-enamelling plants.
- Toxic emissions from process heating, space heating and process chemicals.

Releases to water Liquid effluents from manufacturing plants come mainly from:

- Metal-machining processes using cutting lubricants that often contain chemical additives. From time to time the coolant tanks on machines have to be cleaned out and replenished. The waste-cutting lubricants must not be poured down the drain but must be disposed of safely by a licensed disposal firm.
- Metal-finishing (electroplating) plants use some highly toxic metals and chemical solutions. Great care must be taken to protect operatives in such plants and on no account must the effluent from such plants be poured down the public sewers.
- Metals are often 'pickled' in acids to remove the oxide film left behind after they have been hot worked or heat treated. The spent acid must be neutralised and removed by a licensed waste-disposal firm.
- Process water such as that used in wet (water-washed) spray-paint booths must not be released down the public sewers or into the rivers without first being treated and subjected to analysis. The large-scale use of process water may necessitate the effluent being retained in specially built 'lagoons' where it can be treated and recycled. Care must be taken to ensure that it cannot escape into the ground or adjacent water-courses.
- Only purified water can be released into sewers and rivers and this water must meet stringent requirements to ensure that it does not disturb or destroy the aquatic ecology of the rivers or the bacterial balance of sewage treatment plants into which it is eventually discharged.

Contamination of the land The sites of old factories have become severely polluted over the years by chemical spillage, and badly stored toxic materials. In some cases the toxic substances have been deliberately dumped out of ignorance or to avoid safe disposal costs. This land cannot be reused until it has been decontaminated. As this is a slow and highly expensive process, it is easier and cheaper to let the old sites lie derelict and build on 'green-field' sites on the fringes of towns, but this destroys arable land and erodes the greenbelt around towns adding to urban sprawl.

Energy usage This has already been considered from the point of view of damage to the environment by the generation of electricity and the use of fossil fuels. Although the process of nuclear electricity generation does not produce any direct pollution if properly managed, the disposal and emotional problems associated with nuclear waste renders it commercially uneconomical. Since the provision of energy is destructive and the use of energy is expensive, all new product designs should require manufacture by minimum energy processing.

Outcomes

The main outcomes from any manufacturing process are:

- The required product.
- By-products.
- Waste.

The required product This is the manufactured product that is sold to the customer. It should be 'fit for purpose' environmentally safe and avoid short-term and long-term health hazards. This, however, is not always the case. For example, cigarettes and other tobacco products represent a health hazard despite being the main product for the manufacturer concerned. Further, when the product is no longer required it must be possible to dispose of it simply, cheaply and safely. The materials from which it is made should be recyclable, biodegradable, or chemically safe and stable for disposal in a landfill site.

By-products These are the waste products from the manufacturing process that can be sold on as raw materials for another industry. For example, power station ash is used for the manufacture of insulation blocks for the construction industry.

Waste This is the waste material for which there is no further use. It cannot be used as a by-product. Its safe disposal will depend upon its properties and is the subject of legislation that will now be considered.

12.12.2 *Waste disposal*

No company, large or small, just throws its waste away in a haphazard manner. Local authorities only collect and dispose of domestic waste and light commercial waste; they do not remove industrial waste. Therefore, industry must dispose of its own waste through properly licensed waste-disposal companies. This, however, is costly and is another good reason for keeping waste to a minimum. Waste-disposal companies will grade the waste collected and dispose of it safely. Sometimes they will have to treat it chemically to render it safe or may have to incinerate it to make it safe. Incinerators licensed by the local authority are limited to a throughput of 1 tonne per hour and special licences from the Environment Agency are required for larger plants. Scrap metal is usually sold direct to a scrap-metal dealer for recycling.

From April 1996 – vesting day for the new Environment Agency – the officers of Her Majesty's Inspectorate of Pollution have been absorbed into the new agency and are responsible for *Emissions to the Atmosphere* and the authorisation for section A processes under the Industrial Pollution Act: that is, all incinerators of over 1 tonne per hour throughput, all chemical plant, and all other 'specified' processes. The authorisations issued by the Agency cover:

- Inputs to the process and the nature of the process itself.
- Polluting substances that may be *emitted direct to the atmosphere* and the volume limits per hour of such pollutants.
- Emissions to sewers and any effects on surface-water drainage.
- The nature of solid residues, the need for further treatment before disposal, or the suitability of such residues for disposal directly to landfill sites.

Other industrial processes not covered above may yet fall within the jurisdiction of the Industrial Pollution Control (IPC) officers of the National Rivers Authority (NRA) – now incorporated into the Environment Agency – to provide *control through consent to discharge*. This lists temperature, rate of discharge (at a set volume per hour), pH value, sediment, toxicity (heavy metal content in particular) and bio-oxygen demand.

The Waste Regulatory Authority (WRA) is now incorporated into the Waste Agency and control is exercised:

- By the issue of waste management licences covering the keeping (storing), treating, transport and disposal of waste.
- Through waste regulation (special) waste management licences and associated paperwork. This paperwork giving notice of transport must precede every movement and copies must accompany every movement of the waste. This applies not only to persons actively handling the waste but also to brokers and intermediaries who may not physically handle the waste.
- By registration of carriers of waste. Only the head office needs to be registered and this registration applies to all branch offices throughout the UK. Section 54 places a *general duty of care* on the carrier of controlled wastes. This includes a requirement of the carrier to take such waste only to authorised sites. The main controlling aspects of the duty of care is the adequate description of the waste and any special hazards associated with it. This description must accompany any transfer of waste both physically and in respect of its ownership.

The scale and problems of waste disposal are enormous and Fig. 12.11 puts some facts and figures to this problem. Landfill represents 90 per cent of controlled waste disposal in the UK, and landfill sites must be properly engineered and managed. The officers of the WRA are responsible for ensuring that the well-being of the environment as a whole is not put at risk by operations on the site. For example, no toxic waste must be dumped; no noxious, flammable or explosive gases must be allowed to accumulate; the ground surrounding the site and the water table beneath the site must not be contaminated. The HSE is responsible for ensuring the safety of:

- Employees working on the site.
- Visitors to the site (delivery drivers).
- Persons living and working adjacent to the site.

Landfill operators must not only provide reassurance of minimal impact on local communities during the site's productive life, but also for many years after it has been filled. Filled sites offer opportunities for landscaping and redevelopment of public open spaces in areas of former industrial mining and quarrying dereliction. Restoration is now a key part of landfill management since it turns derelict sites to recreational and agricultural use. Trees are often planted around the perimeter of large modern sites to reduce visual impact and act as a noise barrier, whilst also providing wildlife habitat. Experience has shown that even well-engineered sites are not suitable for building development at least in the short term.

Fig. 12.11 *UK waste industry sector: (a) market size and share; (b) non-hazardous collection (reproduced courtesy of BIFFA)*

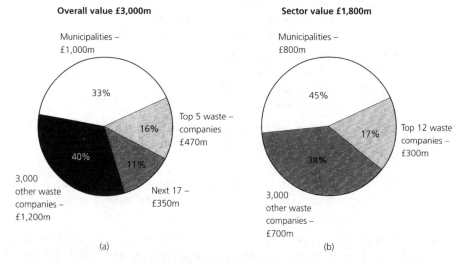

SELF-ASSESSMENT TASK 12.7

1. Discuss how BS 8800 is being implemented or could be implemented at your place of work or within an organisation with which you are familiar.

2. Discuss how BS EN ISO 14000 is being implemented or could be implemented at your place of work or within an organisation with which you are familiar.

3. Explain how the requirements of the above standards could be implemented into an organisation that is already BS EN ISO 9001/2 approved without over-burdening the company with bureaucratic paperwork.

12.1 With reference to the Health and Safety at Work, etc., Act 1974, compare the composition and responsibilities of the Health and Safety Commission with the composition and responsibilities of the Health and Safety Executive.

12.2 Describe the sanctions available to the Health and Safety Inspectorate for enforcing the provisions of the Act.

12.3 Discuss the respective responsibilities of employers and employees under the Health and Safety at Work, etc., Act.

12.4 Explain how the Workplace Regulations 1992 have been implemented at your place of work or within an organisation with which you are familiar.

12.5 Discuss how the European Safety Directive 1992 affects the manufacture, sale, leasing and importing of machinery in the UK.

12.6 Summarise the requirements of the new fire regulations (1989) as set out in the following HMSO publications:
(a) *Guide to Fire Precautions in Places of Work that Require a Fire Certificate*
(b) *Code of Practice for Fire Precautions in Factories, Offices, Shops and Railway Premises Not Required to Have a Fire Certificate*

12.7 Discuss the special guarding and safety requirements of:
(a) grinding machines
(b) horizontal milling machines
(c) power presses

12.8 Draw up a risk assessment form for your place of work or for an organisation with which you are familiar, together with notes of guidance for the completion of such a form.

12.9 Prepare a paper for presentation to senior management, setting out the arguments for the acceptance and implementation of the requirements of BS 8800 within your company.

12.10 Prepare a paper for presentation to senior management setting out the arguments for the acceptance and implementation of the requirements of BS EN ISO 14000 in your company.

Index